In the Shadow of the Moon

Outward Odyssey
A People's History of Spaceflight

Series editor
Colin Burgess

IN THE SHA

Francis French and Colin Burgess

With a foreword by Walter Cunningham

OW OF THE MOON

A Challenging Journey to Tranquility, 1965–1969

UNIVERSITY OF NEBRASKA PRESS • LINCOLN AND LONDON

 Manufactured in the United States of America ¶ ∞ ¶ Library of Congress Cataloging-in-Publication Data ¶ French, Francis. ¶ In the shadow of the moon : a challenging journey to Tranquility, 1965–1969 / Francis French and Colin Burgess ; with a foreword by Walter Cunningham. ¶ p. cm. — ¶ (Outward odyssey : a people's history of spaceflight) ¶ Includes bibliographical references. ¶ ISBN-13: 978-0-8032-1128-5 (cloth : alk. paper) ISBN-10: 0-8032-1128-7 (cloth : alk. paper) ¶ 1. Project Apollo (U.S.) ¶ 2. Space flight to the moon. ¶ 3. Astronautics—United States—History—20th century. ¶ 4. Space race—History. ¶ I. Burgess, Colin. ¶ II. Title. ¶ TL789.8.U6A5337 2007 ¶ 629.45'4—dc22 ¶ 2006103047 ¶ Set in Adobe Garamond and Futura by Bob Reitz. ¶ Designed by R. W. Boeche.

For Dee O'Hara

Spacefarers need more than engineers to support them.

Over the mountains of the moon,
Down the valley of the shadow,
Ride, boldly ride.

Edgar Allen Poe,
Eldorado

Contents

Illustrations

Foreword

Thirty years ago, when I was identified as an astronaut, as often as not the first question I was asked was, "Which one are you?" I was asked that question hundreds of times but was only quick enough to come up with the best answer one time: "I'm this one!"

Today, the same situation elicits "Did you fly in space?" and "What was your mission?"

This happens because, except for John Glenn, Neil Armstrong, and one or two others, the public sees us as generic heroes—a Mark IV, Mod 3 Astronaut. The media made us heroes at a time when we did not take ourselves all that seriously.

When I was selected to be a NASA astronaut in late 1963, it was still the early days of spaceflight. Like it or not, we were instant celebrities or, more accurately, a celebrity's celebrity. We were sought out and "collected" by politicians and Hollywood stars and never thought a whole lot about it. We were not terribly impressed by anyone else, but somehow or other they were frequently impressed to have an astronaut in the crowd.

The public had many misconceptions about astronauts in those days, believing, for example, that there was an "astronaut diet" or an "astronaut physical fitness program." We did have many academic classes the first couple of years of training, but the Mercury astronauts saw to it that they were all "gentleman's courses"—not pressured by grades. The last thing in the world those guys wanted was to be measured side by side with anybody else who arrived later! Without such measurements, they were always at the top of the heap.

There was an order of seniority in the astronaut office because we were all military-trained people. However, that was a whole lot less important than what I call the pecking order. The pecking order pretty much ignored military rank but had everything to do with whether you were a Mercury,

Gemini, or Apollo astronaut. Even though the first thirty of us became pretty well integrated, there was always that pecking order—subject to the occasional exception. When exceptions occurred, it was usually due to a friend's unfortunate death. The top astronauts forgot about the pecking order only rarely.

What else was different in those days? Today's astronauts may seem a bit underpaid for the job they do and the impressive way they do it, but their salaries look pretty good to me. When I went to work as an astronaut, in 1963, I earned a little over $13,000 a year. I once calculated that, during my *Apollo* 7 mission, I had earned the great sum of $660. But we weren't doing it for the money—nobody does a job like that for money. Any one of us would have paid NASA to have the job!

I do recall being just a little chagrined when I first learned what I was going to be paid, and thinking, "You know, if it was so hard to get this job and so many people wanted it, wasn't it worth maybe a bit more than $13,000 a year?"

Something else we did not have and were unable to obtain in those days was life insurance. The second year I was there, we thought we might get coverage when NASA went out for quotes to renew the agency's health and accident policy. They actually solicited two quotes. Insurers had to submit one bid with the thirty astronauts included for death benefits and another bid if we were not covered. The difference was, apparently, quite significant because NASA, bless its sweet little heart, accepted the bid with astronauts *not* included!

None of us planned to collect any life insurance, even though we did not always deserve our reputations as impeccable aviators. It turns out we had just about the same kind of safety record as active-duty military pilots. Most of our accidents were "pilot error"—that is just the way it is with military flying. We lost a lot more astronauts in T-38s back then than we ever did in spacecraft.

By 1965 our program was beginning to roll, but the Soviet Union had already laid claim to many space firsts. In four short years they had launched the first human, Yuri Gagarin, followed by Gherman Titov, the first person to spend a full day in orbit. They also became the first nation to launch simultaneous flights: *Vostok* 3, with Andrian Nikolayev on board, followed a day later by *Vostok* 4, carrying Pavel Popovich into space. The following year Valery Bykovsky and Valentina Tereshkova, the first woman in space, repeated that successful tandem mission.

NASA was hard on their heels. We launched Alan Shepard three weeks after Yuri Gagarin's epic journey, and although Shepard's flight and the following one by Gus Grissom were short, suborbital missions, they got us in the game. It wasn't until John Glenn's flight in February 1962 that the United States began to match the Russians by orbiting humans. The Mercury flights of Scott Carpenter, Wally Schirra, and Gordon Cooper narrowed the gap in what was being called the "space race," before the Soviets once again surprised the world with the first three-person flight. A full four years before the first flight of the three-man Apollo spacecraft, which I would fly, Vladimir Komarov, Konstantin Feoktistov, and Boris Yegorov orbited the Earth together aboard the first Voskhod spacecraft.

Americans seem to rise to the occasion when they find themselves in a race; we just naturally want to win. However, this was not just a race. It was a clash of cultures, systems of government, and a challenge to our way of life. We felt threatened! For the first time in a hundred years, we faced the fact that not only were we not the best at everything in the world, we were not even as good at some things. We were suddenly shocked into doing something.

As 1965 got under way the Soviet Union achieved another notable milestone: a flight featuring the first spacewalk. With the successful conclusion of this mission by Alexei Leonov and Pavel Belyayev, the Russians appeared to be well ahead. However, we were about to launch the first Gemini mission. This impressive and relatively cheap program soon narrowed the gap. As we launched Gemini missions, the Russians were busy creating a whole new generation of spacecraft known as Soyuz. Its birth would not be an easy one.

Sadly, along the way, each nation lost some of its best spacefarers. In my first five years at NASA, we lost seven of the thirty people in our group in aircraft and spacecraft accidents. In the thirty-eight years since, we have only lost five more members of that group. Astronauts Gus Grissom, Ed White, and Roger Chaffee died aboard their *Apollo* 1 spacecraft during a launch-pad fire in January 1967, and the Soviets grieved the loss of cosmonaut Vladimir Komarov when *Soyuz* 1 slammed into the ground following a parachute failure just three months later. Their respective countries mourned them as national heroes. Those of us who knew them as friends and knew their wives and children suffered a far more personal loss. It was my job to fly the next mission into space, a mission with great significance if we were going to reach the moon "before the end of the decade." I felt fortunate to be on the backup crew for *Apollo* 1 and then to be able to fly the first manned Apollo mission. I know that all those getting ready to fly the *next* mission in line feel exactly the same way.

By the time I was ending my tenure as chief of the Skylab Branch of the astronaut office, working on designing America's first space station, the moon race was well and truly over. When Skylab launched in 1973, humans had already visited our moon for the last time that century. I would love to have made that journey myself, but sadly it was not to be.

The space race seems a long time ago, now that we have Russians and Americans orbiting together on a space station. (Now people usually refer to me simply as an "ancient astronaut.") In some ways, the space race between the two nations continues, except now it looks more like a fight for money than national prestige. In the 1960s it was easy to understand the Russians. We truly were in a space race, and we were hollering, "The Russians are coming! The Russians are coming!" We did not know that across the sea, the Russians were saying, "Hey! The Americans are coming! The Americans are coming!" They were doing it for precisely the same reasons: to generate public support and keep the funding up for their program. The Russians were truly in a race with us to land a man on the moon. As we started out behind them getting into space, the Russians made their commitment to go to the moon probably a bit after us. And they did not give up for a long time.

I began to meet and socialize with cosmonauts around 1968, and I have come to know many of them reasonably well. Most of them were just like us, hard-charging guys. Even those who were civilians struck me as having a military background. Even though we always got along with the cosmonauts very well as individuals, I have always been critical of the Russian programs and the people who negotiated the deals with us.

Today, the public pretty much takes space for granted. Most of the time, I really cannot blame them—it's part of the American character. A couple of times a year I go down to the Visitor Center at the Kennedy Space Center and talk to kids. If I ask how many of them want to be an astronaut, I get some hands. In the old days everyone wanted to be an astronaut; everybody would raise his or her hand. It is not like it used to be. But every once in a while something comes along, like the Columbia disaster, and all of a sudden the public is grabbed right by the heart. Death awakens us from our complacency, and we realize again that spaceflight is a very dangerous profession. It is a profession that should be pursued by professionals.

One thing today is as true as it was back in the 1960s—those who fly into space for both nations care less about the politics than they do about getting the job done. As this book shows, we are a rather diverse group of people with very different personal stories. About the only thing we all have in common is that, at one time or another, we each took that walk out to the launch pad, strapped ourselves on top of a powerful rocket, and readied ourselves for the experience of a lifetime. I hope you enjoy getting to know us as individuals in the pages of this book.

Walter Cunningham
NASA *Astronaut*, Apollo 7

Acknowledgments

It would not have been possible to write a book containing so many personal stories without the full assistance and cooperation of a number of key participants in the space program. Many have been interviewed hundreds of times over the decades; others have chosen never to tell their stories in this form before. Yet in each case they sat down with the authors and offered fresh, insightful perspectives into their lives and space careers with openness and patience. In numerous cases they also took the time to invite us into their homes and their confidences despite demanding schedules, and made introductions to others.

Our primary thanks must go to the spacefarers, both Russian and American, who generously gave their time and help so willingly. For their invaluable assistance, our thanks go to Buzz Aldrin, Valery Bykovksy, Mike Collins, Konstantin Feoktistov, Valeri Kubasov, Alexei Leonov, Jim Lovell, Pam Melroy, Ed Mitchell, Pavel Popovich, Jack Schmitt, Valentina Tereshkova, Al Worden, and Alexei Yeliseyev. Dick Gordon also kindly allowed us to fully use an interview we had previously conducted with him.

Some spacefarers went even further in their assistance, in ways that we could never have expected. Bill and Valerie Anders, Scott Carpenter, Gene Cernan, Walt and Dot Cunningham, Charlie Duke, Wally and Jo Schirra, Rusty Schweickart, and Tom Stafford not only provided extensive interviews, but also very helpfully proofread and made valuable amendments to the book. In many cases, this process uncovered stories long forgotten by the spacefarers themselves, which appear here in print for the very first time. Our special thanks to Suzi Cooper and her late husband, Gordon Cooper, who was also very generous with interview time. Sadly, Gordon passed away shortly after receiving some final drafts from this book, which he had helped with and agreed to proofread.

As indicated by this book's dedication, those who stay behind on the ground are often just as vital as those who get to go on a spaceflight, and this proved to be the case when researching and writing the book. Key interviews and some valuable editorial suggestions were made by spaceflight notables and family members Jeannie Bassett, Sam Beddingfield, Howard Benedict, Cece Bibby, W. O. Brown, Harriet Eisele, Susie Eisele Black, Wally Funk, Don Gregory, Gerry Griffin, Paul Haney, Jack King, Gene Kranz, Lola Morrow, Dale Myers, Dee O'Hara, Milt Radimer, Harvey Renshaw Jr., Kris Stoever, Teddy Taylor, Al Tinnirello, Hank Waddell, and Guenter Wendt. In some cases, the events we discussed were controversial, even personally painful at times, and we are most grateful that the participants found this project important enough for the story to be told correctly.

Those who *truly* know their space history are few and far between, so we were most grateful that some expert researchers were so willing to help with this project. Our thanks therefore go to Rick Boos, Tod Bryant, Jim Busby, and Robert Pearlman. Photo researcher Jody Russell at the Johnson Space Center Media Resource Center; Sharon Thomas of the Florida Museum of Natural History; and Shelly Kelly, a University of Houston archivist, were also very helpful in tracking down photographs and long-overlooked documentation. Others who provided much-needed assistance were Kate Doolan, Kerrie Dougherty, Elena Esina, Ed Hengeveld, and David J. Shayler. Erin French provided repeated and meticulous proofreading, and Sonia Lopez assisted in reproducing old photographs.

Particular thanks must go to noted space historians Mike Cassutt, John B. Charles, and former British Interplanetary Society president Rex Hall, MBE. They each gave a good deal of their time to read over each book chapter, providing insights that considerably sharpened the final drafts. We extend our sincere gratitude for their expert help and enduring friendships. Another worthy contributor was spaceflight historian Bert Vis, whose extensive interview archive and helpful advice provided the authors with numerous personal quotations and newly emerging facts from the early days of the Soviet space program.

The authors would also like to acknowledge the skills of copyeditor Paul Bodine, whose subtle changes have sharpened the focus of this work. We also appreciate the support and encouragement of the entire team at the

University of Nebraska Press, particularly that of Director Gary Dunham and former assistant managing editor Linnea Fredrickson.

This book is a tribute to the welcome assistance of all those named above and several other contributors, each of whom lent a willing hand. The authors could not have undertaken nor completed this project without their enthusiastic help and support.

1. Gemini Raises the Bar

With silent, lifting mind I've trod,
the high untrespassed sanctity of space.

John Gillespie Magee

By the end of July 1961 Gus Grissom knew that, barring unforeseen circumstances, he would not fly again in the Mercury program. One of America's original *Mercury 7* astronauts, Grissom had just commanded his country's second manned space mission. The flight went well until the very end, when the hatch of *Liberty Bell 7* unexpectedly blew off, and the spacecraft was lost to the ocean.

In his autobiography, *Schirra's Space*, fellow Mercury astronaut Wally Schirra recalled that Grissom now felt he had something to prove. "He was angry about being blamed for his spacecraft having sunk, and he was fighting to come back out of the pack. Gus was a tiger. He wanted the first Gemini flight, and by God he got it." Gemini, the two-person spacecraft that would follow the solo Mercury missions, was a good place for Grissom to focus. With his astronaut colleagues all waiting their turn to fly, Grissom was now at the back of the line.

"When Gus finished his Mercury flight, he knew he was out of the loop because we had to go through the seven," Schirra reflected. "And he looked at it and said, 'My God, we are not going to have that many flights. I'm going up to St. Louis and play with Gemini.' So it was essentially his spacecraft. He practically had it to himself." Grissom subsequently began working closely with the engineers at McDonnell's St. Louis plant, offering the advice of a flown astronaut as they designed the two-man spacecraft. It was an opportunity they had not been given with the smaller Mercury craft, apart from some suggested modifications, and Grissom threw himself into the task with his usual determination and expertise. "Since we had to fly the beast," he said prior to the flight, "we want one that will do the best possible job." On the whole, the engineers were pleased to have Grissom's input.

On the twentieth anniversary of the attack on Pearl Harbor, Dr. Robert Gilruth had revealed plans for the development of a two-person spacecraft. Just a month later, Alex P. Nagy from NASA headquarters in Washington gave an eminently appropriate name to the project: Gemini.

Nagy's suggested title quickly caught on, and it stuck. In an astrological sense, Gemini was named after a constellation that includes the twin stars of Castor and Pollux, and it is a sign of the zodiac controlled by Mercury. Within its spheres of influence are the qualities of adaptability and mobility—two major objectives of Project Gemini. A circumspect Gus Grissom would later write that "those of us who had become involved with Gemini were beginning to suspect that we'd got, not twins, but a tiger by the tail."

Gemini was seen as a necessary bridge between the pioneering Mercury flights, which primitively tested the adaptability of humans to spaceflight, and the Apollo program, which was designed to send astronauts to the moon and return them to the Earth. Grissom would also draw his own automotive analogy. "Gemini's a Corvette," he reflected. "Mercury was a Volkswagen."

The Gemini astronauts would in effect open the doorway to a moon landing—an event not measured then in decades, but in a mere handful of years. They would not only conduct long-duration missions as a necessary precursor to the lunar voyages, but also experiment with the complex techniques of rendezvous and conducting docking maneuvers in Earth orbit. Over Gemini's twenty-month duration there would be twenty available seats going into orbit. As one Gemini astronaut would describe the program and its list of challenging objectives, it was "test pilot heaven."

Externally, the two-person Gemini vehicle was very similar to the Mercury spacecraft. However, it weighed almost twice as much, and being a foot wider and longer created an overall increase in size of some 20 percent. The most dramatic difference in the spacecraft itself was an increase of 50 percent in its interior volume, which would permit two astronauts to occupy it for flights of up to two weeks' duration.

In every sense it was a pilot's spacecraft. Although 220 dials, switches, and levers surrounded the two astronauts, all of them were within easy reach, while strategically placed mirrors allowed them to view and operate other controls located behind them. The astronauts could also exit the craft while in orbit through two hinged hatches. It was all part of the planning process: the Titan boosters that would carry the three-ton spacecraft aloft used combustible fuel instead of the highly explosive propellant that had fed the Atlas rockets, which allowed designers to dispense with the heavy, rocket-propelled escape towers previously used in Mercury flights. "The weight saved could be put to good use in the spacecraft," Grissom observed in his book, *Gemini*. Having larger hatches was also critically important if a catastrophic situation developed on the launch pad or in early flight. Once a rapid evacuation had been decided upon the hatches would fly open in milliseconds, while ejection seats would simultaneously blast the two crew members out and away from their spacecraft.

Another feature of Gemini was the white, one-ton adapter section located at the rear of the spacecraft. Essentially a harmonizing collar between the craft and the Titan booster, the section also housed sixteen small rockets for orbital maneuvering, containers of oxygen, and batteries for electrical power. This adapter section, its job done, would be jettisoned just prior to reentry, exposing the spacecraft's retrorockets and heat shield.

In discussing developing plans for Project Gemini at a press conference, NASA's Director of Manned Space Flight, Dr. H. Brainerd Holmes, explained that the changes were important, "for they will allow us to put manned space flight on more of an operational basis, as opposed to the research and development effort that is involved in Project Mercury. They are the same changes that are generally found when we compare a first prototype aircraft with its later production version."

The Martin Company's Titan II had essentially been designed as a ballistic missile; in fact, it was the most powerful Cold War booster ever developed by the United States, capable of carrying an enormous warhead measured in tens of megatons of thermonuclear energy. This mighty booster's first stage would develop around 430,000 pounds of thrust—some 65,000 pounds more than the Atlases that would launch four American astronauts into orbit during Project Mercury—while the second stage would produce an additional 100,000 pounds of thrust.

Unlike the Atlas and other boosters of that era, the Titan II did not burn liquid oxygen or kerosene. Instead, it was powered by highly toxic hypergolic propellants: room-temperature fuels and oxidizers that spontaneously and violently ignited on contact with each other in the rocket's combustion chamber.

The one truly strange thing that happened to the Gemini craft and its interior fittings during their development was that they began to be fashioned around Gus Grissom—one of the smallest of the astronauts. Later, taller astronauts would have problems squeezing into the same space and using the same apparatus designed for Grissom's compact frame. He told *Life* magazine that "When the other [astronauts] started looking at the Gemini mock-up, it was pretty clear it was designed around me." In fact, this became so evident that the astronauts began to call the Gemini spacecraft "The Gusmobile."

Fellow astronaut John Young, who would be assigned to the first mission with Grissom, reflected the feeling of the others: "Gus really had a big hand in everything," he said, "from the way the cockpit was laid out to what instruments went where. It was his baby." Grissom demanded a lot of himself, especially as Gemini came with the accountability of command. "I was responsible for my own skin in my Mercury flight," he said at the time. "But now that I'm going up for my second flight and have John in that co-pilot's seat, I'm responsible for two. This will mean some of the decisions may come a little harder, but I've asked for the responsibility and I've got it."

In most of John Young's photos, past and present, he generally looks detached and unsmiling. Yet those who know him say that although the former navy pilot might seem aloof, he actually possesses a lively, very dry wit. Blink and you might miss it, but it is there. At various times he has been described as a loner, an adventurer, and a mechanical genius, with a curious dislike for the great outdoors.

"John is an amusing bundle of such contradictions," his first wife Barbara once commented. Fellow astronaut Michael Collins would also describe his colleague as "Mysterious. The epitome of the non-hero, with a country boy's 'aw shucks—t'ain't nothing' demeanor, which masks a delightful wit and a keen engineer's mind." NASA managers, in their first technical meetings with the rookie astronaut, were regularly thrown by his demeanor. At first Young would sit in silence, absorbing the opinions being offered, and the others attending would assume he had nothing of importance to add. Then, toward the end of the meeting, Young would gently make a point so devastatingly precise that they would realize he was ahead of them all. He would have summarized all the key points and seen another that they had all missed. He was rarely underestimated twice.

Just a week shy of his thirty-second birthday and then a lieutenant commander in the U.S. Navy, Young was named as one of NASA's second group of astronauts. He had not been eligible for consideration in the first Mercury group, as he was still a student at test-pilot school. John Watts Young was born in San Francisco on 24 September 1930 to parents William Hugh and Wanda (née Howland). He was less than two years old when the family moved east—first to his father's birthplace of Cartersville, Georgia, where his younger brother, Hugh, was born in 1933. The future astronaut began his education at Cherokee Avenue School in 1935, and while his father looked for work he and his brother lived with their aunt, who remembered his childhood fascination with airplanes and model trains. She would later write of her nephew that "He has a brilliant mind with amazing concentration. John is a very reserved and private person completely absorbed and happy in his space work."

The following year his father found a new job in Orlando, Florida, but it would be some three years before the family moved there permanently, and into their own home. Young, now nine years old, and his brother were enrolled at Princeton Elementary School. Then, during World War II, his father served as a navy commander, flying patrols over the Pacific. Young's interest in aviation continued through his high school years, both at Memorial Junior High School and later at Orlando High. According to his father, he also "used to draw pictures of airplanes and rockets all the time." One of Young's former high school classmates recalls that "in the eleventh grade everybody had to make a talk on something. . . . John chose rockets."

Academically, he had little difficulty in high school, breezing through with straight As. Football proved a handy release from the strictures of study, and Young also became a promising track athlete. A member of the National Honor Society, he graduated in 1948 and was awarded the faculty's highest senior class honor, the Guernsey Good Citizenship Cup. Young received an ROTC scholarship and enrolled at the Georgia Institute of Technology. He subsequently earned a magna cum laude degree in aeronautical engineering from the institute in 1952, finishing second in his class, and received a commission as an ensign in the U.S. Navy.

A temporary assignment as fire control officer aboard the Fletcher Class destroyer USS *Laws* followed, and even though he bore the fledgling rank of ensign Young seems to have made a lasting impression on people. One of those who served with him, Joseph LaMantia, recalls Young as being "the most respected officer on the ship."

His next assignment was to attend basic flight training school in Pensacola, Florida, and in 1954 he won his wings. An advanced six-month flying course was followed by four years' duty on the carriers *Coral Sea* and *Forrestal*, flying TF-9D Cougars and F-8D Crusaders with the 103rd Fighter Squadron. During this exciting period Young married Barbara White of Savannah, Georgia, with whom he would eventually have two children—daughter Sandy, born in 1957, and a son, also named John, born two years later.

"John started out in the black-shoe Navy," Grissom wrote with dry humor in his autobiography, using the term for navy personnel who were not naval aviators, "and I never let him forget his destroyer days. After switching to flying, he set a world time-to-climb record in 1953, during Project High Jump." In fact, as a test pilot, Young set world "time-to-climb" records for both three thousand and twenty-five thousand feet in a navy F-4H-1 Phantom II fighter aircraft. Prior to his astronaut selection in 1962, he had accumulated an impressive record as a navy test pilot, instructor, and project evaluation officer, with around thirty-two hundred flying hours to his credit, including nearly twenty-seven hundred in jet airplanes. As Young told the *Washington Evening Star* newspaper in 1965, "I left the best job in the Navy to get the best job in the world."

"John's not what you'd call the talkative type," Grissom later recalled of his Gemini copilot, "but he's got a good solid sense of humor, which is a prerequisite in this space business. It's a bit unusual and takes you a while to catch on to, but he knows how to ease the strain. Without humor, you're in a bad way when the glitches come." Neither Grissom nor Young were known for wasting words, and as they continued their mission training they were regarded as well matched. As the first non-Mercury astronaut to get a confirmed prime crew assignment, Young seemed in fact to imitate his commander down to the tiniest detail. Both quietly absorbed themselves in the engineering details of the mission. When reporters tried to ask them questions about themselves, they would get a reply about machinery instead. It was the way the two astronauts wanted it. Neither man cared much for the publicity, and would rather have been quietly discussing the latest simulator run together than responding to reporters' questions.

In one aspect of the mission the crew seemed almost willfully intent on derailing NASA's publicity plans. Tradition demanded that the spacecraft receive a name, and this privilege fell to Grissom as commander of the vehicle. His first idea was to name it after an Indian tribe from the area in Indiana where he was born, and he and John agreed that the name Wapasha was appealing. Then someone suggested that the press would probably tongue in cheek call it the "Wabash Cannon Ball," after a popular rail-themed country song. At the time, Grissom's father was working for the Baltimore and Ohio Railroad, and he didn't want to impose that burden on him. "How would he explain that one to his pals on the B&O?" Grissom later reflected.

Then he had a devilish notion, inspired by the name of the popular Broadway show and movie *The Unsinkable Molly Brown*. Harking back to the loss of his Mercury spacecraft *Liberty Bell 7*, Grissom began to feel *Molly Brown* might be a perfect choice. Once again, John Young agreed, so their Gemini craft was named after the *Titanic* survivor portrayed in the musical. "Some of my bosses were amused;" Grissom recalled, "some weren't." A couple of those who didn't like the name asked the unrepentant astronaut what his second choice might be, thinking he may have a better option. His second choice, he told them, was *Titanic*. "Nobody was amused, so *Molly Brown* it was."

In hindsight, Deke Slayton could hardly have selected a better crew for *Gemini 3*. It was essentially a "shakedown" flight, as pilots liked to call it, which called for the crew to exercise the spacecraft's orbital attitude and maneuvering system. This system was crucial to the planned rendezvous and docking flights that would later be attempted. Another of their vital tasks was to attempt a controlled reentry using an on-board computer. "In comparison with some of the later missions," Grissom once wrote, "ours was what John Young would describe as 'a piece of cake.' But John and I had practically lived with our spacecraft since the first rivet was put in it at the McDonnell plant. We had studied every one of its systems as each was installed, and sweated out the glitches along with the McDonnell engineers. So we had the vital ingredient of confidence going for us all the way."

Originally, the first manned Gemini mission was scheduled for December 1964—America's only manned flight planned for that year. Then the postponement of a critical test flight and the forces of nature brought about a lengthy delay. The second precursor flight, designated *Gemini 2* and scheduled for 6 October, involved launching an unmanned Gemini craft out over the Atlantic to demonstrate reentry techniques and recovery systems. On 17 August the Titan II booster being readied for the task was damaged by lightning as it stood on the launch pad. Later, as a precautionary measure, the second stage was removed and placed under shelter when Hurricane Cleo began advancing on the Cape. The first stage remained on the pad but was firmly battened down. After Cleo, another two hurricanes began moving in, so the first stage was also lifted from the pad and moved to a safe area. NASA's associate administrator for manned spaceflight, Dr. George Mueller, announced that these factors had combined to eliminate the slim possibility of a manned launch that year, and he confirmed that the launch would take place no sooner than February or March.

The Titan II test flight eventually left the pad on 19 January, culminating in the recovery of the automatically piloted spacecraft by crews from the aircraft carrier *Lake Champlain* an hour and forty-four minutes after liftoff. The craft splashed down only twenty-four miles short of its intended target after a journey of 2,126 miles. A jubilant Gus Grissom said the almost-faultless trial had left a "clear road ahead" to the flight he and John Young would conduct just a few weeks later. At a news conference held in a giant assembly room of the Martin Company plant in Baltimore, Grissom was full of praise for the specific rocket that would carry them into space. "This Titan is the cleanest booster that has gone through all its tests—cleaner than any booster flown to date," he said with conviction. "Very definitely, Titan is *go*!"

One aspect of their spacecraft that never really sat well with the crew was the ejection system, intended to save them in the event of a booster catastrophe at launch or in early flight. As Young watched one test of the system, his hope turned to dismay when the hatch failed to open as planned. The ejection seat fired and plowed straight into the hatch. Such an event would have proved fatal, or, as Young dryly put it later, "One hell of a headache—but a short one!"

On 5 February, *Molly Brown* was slowly trundled out to Launch Complex 19 and carefully hoisted into position atop the ninety-foot Titan II launch vehicle. Three days later, in yet another positive move for the Gemini program, NASA announced that Mercury astronaut Gordon Cooper and rookie Charles "Pete" Conrad had been selected to fly the *Gemini 5* mission, then planned for September. With a provisional launch date of 23 March to work toward, the two *Gemini 3* astronauts stepped up their training. They worked well together. Young admired his commander's experience and dedication to the task at hand, while Grissom took a liking to his copilot's pragmatic enthusiasm, engineering skills, and wry humor. "Until we joined up for this flight," Grissom said of him, "I didn't know John Young any better than any of the others in their group. They're all talented. In fact, when one of them comes up with a new answer for some problem, I think they are a lot smarter than our original group of seven. By launch time John and I will know each other pretty well. We take a note book with us into the flight simulator . . . and jot down all the problem areas. Then at night we sit down . . . and talk things over, deciding who does what."

Nothing was going to prevent Gus Grissom from making this flight—not even a possible fracture in his arm. Within weeks of launch, he hit a door with his hand during a relaxed function in the Bahamas, and according to his wife, Betty, "apparently broke a bone in his left wrist." Almost immediately the wrist began to swell up, and he had to strap it tightly with a plastic bandage. Somehow he managed to keep it a secret and worked privately on the wrist to bring the swelling down before launch date. He knew very well that if anyone found out it could easily have caused him, or both crew members, to be replaced. "Gus was never entirely sure the wrist was broken," Betty Grissom later revealed in her book, *Starfall*, "because he did not consult a doctor, at NASA or anywhere."

As launch day approached, the crew's training included a full seven-hour countdown simulation inside their spacecraft on the launch pad. The test was deliberately halted just one minute before the theoretical liftoff. Even though several minor problems had emerged, officials declared the trial launch a success. Then, on 17 March, just six days before *Molly Brown* was due to fly, the Soviet Union dropped a sickening bombshell. *Voskhod 2* was launched into orbit, and cosmonaut Alexei Leonov made a historic spacewalk. The timing was clearly aimed at stealing NASA's thunder, and it gave the space agency quite a jolt. When asked the seemingly inescapable question about where this latest feat placed NASA in the space race, officials calmly reacted by talking about Gemini's elaborate and structured preparations, while politely downplaying what they claimed was a very obvious (and successful) instance of Soviet propaganda.

At 9:24 a.m. on 23 March 1965, after a minor delay, the Titan II's first stage erupted into life; NASA's first manned Gemini mission had begun. The launch was later said to have been so smooth and relatively free of noise that neither Grissom nor Young realized the Titan had left the pad until CapCom Gordon Cooper reported "Bolts and liftoff!" Then they noticed that the mission elapsed-time clock was running. Less than half a minute after launch the two astronauts once again heard Cooper's calm Oklahoma twang. "NASA never approved the name for the spacecraft," Cooper later recalled, "but I gave it my official approval by calling to Gus and John, 'You're on your way, *Molly Brown.*'" Grissom responded with an enthusiastic, "Yeah, man!"

Spectators on the ground watched as the Titan II rocket arced into a startlingly blue sky with a deep, thunderous roar. Witnesses to the launch were amazed that there was no fiery contrail as in earlier Atlas liftoffs due to the use of hypergolic fuel in the booster. Two minutes after liftoff, Flight Control reported, "everything looking well" as the Titan reached a speed of 3,000 miles an hour and the crew prepared for staging. "Staging is really something," John Young said of the event. "It's called 'fire in the hole' because you fire the second stage engine before you get rid of the first stage. [It] blew out everything and fire came all around the vehicle and you could see it. That was a surprise to me. But it is only momentary, and with the second stage firing you get right out of there."

As the second stage kicked in and the spent first stage dropped away, acceleration increased dramatically. Flight Director Chris Kraft gave the crew a tentative "go." Soon the spacecraft was approaching an orbital speed of 17,500 miles an hour. Flight Control continued to report that all systems were functioning well, while Grissom confirmed "everything fine" within the spacecraft. He had good reason to feel fine himself, as he had just become the first person ever to fly into space a second time. It was an occasion not lost on him.

At 9:30 a.m., just six minutes after liftoff from the Cape, the Titan's empty second stage fell away, and *Molly Brown* slid gracefully into orbit over Bermuda. The first manned Gemini craft had achieved an elliptical orbit of 87 by 125 miles above the Earth. This was slightly higher than planned, and resulted from a better-than-expected performance of the Titan II booster. Shortly after, while Grissom was busy verifying the parameters of their orbit with Mission Control, the crew was suddenly faced with indications of a potentially life-threatening problem. Grissom would later describe this situation by saying "all hell broke loose." Young, however, was straight onto it and calmly reported that the fault was in the oxygen system, causing mixed instrument readings. He knew his spacecraft backward and forward, and much to Grissom's relief his copilot had soon diagnosed the problem as an electrical failure. Young promptly switched to a backup unit and reported that the instrument readings had returned to normal. Grissom, who had hastily lowered the visor on his helmet at the first sign of trouble, flipped it back up and resumed his duties. Grissom also reported that the spacecraft was experiencing a slight yaw drift to the left, but it was never considered a serious anomaly, and was eventually found to have been caused by a venting water boiler.

Seated at his CapCom console in Muchea, Western Australia, Pete Conrad relayed some happy news to his fellow astronauts: Kraft had given them the okay to continue for at least a second orbit. At 10:58 a.m., just before *Molly Brown* entered that second orbit, Grissom fired the two forward thrusting Orbit Attitude and Maneuvering System (OAMS) rockets, which effectively reduced the elliptical orbit to an almost circular one of 105 by 97 miles. Constantly issuing updated bulletins on the mission, public affairs officer Paul Haney announced news of the operation. "This is Gemini Control," he reported. "Within the last five seconds the *Molly Brown*, Gemini 3, spacecraft has completed the transitional maneuver by firing their forward-firing thrusters, eighty-five pounds each, for some seventy-five seconds. The maneuver has apparently been a successful one, demonstrating the extraordinary steering capability of this spacecraft."

It was more than just "successful"—for NASA it was an outstanding triumph. For the first time ever the flight path of a spacecraft had been manually altered by a pilot in orbit. As Flight Director Chris Kraft later put it, "We'd just crossed a major milestone on the way to the moon." And there was more to come. Young, as copilot, had not only operated the first computer on a manned spacecraft, but had also been entrusted with several experiments during the four-plus hours of the mission, with Grissom assisting. Samples of human blood needed to be irradiated, and sea urchin eggs were meant to be fertilized to gauge their reaction to weightlessness. The latter experiment did not quite go as planned, however. As Grissom later related in his book *Gemini*, it was a "pathetically simple" experiment, and all he had to do "was turn a knob, which would actuate a mechanism" to fertilize the eggs. "Maybe, after our oxygen scare, I had too much adrenalin pumping," he wrote, "but I twisted that handle so hard I broke it off." Much to his amusement he would later learn that a ground controller, who was simultaneously duplicating the experiment on the ground, had snapped off his handle in exactly the same way.

Young, meanwhile, was testing a new array of water-reconstituted food items such as dried beef pot roast, orange and grapefruit juice, brownies, chicken, and apple sauce. With a few minutes to spare before the next maneuver—the most critical of their flight—Young sprang a small surprise on his commander when he slipped out a carefully wrapped corned-beef-on-rye sandwich from a pocket in the leg of his spacesuit. It was a parting gift that that Wally Schirra had furtively handed to him with a sly wink just before hatch closure. "I catered that," Schirra has since admitted. "I got the sandwich at Wolfie's, a deli near the Cape, the day before. It was refrigerated all night so it was perfectly safe; we ate that food all the time." Young offered the sandwich to Grissom, who was feeling hungry, and he took it with a big smile.

"It was no big deal," Young later said of the incident. "The horizon sensors weren't working right so I gave this sandwich to Gus so he could relax. There was nothing he could do in the dark to make that thing work until we got back into the daylight." However, after taking two bites from the highly aromatic sandwich, Grissom realized that some loose crumbs (and the strong odor of corned beef) had begun drifting around the cabin. He quickly rewrapped and packed it away in his pocket, and the two men soon forgot all about it. But others did not. That small prank would later cause NASA and the two astronauts a lot of grief when enraged congressional representatives came down hard on the space agency for allowing crew members to engage in activities that contravened strict health and safety regulations aboard a U.S. manned spacecraft.

After lowering and circularizing their orbit, the crew of *Molly Brown* finally began the planned task of altering the spacecraft's orbital plane. In other words, they would change the angle of their orbit compared to the Earth below by nudging the spacecraft laterally, using measured bursts from their small maneuvering rockets. It was not performed simply as an exercise in orbital aerobatics, however—it was a maneuver critical to future rendezvous missions between two spacecraft. The Mercury spacecraft could not do it, nor could Vostok and Voskhod—despite the Russians' desire to give the public the impression that Soviet spacecraft had this ability. These primitive spacecraft, hurled into space by their boosters, could only turn to face in different directions before firing retrorockets to begin reentry. Their orbits were based solely on the accuracy of the booster that carried them into space, not by anything the spacecraft could do. It would be impossible to travel to the moon—or even to complete the remainder of the Gemini program—without being able to change orbit, and doing so was *Gemini 3*'s most vital task.

The maneuver was begun by feeding data into the ship's computer, which instructed the attitude control rockets to fire and rotate the spacecraft broadside to its flight path. Once this had been achieved Grissom fired the forward thrusters, propelling *Molly Brown* about a mile to the south, the craft's orbit now canted slightly to the original flight path. Chris Kraft in Mission Control even allowed himself a smile. "All that theory about rendezvous was being converted to fact," he later wrote of the accomplishment. A third orbital maneuver brought the low point or perigee of their orbit down to just fifty miles above the Earth, at which time Grissom jettisoned the adapter section and then fired the retrorockets to bring *Molly Brown* home.

It had been a good day. Though the spacecraft had certainly incurred some small problems during its orbital checkouts—yaw drifts, oxygen readings—these were quite minor compared to the overall success of the first Gemini mission. The first true space vehicle, one that could truly travel in space rather than just pass through it, had all but completed its vigorous maiden voyage. "In the *Liberty Bell 7* I was a man in a can, just along for the ride," Grissom would later say. "*Molly Brown*, bless her heart, was a machine I could maneuver." In fact, Grissom was so elated by the smooth performance of the spacecraft that he even permitted Young to fly it for a few minutes—something that most commanders allowed only with great reluctance. "It was a really good mission," Young told a later press conference. "Gus performed more than twelve different experiments in the three orbits—he did a really great job. I don't think he really got enough credit for the great job he did. He proved that the vehicle would do all the things needed to stay up there for fourteen days."

As reentry began, Grissom noticed that the onboard computer was indicating that they would splash down a little short of their target, so he manually banked *Molly Brown*, first to the left and then the right, in order to bring the spacecraft down a little closer to the waiting fleet of recovery ships. Hurtling backward toward Earth, John Young was entranced by the vividness of what he was seeing. "The first thing you notice is at about six and one-half minutes after retro fire a slight orange haze that envelops the spacecraft. And this haze layer increases and changes colors to a dark green. It's a very beautiful thing. And then orange sparks of ablative material start flying forward." He also reported at the postflight news conference that their reentry through the atmosphere was so fiery that "at one point over Florida I could see the path we left as we came down; . . . it was long, funnel shaped." It would later be discovered that the seventeen hundred degrees of heat encountered in this phase had actually melted away one of the layers of the protective heat shield.

There was a bad surprise in store for the two crew members when the main parachute deployed. It snapped open so violently that they were both thrown against their hatch windows. Grissom hit his helmet's faceplate so hard on a windshield mounting bracket that it was punctured, while Young's faceplate was scratched. Just as they were recovering from that, the spacecraft plunged into the Atlantic. *Molly Brown* splashed down four hours and forty-five minutes after launch, having completed three Earth

John Young (*left*) and Gus Grissom on the recovery helicopter transporting them to the carrier USS *Intrepid* after their successful *Gemini 3* mission. Courtesy NASA.

orbits. Grissom was naturally keeping a close watch on his window and wondering why he could see blue water. For a few moments memories of *Liberty Bell 7* came flooding back to him, and he later revealed that his first thought was "Oh my God, here we go again!" Then he realized he had not severed the parachute, which was acting like a large sea anchor and dragging them across the water. He reached out and triggered the quick-release mechanism, and *Molly Brown* suddenly swung upright, just as planned.

As they waited for the recovery team from the USS *Intrepid* to pick them up, the inside of the craft, which was pitching and tossing in a heavy, five-foot swell, quickly became extremely hot and stuffy. Both men were becoming increasingly queasy with the rolling motion. After enduring this for a short time they elected to remove their spacesuits—not an easy task in the confines of the spacecraft. Then, to his great consternation, Grissom became physically ill. Somehow Young was able to keep himself from following suit. "Gemini may be a good spacecraft," Young would later grouse, "but she's a lousy ship!"

Helicopters from *Intrepid* had arrived on the scene only three minutes after splashdown, but the carrier itself was some fifty miles away. Grissom and Young elected to remain in *Molly Brown* until the *Intrepid* was much closer. Meanwhile, a pararescue team had been dropped nearby in order to secure a flotation collar around the spacecraft. An hour after splashdown Grissom finally opened his hatch and scrambled out into a waiting life raft. He would say that he was the first to depart "because my hatch was the only one fully out of the water and could be opened without danger of flooding the cabin." John Young followed, and would later wryly observe that it was the first time he'd ever seen a captain leave his ship first. The two men were then hauled up into a recovery helicopter and transported to the *Intrepid*.

Ninety minutes after Grissom and Young arrived safely aboard *Intrepid*, another helicopter gently lowered *Molly Brown* onto the deck of the carrier. The astronauts quickly found that they had hit the water some fifty-eight miles from the planned area, but this was later attributed to the spacecraft's failure to develop quite as much lift as expected while Grissom was flying it back to Earth. It also seems Grissom's injured wrist had not caused him any problems, as both men quickly passed the postflight medical checkup.

Despite their post-splashdown nausea, the two astronauts were ecstatic over their flight. Grissom would later say, "I do know that if NASA had asked John and me to take *Molly Brown* back into space the day after splashdown, we would have done it with pleasure. She flew like a queen, did our unsinkable *Molly Brown*, and we were absolutely sure that her sister craft would perform as well."

Even the normally taciturn Young found it hard to describe his first experience of spaceflight during the postflight news conference. "It was incredible; it was unbelievable," he enthused. "There are no words in the English language to describe the beauty of it. Man—I was impressed!" Reporters commented that the formerly gruff astronauts were now ad-libbing like a pair of comedians. "I think zero-g flight could make an extrovert out of anybody," Young conceded.

The flight of *Gemini 3* also marked something of a milestone in America's space program, as the last flight controlled from the Cape. Virtually from the time *Molly Brown* left the pad, new mission facilities at Houston's Manned Spacecraft Center had assumed control of the flight, as they would for those that followed.

For Young, one unfortunate legacy of his first spaceflight has been enduring questions about *that* corned beef sandwich. He has always regarded the incident as an innocent prank. Decades after the flight he still shakes his head and frowns in disbelief whenever the subject is raised by reporters in lieu of a legitimate, technical question. He was certainly hauled over the coals at the time, he admits, but it did not affect or hurt his astronaut career in any way. It was not considered a harmless prank at the time, however. As Paul Haney recalls, "Jim Webb was furious. He held up a promotion Grissom had coming, and minimized the crew's post-flight Washington celebration to a quick visit to Vice President Hubert Humphrey's office in the Capitol. Not so much as a drive-by of the White House."

Grissom also typically shrugged off the incident. "After the flight our superiors at NASA let us know in no uncertain terms that any non-man-rated corned beef sandwiches were out for future space missions. But John's deadpan offer of this strictly non-authorized goodie remains one of the highlights of the flight for me." In fact, after the flight, John Young cheekily enshrined the remains of the notorious sandwich in a small plastic container, which he kept on his desk, alongside the official letter of reprimand.

On a far more positive note, the flight of *Molly Brown* gave Project Gemini the start it needed: a near-perfect test flight that admirably demonstrated the capabilities of the new generation of American spacecraft. Both astronauts were feted for their accomplishments on the short, low-orbit flight, and Grissom accepted this praise with good grace. "I think I can understand the outpouring of good will John and I received for our relatively easy flight," he said after their mission. "After all the Russian spectaculars, the United States was back in the manned space flight business with probably the most sophisticated spacecraft in the world."

Following their successful *Gemini 3* mission, Grissom began looking ahead to yet another generation of American spacecraft—Apollo. Young, meantime, was not finished with Project Gemini. He would soon be assigned to a far more complex mission on *Gemini 10*, this time with Group 3 astronaut Michael Collins. For many years, in fact decades, John Young would actually be NASA's most experienced astronaut, with a six-mission record that was not broken until 2002 by shuttle astronaut Jerry Ross. John

Young once summed up his lengthy astronaut career very succinctly: "I can't think of a single job I'd rather have—in this world or out of it."

There were moments of levity at the preflight press conference for the crew of *Gemini 4*, Jim McDivitt and Ed White. Amid a flurry of questions, one reporter asked McDivitt: "Have you selected a name for the spacecraft?" and he responded with a wide smile, "I don't know—what's playing on Broadway these days?" In fact, if the crew had been given permission to do so, they would have demonstrated their shared patriotism by naming the *Gemini 4* spacecraft *American Eagle*. There was even some back-room chatter that the lesser-known, alternate name for *Gemini 4* had once been *Little Eva*. But when Gus Grissom impudently chose to call his vehicle *Molly Brown* it had caused such a furor within NASA management circles that the naming concession was immediately revoked.

Humorous wordplays aside, the flight was once planned to feature a "little" EVA, or limited exposure of an astronaut to space. Ed White, one of two space rookies on the crew, had begun training to open his hatch in space, stand up on his seat, and poke his head and shoulders out of the spacecraft. When the Soviet Union executed the first EVA, these plans changed dramatically.

The crew commander, James Alton McDivitt, was born in Chicago, Illinois, on 10 June 1929. His family later lived for five years in Kalamazoo, Michigan, during which time he attended Central High School. Rosemary Sheldon shared a year in biology class with McDivitt at high school. She remembers him as "easygoing, quiet, very pleasant and always smiling about something, . . . neither outstanding for his doings one way or another. To me, he was just plain Jim, a very ordinary fellow, dissecting his frog just like the rest of us."

After high school McDivitt had no definite ideas about a future career, and for a time he took on temporary work as a water boiler repairman. Then in 1948 his family moved once again, to nearby Jackson, where McDivitt entered Jackson Junior College. He graduated in June of 1950, the same month that the Korean War erupted. He was twenty years old, and rather than be drafted he decided he wanted to serve his country as a fighter pilot. "I'd never been in an airplane," he told interviewer Doug

Ward in 1999. "Never been off the ground. I'd already joined the Air Force, was in the Air Force, and was accepted for pilot training before I had my first ride. So, fortunately I liked it!"

McDivitt entered the air force as an aviation cadet in January 1951, gaining his pilots' wings and a commission as a second lieutenant the following May at Williams AFB in Arizona. He completed combat crew training at Luke AFB six months later and was shipped over to Korea, where he took to the skies in Lockheed F-80s and swept-wing North American F-86 Sabres. Flying for the 35th Bombardment Squadron, McDivitt carried out the required one hundred combat missions and stayed on for forty-five more, because "there was a shortage of experienced pilots," he later said, and because "I thought I could do a good job." In fact, he did such an outstanding job that he was decorated with the Air Force's Distinguished Service Medal with oak leaf cluster, the Distinguished Flying Cross with three oak leaf clusters, and the Air Medal with four oak leaf clusters. "I flew my last [combat mission] two hours after the armistice was signed—with a lot of trepidation," he told Ward. "Nobody wants to get shot down after the war is over!"

Following his return to the United States, where he decided to stay in the air force, McDivitt met his first wife, Patricia Ann Haas, and they were married in June 1956. The following year he was promoted to the rank of captain, and the air force recommended his enrollment at the University of Michigan under the Air Force Institute of Technology program, for further studies in aeronautical engineering. In 1959 he was awarded a bachelor of science degree after ranking at the top of his class of 607 students with a straight-A record. At university he had also become great friends with one of his classmates—a fellow pilot named Ed White.

Edward Higgins White II was born into a family of career military officers in San Antonio, Texas, on 14 November 1930. At the time, his father, who had earlier been one of the U.S. Army's pioneering balloonists and aviators, was stationed at nearby Kelly AFB. Although he was only eight months old when his family moved on, to an air force base in Hawaii—one in a long string of such assignments for his father—White always considered San Antonio his hometown, and was proud of his Texas heritage. Following in his father's footsteps, he received an appointment to the military academy at West Point and was subsequently commissioned in the air force after his graduation in 1952.

On completing flight school he married Patricia Finegan, a girl he had met while at West Point, giving both he and McDivitt something else in common—wives named Pat. He then served as a tactical fighter pilot in Germany until 1957, when he was also selected by the air force to attend the University of Michigan and study for his master's degree in aeronautical engineering. Like McDivitt, White achieved the rank of captain while at university, graduating in 1959. While at Michigan McDivitt and White were not only good friends; they also lived just down the street from each other, and would often discuss what each man might do in his future career. At one time, McDivitt had thought about earning a master's degree through the air force, but his application was turned down. He was also offered an assignment as a Project Office engineer at Wright-Patterson AFB, but it was a nonflying job. He said that was "about the last thing I wanted to do."

Ed White, on the other hand, knew exactly where he was heading. "He was going on to the Test Pilot School," McDivitt recalled. "He'd applied for it and been accepted. I thought, 'Heck, I'll do that!' . . . and so I put in my application. I heard back in about two weeks that I'd been accepted, which was a whole different route for my career." McDivitt and White then attended the USAF Experimental Flight Test School (EFTS) at Edwards AFB as part of Class 59-C, graduating in 1960. Not only did McDivitt receive the prestigious A. B. Honts Award as the outstanding graduate of his class, but he was also honored with other awards for best all-round flying performance and best academic student—making him the first EFTS student to win all three awards. He then elected to stay on at Edwards as a test pilot in Flight Test Operations. "And it was a great job," he said. "Probably the best job I've ever had." Ed White, meanwhile, was sent to Wright-Patterson AFB in Ohio, where he flew as an experimental test pilot with the Aeronautical Systems Division.

"I wanted to fly airplanes and continue to do what I was doing," McDivitt said of that exciting time. It was therefore with a measure of reluctance that he attended the USAF Aerospace Research Pilot School, which was preparing pilots for space programs such as Dyna-Soar. One of his instructors at the school was Frank Borman, also destined to become a NASA astronaut and who, ironically, would serve as his backup pilot for *Gemini 4*.

After graduating from the aerospace school in December 1961, things happened in a rush. Late in 1962, X-15 pilot Robert (Bob) White said he would be leaving the program for a posting in Germany. McDivitt was asked if he would like to step in as a replacement and fly the mighty research airplane. "Of course, I jumped at that!" he recalled. He was also offered the role of project pilot on the F-4 program, "a cushy, wonderful job," and then one of his bosses asked if he would like to be assigned to the Dyna-Soar program. After flying to the Boeing plant and discussing the project with those developing the winged spacecraft, McDivitt concluded that "the Dyna-Soar Program would never fly. It was a lot more screwed up than anything else I'd ever seen. The spacecraft was wrong. They were trying to make it a military weapon system. It wasn't. It was too heavy. There wasn't a rocket that could lift it. So I went back and told my boss I really wasn't interested in the Dyna-Soar Program." With that, he was assigned to both the X-15 and F-4 programs. Soon after, his commanding officer happened to mention that NASA was looking for a second group of astronauts, but with all that was going on McDivitt was not interested. "I want to fly the X-15," he recalls saying that day. "I want to stay here and fly airplanes."

Some time later he began to have a change of heart, and discussed his future with the retiring X-15 pilot, Bob White. White told McDivitt what he'd already suspected, that he should use his expertise in the way that would serve his country best, and if that meant being an astronaut rather than an X-15 pilot, then so be it. "That sounds kind of dorky today," McDivitt said of his youthful patriotism, "but that's the way I felt." There was only one small problem—astronaut applications had closed a few weeks earlier. However, despite being only a lieutenant colonel, Bob White was a man of considerable influence. He soon persuaded some people within the air force hierarchy to submit a late application nominating McDivitt.

Early the next day, a livid commanding officer stormed into McDivitt's office, "threw my application on the floor, and called me a traitor." He demanded that the application to join NASA be immediately withdrawn. When McDivitt refused, he yanked him from the X-15 program, and venomously added, "Your career is *over*!" It was a very difficult time for the young air force officer, but then to McDivitt's relief the cards fell neatly into place once again. "Fortunately," he reflected, "I got selected!" Not only had he become a member of the second astronaut group, but his best friend, Ed White, had also been successful. As McDivitt told Ward, he

The newly selected crew of *Gemini 4*: space "rookies" Ed White (*left*) and Jim McDivitt. Courtesy NASA.

had received a welcome surprise while attending an initial NASA selection interview in Washington DC. "I walked in the room in the Pentagon and Ed was already there . . . and he says, 'I knew you'd be here!' And I said, 'I knew you'd be here, too!' "

One of their former Edwards instructors, Tom Stafford, was also selected in the group of nine new astronauts, and he would later say of the two friends, "I knew they were outstanding choices." McDivitt's selection as an astronaut in the civilian space agency actually led to a small but unexpected problem for a person who'd basically worn nothing but a variety of military uniforms for many years. "All of a sudden," he said, "ninety-five percent of your wardrobe is extinct!" The only civilian clothing he possessed was "one or two obnoxious-looking sport coats and maybe two pairs of slacks." He and Pat had to hurriedly add a little clothes shopping to their list of things to be done.

As an astronaut, McDivitt quickly displayed his exceptional leadership potential. Conservative but incisive, deeply religious with a fine and thorough military orderliness, he began to impress the right people. "Jim didn't talk much," according to Gene Cernan. "He let his work speak for him." One reporter described McDivitt as "a deceptively easy-going sort who combined the rare ability of getting speedily ahead with the even rarer trait of winning friends on the way."

When the two friends stood on a Houston stage in September 1962 to be introduced as members of NASA's second intake of astronauts, it was White who said in his brief speech that "Jim and I have been following right along together." He then added with a smile, "It seems that every time we got together, we were taking examinations of some kind."

Once their group had settled in at NASA, the two friends were given engineering assignments under the mentorship and guidance of the Mercury astronauts. McDivitt's work related to guidance and navigation, while White concentrated on flight controls. During this period, and before moving to the new Manned Spacecraft Center (MSC), they also shared an office in downtown Houston. "We had a very close career all through that time," McDivitt recalled, then added, "He was the best friend I ever had."

Ed White was a superbly fit athlete. In fact, he had narrowly missed being selected for the 1952 U.S. Olympic track team, failing to qualify in the 400-meter hurdles by just four-tenths of a second. As West Point's *Assembly* magazine once noted of the former cadet, "He had inherited the qualities of leadership, devotion and perseverance from his parents . . . added his own inborn attributes and then forcefully began to make his way through life. His devotion to Duty, his Country and to his classmates provided Ed, from the very beginning, with a natural trend towards greatness." Indeed, the 1952 West Point yearbook, *Howitzer*, had said of White that he was "craving excitement and adventure and seldom passing up the chance to do something out of the ordinary." He soon proved the editors right.

At NASA, White continued to impress with his physical fitness, keeping in shape by running long distances, supplemented by weightlifting, squash, handball, and water-skiing. Dr. Charles Berry, the astronauts' physician, stated that "He could run the mile, fight a ten-round boxing match, swim twenty miles and do anything that would require physical endurance." White also became extremely popular with the media, who appreciated his exuberant, talkative attitude.

McDivitt and White were publicly named to the flight of *Gemini 4* on 27 July 1964, with Frank Borman and Jim Lovell as their backups. Compatibility was important on Gemini, especially on the planned long-duration missions, and Deke Slayton took this into account when selecting his crews. He already knew McDivitt would prove an outstanding spacecraft pilot, while the two men were buddies from way back.

At first, the mission objectives were hardly those to attract much media attention. Under the command of McDivitt, the crew would conduct a four-day flight around the Earth sometime in 1965, during which they would photograph the Earth's weather and terrain, measure radiation, and study electrical discharges. Or, as McDivitt characterized it, "Medical experiments, tests and other assorted junk."

Then, things changed. A NASA document circulated on 12 March 1965 suggested that one of the *Gemini 4* crew members might open the hatch of their spacecraft, and stand with his head and upper torso exposed to raw space. Deputy Manager of Project Gemini Kenneth Kleinknecht hinted to reporters that NASA was studying the exciting possibility. Initial plans had McDivitt penciled in for this task, and for a time both men involved themselves in early EVA training. Then McDivitt concluded that being commander was a full-time job in itself, and after lengthy consultations with the decision-makers Ed White was officially given the job.

Whenever a reporter asked about those early plans for a limited EVA, NASA officials would stress that they hadn't made any firm decision. It was something of a smokescreen, perhaps to appear noncommittal in light of the Soviet Union's track record for beating NASA's announced goals. But while NASA procrastinated, the two astronauts were in serious but clandestine EVA training. Events were also moving fast on the other side of the world, and within the Cosmonaut Training Center Soviet spacefarers were deep in training for yet another space spectacular.

Meanwhile, McDivitt and White watched and worked with McDonnell's engineers as their spacecraft was being built at the plant in St. Louis. They also practiced for hundreds of hours in the simulator replica at Houston, until they were largely in harmony with its environment and systems. One of the first real problems McDivitt and White encountered was a distinct lack of headroom in their spacecraft. Designed around Gus Grissom's smaller frame, the taller astronauts quickly discovered that the tops of their helmets would be jammed against the hatch for the duration of their flight. Accordingly, McDonnell's engineers decreased the amount of padding on the inside of the hatches, but for the almost six-foot-tall McDivitt and White it still proved a tight squeeze.

A launch target date of 3 June was announced for *Gemini 4*. Then, in the middle of March, Alexei Leonov squirmed out of his air lock to conduct history's first spacewalk, and once again a frustrated NASA had been beaten to the punch. Something had to be done—and quickly—to prove that American space technology was equivalent to, if not better than, that of the Soviet Union. An American EVA would provide the obvious response. According to Frank Borman, "The general feeling was that we weren't quite ready for EVA at this stage of the program, but the Soviet walk in space changed that thinking in a hurry." Soon after, acting under new mission guidelines, the two astronauts went into training for a full EVA outside of their spacecraft.

Other objectives evolved. Now there were plans for McDivitt to perform some formation flying with the second stage of their Titan booster after reaching orbit and for White to somehow make his way over to it on his spacewalk. It was a highly ambitious and audacious plan. Their training intensified, and the two men secretly practiced EVA techniques until late in the evening with the special equipment they would use on the flight. A propulsion device was badly needed if White was to move around outside the spacecraft, and one was quickly devised. It was a small, gas-propelled unit, which was tested with the aid of an air-bearing table at MSC; but it was extremely difficult to replicate the way in which White would move around in weightlessness.

For the flight, McDivitt would be wearing the new, standard Gemini suit, weighing twenty-six pounds, but White's had extra measures built in. His suit, five pounds heavier than McDivitt's, featured eighteen layers of material—felt, Dacron, and aluminized Mylar plastic—to protect him against possible micrometeorite impacts and the extreme temperatures he would face outside the spacecraft. He would also have a small chest-mounted strap unit containing an emergency oxygen supply. As well as an inner visor to maintain the suit's internal pressure, White's helmet also featured a gold-coated middle visor of Lexan plastic, which would protect his eyes by allowing just 10 percent of the sunlight to reach his face, and finally a blue-green outer visor made of Merlon, thirty times tougher than Plexiglas. Additionally, both spacesuits featured something new to U.S.-manned spaceflight: small American flags sewn onto the left shoulders.

On 24 May an important milestone for the flight was achieved when qualification tests for all of the EVA equipment were successfully completed. By then White had spent sixty hours rehearsing his EVA in an altitude chamber set at 180,000 feet, even though the spacewalk would take place at nearly three times that altitude. Included in this training were 110 rehearsals that involved climbing out of the hatch and then returning to the cabin.

On 3 June 1965 the two astronauts were awakened at 4:10 a.m., fifty minutes before the final countdown began. They had a brief physical examination and then ate a hearty breakfast of sirloin steak and eggs. With the flight of *Gemini 4*, Jim McDivitt would also become the first Roman Catholic to fly into space, so before he and White left the crew quarters he took part in a small communion ceremony with two priests who had joined the crew for breakfast.

Unlike the earlier Mercury flights, the two astronauts were dressed in casual clothing as they were driven out to Launch Pad 16 in the transfer van. Here they would spend the next hour and a half having sensors attached to their bodies and donning their spacesuits. When ready, they would ride up the elevator and enter their spacecraft a little under two hours prior to the scheduled launch time.

Thirty-four minutes before launch there was a hitch in the countdown when a faulty electric motor had to be replaced in the erector. This was the cranelike gantry that cradled the ninety-foot Titan II rocket, and was normally lowered during the last phases of the launch. Before the problem was fixed, McDivitt was a little concerned that the erector might go haywire and hit their Titan rocket, which would probably have resulted in the crew being forced to eject. But when the repairs seemed to be going well he settled back, and both men even fell asleep. A while later they were awakened and asked if they'd like to speak to their wives while work continued on the faulty erector, which proved a welcome diversion before their launch. Once the motor had been installed and tested the countdown was resumed, and at T-30 minutes all the technicians were ordered to clear the pad area.

At 10:16 a.m., an hour and sixteen minutes late, the Titan's engines roared into life, and the rocket trembled as it waited to be unleashed. Then the hold-down clamps were released, and *Gemini 4* was on its way. Soon after, McDivitt reported, "Roll complete."

"*Roger*," was the response.

Just five minutes and thirty-nine seconds after launch, *Gemini 4* was in orbit. It now became the first manned mission to be controlled by the new Mission Control Center in Houston. "It looks great up here," McDivitt reported. After separating from the second stage of the booster rocket, which had burned all its fuel, he fired two of the spacecraft's aft thrusters to add thirty feet per second to *Gemini 4*'s velocity and pull them away from the tumbling second stage.

McDivitt turned their spacecraft around so the blunt end was facing forward and reported that *Gemini 4* was established in orbit. He also said that the booster's 27-foot second stage was approximately 500 feet away, but in a slightly lower orbit. Soon after, as they flew into the darkness of the night side of the planet, the crew began to prepare for the first major task of their mission. McDivitt would attempt to fly formation with the drifting upper stage and maneuver in as close as he could safely manage. Plans then called for White to commence his EVA and float over to the booster with the aid of the propulsion gun. Once there, he would remove a piece of material that had been placed on the outside to study the effects of launch heating. If they could accomplish this, the bold plan would certainly result in some truly breathtaking photographs.

McDivitt would later state that a number of unexpected difficulties then arose. In the hurried mission planning, no one had taken into account the fact that the booster's tanks would normally go into a vent condition once it had achieved orbit. This meant that the tanks were still expelling residual propellants, resulting in the same effect as a small rocket engine. In the vacuum of space, this caused the weightless booster to tumble, slowly but continually rotating around its center of mass. Maneuvering close to it was now a high-risk proposition.

McDivitt not only had this to contend with, but as *Gemini 4* swept around in the darkness of Earth's shadow another problem also became apparent. To facilitate the rendezvous, the Martin Company had installed a strobe light on either side of the Titan II booster. The bright flashing lights on the booster were dazzling, and McDivitt also quickly discovered it was almost impossible to align *Gemini 4* in such a way that he could see both lights simultaneously. Most of the time, due to the slow rotation of the booster, only one of the lights could be seen, meaning he could get no depth perception at all on the target.

As they emerged into sunlight once again, things only got worse. Being in a slightly lower orbit, the booster stage was traveling faster relative to the Earth than the manned spacecraft, and McDivitt experienced great difficulty catching up. Because no simulator had been available for ground training in this difficult maneuver, much of what he attempted was purely instinctive. He also did not have the luxury of an onboard radar or computer, and had to make all his calculations by peering out through a small triangular window, with nothing more than his own eyes and judgment to guide him. "I knew what would happen in theory," he said later. "But we had no practical training." Eventually, after half an hour and despite his best efforts, the gap between them had widened to half a mile.

As they flew over Mexico, McDivitt reported that "it is taking more fuel than we estimated. Shall we make a major effort to close with it, or save the fuel?" Gus Grissom was on CapCom duty at the time, and he looked over at Flight Director Chris Kraft. "Give it up," Kraft sighed. "Tell him to save the fuel." In fact, McDivitt had already expended nearly 50 percent of his fuel supply in the futile chase, and the remainder would be needed for the spacewalk and reentry. Grissom passed the message to McDivitt, who concurred.

As Grissom would later point out, "the effort did point out that the orbital mechanics involved in rendezvous were a little more involved than had been expected."

Their disappointment did not last long. Just seconds later Grissom relayed a welcome message from Chris Kraft to the crew: "Okay, we're giving you a GO for your EVA at this time." Though the EVA was originally planned for the second orbit, it took McDivitt and White so long to go through the forty-item checklist that Mission Control wisely decided to delay until the third orbit. McDivitt also wanted White to relax and cool down a little, as he had noticed that his copilot looked tired and hot. "It sounds like you've been awfully busy," Gus Grissom commented to the crew as *Gemini 4* slipped over Florida at the end of its second orbit.

"Yeah, I finally got a chance to look out the window," McDivitt replied. "It's really nice."

The revised EVA schedule meant that the egress would take place over Bermuda in full daylight. Chris Kraft harbored concerns about trying to get White back inside the spacecraft before they flew into darkness on the night side, so he wanted the spacewalk kept short, as planned. In particular, Kraft did not want the crew "fooling around in the dark." As he continued to monitor the flight, the two astronauts checked their pressure suits, sealed their helmets, and began purging the air from *Gemini 4*'s cabin as they neared Carnarvon in Western Australia.

Robbed of the chance to do an EVA that involved the booster, White now began to prepare for a far less hazardous spacewalk, in which he would simply exit the spacecraft and float around for a few minutes, maneuvering with the aid of short bursts of compressed nitrogen from the tiny space gun. This seven-pound device, slightly resembling handlebars on a bicycle, was known officially as the Hand Held Maneuvering Unit, or HHMU, but it had quickly become known to everyone as the "Zot" gun. Its chief drawback was that it carried only a miniscule amount of fuel, but McDivitt later remarked that it would not have made much difference anyway. "It was a hopeless device. There was no way you could really control yourself in six degrees of freedom with just that. The gun was useless."

When they were both ready, White reached overhead and tried to unlock his hatch, but there was a mechanical malfunction. A spring inside the hatch had failed to compress. The two began prodding and poking around the hatch ratchet, and to their relief the hatch suddenly cracked open. Even though it proved surprisingly difficult for White to push outward, even in zero-g, the hatch was soon fully open.

As White peered out through the open hatch, both men would have felt a lingering concern. If the hatch had proved tricky to open it might be just as difficult to close, and if they reported the problem to Mission Control, the EVA they trained so long and hard for might be canceled. Either way, the hatch would still have to be closed sometime, so they figured they might as well go ahead with the EVA and not tell the ground. "I mean, there was nothing they could do," McDivitt later said of their predicament. He then coyly admitted, "and they probably would have said no."

One hundred and thirty-five miles above the Earth, White carefully stood up on his couch, mounted a camera to the exterior of the spacecraft, and snapped off a few photos. McDivitt reported that everything looked good, so Kraft gave White permission to proceed with his EVA. Shortly afterward, White pulled himself outside into raw space, followed by a glove spinning loose from the cabin. He was attached to *Gemini 4* by a 25-foot-long, one-inch-thick lifeline wrapped in plastic tape coated with gold. Known as White's "umbilical cord," it carried an oxygen supply and communications wires. He then used his propulsion gun to move to the front of the spacecraft so McDivitt could take some photographs. White quickly attached a 16-millimeter movie camera to a brace at the rear of the spacecraft to record some of his actions, and used a still camera attached to his HHMU to photograph the view below. At the postflight press conference, White would recall his impressions as the first American astronaut to walk in space: "There was absolutely no sense of falling. There was very little sensation of speed, other than the same type of sensation that we had in the capsule, and I would say it would be very similar to flying over the Earth from about 20,000 feet. I think as I stepped out, I thought probably the biggest thing was a feeling of accomplishment as one of the goals of the Gemini 4 mission. I think that was probably in my mind."

All too soon, the Zot gun ran out of fuel, and White had to maneuver himself by manipulating and tugging on his tether. He made his way to the nose of the spacecraft once again, where he thought he might be able to hold onto something, while McDivitt performed a small attitude correction using the OAMS thrusters. In fact, whenever White impacted with the spacecraft it moved, just as the Soviets had found on their EVA mission. However, McDivitt was having very little difficulty controlling the Gemini craft and keeping it stable. The only thing White could see to hang onto as he moved up the spacecraft's exterior was a communication antenna, but he decided against it. After the flight, he said, "I took one look at the stub antenna, which was our connection with radio back to Earth, and I felt that this wasn't any place to play around with, and so I didn't do any work around the nose of the spacecraft."

During his spacewalk three months earlier, Alexei Leonov was said to have reported feelings of disorientation, so doctors on the ground were anxiously waiting for White to comment on his condition. This came as *Gemini 4* was approaching the coast of California. "There is absolutely no disorientation associated with this thing," White happily reported, as he pushed himself off the spacecraft and practiced rotating around, using the tether. He then hauled himself back to *Gemini 4* and by tugging on the lifeline even managed to walk on the corrugated spacecraft. "There's no difficulty in recontacting the spacecraft," he added. "I'm very thankful in having the experience to be first. Right now I'm actually walking across the top of the spacecraft. I'm on top of the window." Meanwhile, McDivitt had been snapping away with his camera, busily recording the historic event. "Hey," he jokingly protested, as White brushed against his window. "You just smeared up my windscreen, you dirty dog!"

The banter continued as *Gemini 4* flew over Texas, with White describing some of the spectacular sights below, but also reporting on the effectiveness of his spacesuit, helmet, and other equipment. But while they were conversing between themselves they were effectively unable to receive voice communications from the ground. In order to talk to Houston and receive messages, McDivitt first had to flip a switch. Flight Director Chris Kraft would later say that this was a monumental error, "a flaw in our thinking that I made certain never happened again." At last, McDivitt told White he should see if Houston had anything for them, and flipped the communication switch. Immediately, Gus Grissom said: "The flight director wants for White to get back in." Unfortunately, they still did not hear this transmission, as the astronauts were using an overriding voice-activated internal communication system that made it difficult to hear ground-based transmissions. So as White continued his twists and tumbles, Grissom kept repeating, "Gemini 4, Gemini 4 . . . "

Eventually, McDivitt heard the transmission and responded. "Hey Gus, this is Jim. Got any messages for us?" For the first time on the flight, and to Grissom's surprise, the flight director, concerned with White's protracted stroll, unexpectedly assumed control of communications with the spacecraft by engaging an override switch. In doing so he had violated one of his own strict rules on ground-to-spacecraft communications, but he wanted

White in, and said so. "Yes!" he barked in response to McDivitt's inquiry. "Tell him to get back in!"

The message was immediately relayed to White, whose breezy comment was "This is fun." By now, he had been floating freely in space for nearly twenty minutes, and had conducted his spacewalk from Hawaii clear across to Florida. McDivitt now began coaxing his colleague in with increasing urgency as the darkness of Earth's night side loomed. "Come on, Ed," he finally called, his voice just a little sharper. "Let's get back in here before it gets dark."

"It's the saddest moment of my life," White replied with obvious reluctance, as he made his way back into the spacecraft. Then it was time to try closing the hatch, and both men prayed that they could accomplish this. Despite the difficulty they had encountered in getting it unlatched, McDivitt felt confident they could secure the hatch once White was back inside. Since there was no way they could reenter Earth's atmosphere with the hatch not fully closed and latched, McDivitt was also pressing White to get back in while there was still some daylight to work by. Unfortunately, it was already dark again by the time White and all of his equipment were safely inside the cabin. They closed the hatch, and then their worst fears were realized. It wouldn't lock. If they could not find some way to rectify this, *Gemini 4* and its occupants could well be incinerated during reentry.

Fortunately, McDivitt had taken careful note of something a spacecraft technician had demonstrated to him after an EVA simulation in the altitude chamber in St. Louis. On that occasion, they had also been unable to lock the hatch down again, so after they had showered and changed McDivitt wandered back to see what the problem was. As he watched with interest, the technician placed his finger into a slot in the locking mechanism, located a small cog that hadn't engaged properly, and nudged it back into place. After that, the hatch locked down without any further problem.

It was not quite as easy for McDivitt, as he was working in the dark. He also had to be very careful not to puncture the tips of his gloves as he slipped his finger into the slot and fiddled around trying to locate the small cog. Finally, he felt what he was after, and pressed it into place. The hatch locked down securely, and both men breathed a deep sigh of relief. As Frank Borman later observed in his book *Countdown*, "Making it even tougher were the awkward, rather primitive spacesuits White and McDivitt were wearing—performing any manual task with those bulky gloves was like trying to open a can with handcuffs on your wrists." Because of their problems with the hatch the crew decided to cancel a scheduled second hatch opening, which would have allowed them to jettison some bulky equipment that had only been needed for the spacewalk and would clutter up the cabin. Despite the overcrowded conditions, they would keep all their gear in the cabin until they landed.

Once they had secured and repressurized the spacecraft, the two astronauts were finally able to relax. Both were overheated and physically exhausted after their efforts, particularly White, who took a couple of hours to fully recover in his much heavier spacesuit. Beads of sweat covered their faces, and the inside of their helmets had fogged up. However, once McDivitt had powered down the spacecraft things finally settled down, and both men eventually fell asleep.

White's spacewalk, for all its problems, had happily resolved the fearful issue of whether an astronaut would become disoriented on EVA. The crew had proved that astronauts could move freely outside of their spacecraft, which allowed future missions to include many external functions and work in space—a vital step on the way to the moon.

There was, however, a hint of the problems that would soon affect other American spacewalks. At their postflight press conference, McDivitt stated that he had noticed White was breathing quite heavily with exertion after the relatively simple task of mounting the external movie camera. White responded that he hadn't realized he was tiring himself so much, "but I did notice that I was putting out energy for the first time during the EVA exercise." The superfit White had made his walk in space look so apparently easy and effortless that it induced some unfortunate complacency among planners, who set even harder and more complex tasks for subsequent EVA missions.

Relaxed and refreshed after several hours of drifting flight, the astronauts spent their second day in space coping with "garbage all over the place." They had a small well under their feet for stowing used equipment, but with White's spacewalk gear still aboard they had precious little room in the cabin. In fact, White said they were so short of space that he was carrying the bulky umbilical cord nestled in his lap. Later that day the crew talked with their excited families at the MSC in Houston, and were congratulated on the spacewalk by their wives. White responded, "Quite a time we had. Quite a time." Physicians had expressed a little concern about the crew's lack of water intake, so Pat McDivitt and Pat White told them in wifely, no-nonsense terms to drink more water and get enough rest.

One of the flight's chief objectives had been to test the effects of prolonged weightlessness and confinement under conditions of possible stress on the mind and body. So much remained just theory. Some doctors had even speculated that being in zero-g for a hundred hours might prove too much for the human body, and the astronauts might even die when they landed. In reality, however, McDivitt and White felt good, and the drifting tumbling of their spacecraft did not cause them any concerns or illness. They knew that even their failure to make a rendezvous with the booster rocket would eventually provide useful lessons and information for the future.

On their final day in space, McDivitt and White took star and horizon readings, took photographs for meteorologists of a bank of eddy clouds over the Canary and Madeira Islands, and performed further physical exercises for the flight surgeons by using a strong bungee cord they could attach to their feet. Then, as the crew began receiving its landing instructions, McDivitt told controllers that they were trying to stow as much gear as they could for the reentry. Ripples of laughter erupted at MSC when he commented that the cabin was "so full of things I can barely see out the window."

Despite their cramped conditions the crew was in good spirits. When they were asked, "You two ready to come home yet?" McDivitt joked of his traveling companion, "I'm sure tired of looking at this ugly face. It needs a shave." To a light-hearted suggestion from MCS that "we are thinking of extending the mission about a week," White—a man with a famously huge appetite—chimed in, "Well, if you do, you'd better send up some more food."

In preparation for landing, McDivitt began firing control rockets over Hawaii as *Gemini 4* approached the end of its sixty-second orbit, and jettisoned the adapter module. An IBM computer had been installed in *Gemini 4* to help McDivitt's landing maneuvers, but when he had earlier tried to shut it down to save battery power it had refused to comply. Eventually, it was decided to disconnect and reboot the computer, but when it was switched on again it no longer functioned. The crew would now have to rely on the system used by the Mercury astronauts, which essentially allowed gravity to do the work once they had begun reentry. The maneuvering rockets had eased *Gemini 4* down to a new orbit with an apogee of around ninety-eight miles. Then, on the command "Start burn," as they passed over Guaymas, Mexico, McDivitt fired the four retrorockets. "Affirmative," he reported. "Am firing." *Gemini 4* then began its swift descent back to Earth.

McDivitt brought the spacecraft down to a splashdown in the Atlantic Ocean, southwest of Bermuda. Despite the failure of its landing system computer they were only fifty-six miles from the waiting carrier USS *Wasp*, and two minutes ahead of schedule.

As they waited for the recovery force to move in, McDivitt couldn't help but recall the grim preflight predictions of some alarmist doctors, who feared that the astronauts' bodies might not be able to cope with protracted spaceflight and might even succumb to the stresses of reentry. Despite their worries, he felt good. He looked over at White and asked, "How do you feel, Ed?" White grinned and said, "Just fine." With a chuckle in his voice, McDivitt then commented, "I guess we aren't going to die after all!" White's weak stomach eventually let him down, and he became seasick in the bobbing spacecraft before a helicopter managed to pluck the two men from the sea. They were rapidly transported across to the waiting carrier, where a red-carpet reception was waiting for them.

Though the bearded astronauts were pale and exhausted after their space marathon, they happily waved, gave the thumbs-up to photographers, and saluted with gusto. White even surprised everyone, particularly the doctors, by performing a little jig on the flight deck. Following this, the two men then endured five and a half hours of medical tests. During the tests they were able to speak with their wives by telephone and also took a call from an ecstatic President Johnson, who told them: "You have written your names in the history books and in our hearts. What you have done will never be forgotten." He then invited them down to his Texas ranch the following weekend and added enticingly, "I have been saving a little something for you." Once their tests were completed, the weary duo visited the captain's cabin, where they showered and shaved. Afterward they enjoyed a light meal with the recovery task group commander, Adm. William McCormack, before crawling into their beds and falling sound asleep. Both then slept for nine and a half hours. Meanwhile, initial medical reports said both crew members were in fine physical condition after more than four days of weightlessness.

The next day, White saw a few Marines and midshipmen enjoying a hearty round of tug of war. To everyone's amazement, he grabbed the rope and joined in. In Houston, George Mueller summed up the flight by saying: "We are all tremendously pleased with the results of the *Gemini 4* flight. It was one of the most successful missions in our manned spaceflight program." The astronauts' physician, Dr. Charles Berry, predicted that medical data from the record-breaking flight would "knock down an awful lot of straw men" concerning the human ability to live and work effectively in space for long periods. "For one thing, McDivitt and White exhibited no symptoms of lightheadedness when they came back to Earth," he said. "They are in better shape after four days than Gordon Cooper was after thirty-four hours in space two years ago. Possibly because they had exercise equipment aboard. One straw man that *was* knocked was the theory that we might have a couple of unconscious astronauts on our hands."

McDivitt and White never did make it down to LBJ's Texas ranch, but a few days after their homecoming they attended an official ceremony at the Manned Spacecraft Center, attended by President Johnson. The president's promised "little something" turned out to be presidential nominations for immediate promotions from major to lieutenant colonel, which he said was "for their spectacular achievements." He then handed over brand-new silver maple leaves to two very surprised astronauts. For good measure, and since the furor over the previous mission's corned-beef sandwich incident had died down, the president finally promoted Gus Grissom to lieutenant colonel, and added that Gordon Cooper would be elevated to the same rank on 15 July. In his congratulatory speech to the crew of *Gemini 4*, Johnson remarked: "What impresses me and gratifies me most about these two heroes—and all the other astronauts—is the quality of personal modesty and humility." He then added, "I haven't yet met a man who has not come down from space wanting to give more credit to the men and women on the ground than he will accept for himself up there."

For their part, McDivitt and White presented Lyndon Johnson with—as McDivitt called it—"a picture of Ed." It was one of the magnificent shots the commander had taken of White during his history-making spacewalk, which had just been released to the public.

Following *Gemini 4*, Ed White joined Michael Collins on the *Gemini 7* backup crew, which should have automatically placed them as the prime crew for *Gemini 10*. Instead, White told Collins that he had discussed his future options with Deke Slayton, the Mercury astronaut who was now Director of Flight Crew Operations. White had agreed to forgo *Gemini 10* and move straight into Apollo training. Slayton was also keen for McDivitt to move into Apollo; it was known that he wanted McDivitt to eventually command a lunar landing mission. In fact, McDivitt would go on to command *Apollo 9* and later become one of the youngest air force officers ever selected for promotion to general. Amiable without going overboard, he was always highly regarded by his peers in the astronaut office.

In his *The All-American Boys* Walt Cunningham offers this characterization: "No one was more deliberate or meticulous than Jim McDivitt; no detail was too small if it affected his mission. He eliminated problems by working them to death. When he presented his case he had the facts. . . . Most of Jim's contemporaries considered him the 'anointed one' of the nine Gemini astros. Looking back it does appear he was given every chance, break, prerogative, and option, beginning with command of the second Gemini mission."

It was now increasingly evident that the two successful Gemini missions had helped NASA bridge the gap with the Soviet Union in the race to the moon, and there was the tantalizing promise of more outstanding missions and accomplishments to come.

As pad leader Guenter Wendt would later write, "We were ready to get on with the real business of spaceflight: long-term missions, and orbital rendezvous and docking. Everything else had been done just to get to this point. Now, the real work began."

Gordon Cooper's masterful flight aboard *Faith 7* had brought the Mercury program to a successful conclusion. Once the postflight celebrations had died down he was keen to pick up a Gemini mission assignment, but his boss, Deke Slayton, was a little more cautious. Slayton would later say that, despite Cooper's outstanding performance on the final Mercury flight, he was still regarded as "a question mark" by management. For a while, therefore, he was unsure whether to assign him to another flight.

The Mercury astronaut's talent and skills eventually prevailed over his "hot shot" reputation with the NASA hierarchy. Cooper apparently remained unaware there had ever been a problem. Referring to the pre-Mercury flight tension between him and the head of flight operations, Walt Williams, Cooper told the authors that the two of them "became very good friends after my first flight, when he came to me and said, 'I just wanted to tell you I was wrong.' So I had no idea. I never got that feedback from them." Slayton finally penciled Cooper in for the long-duration *Gemini 5* mission in the middle of 1963, although the assignment was not officially announced until 8 January 1965. The acclaimed Mercury astronaut was teamed with the fun-loving but highly regarded space "rookie" Pete Conrad. They had formed a close working friendship during their jungle survival training early in Pete's astronaut career.

This mission would not only beat the endurance record then held by the Russians, but would hopefully also provide evidence that people could live and work in space for several days. In fact, the *Gemini 5* flight would last for the length of time it would take an Apollo crew to reach the moon and return. Conrad was naturally elated to have picked up an early flight, especially as he was paired with the experienced Cooper. "I feel a lot better about sitting up in the right-hand seat on my first ride with somebody that's been there before," he said of his mission commander. In his book, *Leap of Faith*, Cooper described Conrad as a "go-getter-type guy . . . the shortest of the astronauts, at five feet six inches, and also one of the funniest and most irrepressible with his lively sense of humor and gap-toothed grin." Outspoken, funny, and feisty, Conrad never worried about what others might think of him, instead giving his opinions in a free-spirited way that kept his colleagues in stitches. Indeed, some astronauts who worked with him later commented that it was hard to be close to him because it seemed everyone knew him and liked him. A cool, snazzy dresser, Conrad personified to many the hotshot astronaut of the era.

Conrad was not all wisecracks, however. It was also very evident to his colleagues that he was one of the best, brightest aviators and engineers they had ever met. His combination of wit and sharp ideas transmitted enthusiasm to everyone around him. His colleagues would work hard to achieve his ideas and ideals, as they did not want to disappoint him. As Guenter Wendt wrote: "Pete Conrad was what everyone thought an astronaut should be."

At first, the projected third two-person mission was to have included the first EVA, to be carried out by Pete Conrad. When Alexei Leonov floated free of *Voskhod 2* in March 1965, however, things changed, and NASA planners moved EVA plans forward to *Gemini 4*. The mission focus of *Gemini 5* now switched from EVA to one of extended duration. When asked at a news conference to comment on Leonov's shock spacewalk the previous week, Conrad was both philosophical and gracious. "I have nothing but admiration," he observed. "Certainly, I would have liked the opportunity of being first, but I don't think it really matters to me. I have no sour grapes about the Russian space program. I feel they have a program which is just as aggressive as ours."

Cooper had been looking forward to their EVA flight. However, the turnaround in mission objectives, while a considerable nuisance, would actually prove to be quite providential. "I suppose a long-duration flight was fine," he recounted for the authors. "It gave me the opportunity to do a lot of things. Actually, although we wanted to do the EVA, it kind of relieved us of some problems. We were running out of room with all the experiments we were carrying, all the things for a long-duration flight. It was going to be pretty nip-and-tuck, being able to carry all of the EVA equipment as well."

Despite the mission change, Cooper later said that he and Conrad were told they still had to wear bulky, uncomfortable EVA suits for the entire eight-day mission. "We felt that Gemini would have been proven well enough by Gemini 5 that we could wear regular flight suits," he mused. "We made a major project of trying to convince NASA of this, and almost made it. But there were just one or two 'weak sisters'—people who didn't have the courage to make the decision and let us go with them. Since we weren't opening the hatch, and had a very reliable pressurization system for that period of time, we felt like we could wear just a regular airplane flying suit and helmet."

Charles Conrad Jr., selected in the second group of astronauts, went by the nickname "Pete." Though he was named after his father, a World War I balloon pilot, Conrad's mother wanted him called Peter. She may have lost the argument then, but he quickly became known as "Pete" anyway, and the name stuck, though later he was also known as "Tweety Bird" to many of his fellow pilots. Married to the former Jane DuBose, he had four young sons when he joined NASA.

Conrad, born 2 June 1930, had wanted to be a test pilot from the age of six. Growing up in a middle-class family, he developed an early passion for flying through his father. At the age of nine he was building simulated cockpits "out of dining room chairs and soapboxes" in his bedroom, which provided endless hours of enjoyment. "I'd be sitting in the Spirit of St. Louis for hours, right alongside Lindbergh," he recalled. By the time he had turned fifteen, Conrad was working as an apprentice mechanic in an airfield machine shop near his home in Haverford, Pennsylvania. He was also learning to fly, and went solo for the first time at the age of sixteen. He gained his pilot's license soon after, and then went to Princeton University, a prestigious "Ivy League" center of higher learning, on a U.S. Navy scholarship. While there he completed the regular ROTC program, and when he graduated from Princeton with a bachelor's degree in aeronautical engineering in 1953, Conrad also received his commission. He immediately applied for flight training and became a naval aviator, later serving at the Jacksonville Naval Air Station. In 1957 he entered the Naval Test Pilot School at Patuxent River.

Some time in his early teens, during the war, Conrad acquired a tattoo—a blue anchor surrounded by stars with his initials underneath. Unless any of the early air force cosmonauts had one, it is likely that Pete Conrad became the first person to fly into space with a decorative tattoo.

Having completed his training, Conrad served as a project test pilot with the armaments test division at Patuxent River Naval Air Station (NAS), working with such high-powered airplanes as the F-8 Crusader. His next assignment came in 1960, as a pilot instructor and performance engineer at the Patuxent River Test Pilot School. Conrad had been interviewed in 1959 for a place in the first group of astronauts but did not make the cut. Although he possessed all the necessary skills and qualifications, his carefree, even flippant attitude during the medical and psychological examina-

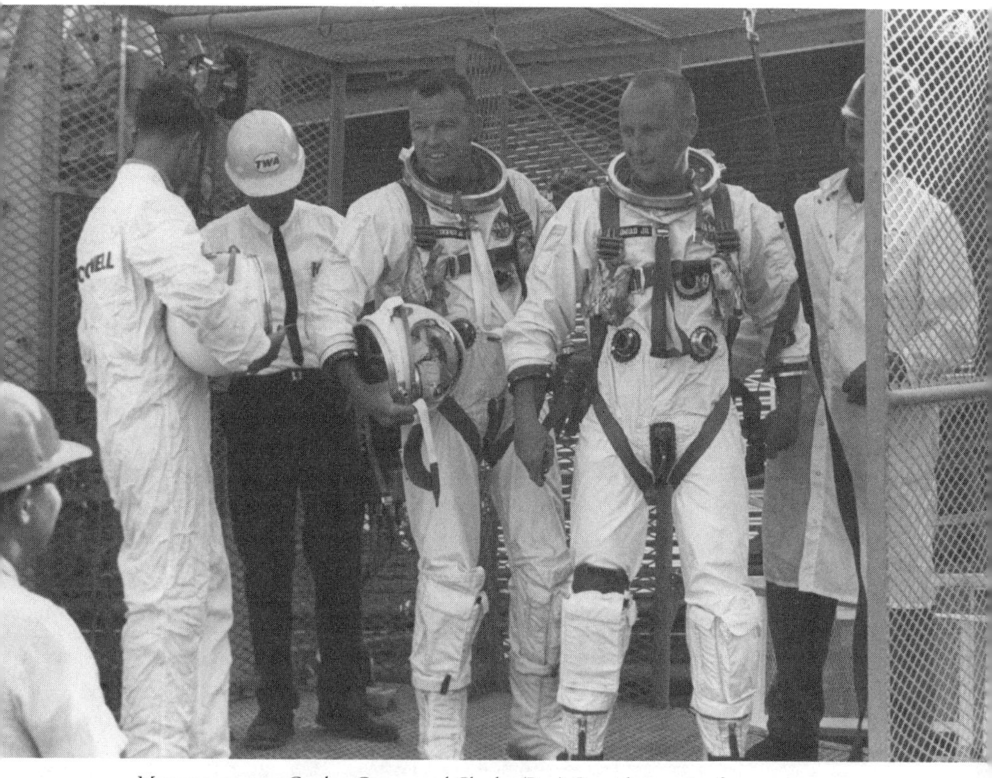

Mercury astronaut Gordon Cooper and Charles (Pete) Conrad preparing for their long-duration *Gemini 5* mission. Courtesy NASA.

tions (he even told the proctologist he'd been unnecessarily rough) did not sit well with the selection panels, and he was rejected. According to Frank Borman, "Pete Conrad's advice on handling interviews was, 'If you can't be good, be colorful,' but not many of us could be both good *and* colorful as Pete was—he had a wonderful sense of humor that he erected as a façade over his exceptional ability."

Wally Schirra agreed, telling the authors that he'd also met his convivial match: "Conrad, they felt, wasn't able to live alone in space, or endure in space. The shrinks pretty well screwed up on that one. He was considered one smart ass-tronaut. I held the title until Conrad came along. There was no way I could compete with him." Three years later Conrad was asked to volunteer again, and this time he was accepted. "I didn't regard it as something new," he said of applying to join NASA. "It was very straightforward—simply a logical step for a test pilot. We had reached as far as we could go in the atmosphere. Space was the obvious follow-on. The problems are not all that much different for a test pilot."

Provisionally, the flight of *Gemini 5* was targeted for the first day or two of August. Cooper and Conrad had been in training since February, giving them six months to prepare for the longest space mission yet attempted. The June flight of *Gemini 4* resulted in a colossal amount of data that had to be examined, absorbed, and resolved, and this only served to complicate matters for both the crew and the flight control team. Cooper found himself struggling to keep up. According to Tom Stafford in *We Have Capture*, "Training for the flight didn't go smoothly. Gordo also had a fairly casual attitude toward training, operating on the assumption that he could show up, kick the tires, and go, the way he did with aircraft and fast cars."

As precious weeks ticked by, Deke Slayton began to realize the crew would never be ready in time. "I got on the plane and went to see George Mueller [associate administrator for the Office of Manned Space Flight at NASA headquarters] to ask for help and he delayed the launch by two weeks."

As well as carrying twice the fuel capacity of *Gemini 4* for maneuvering and altitude control, *Gemini 5* also had the primary objective of testing a newly developed, sophisticated device called a fuel cell. These cells were designed to replace heavy, conventional batteries, and the long-duration flight presented an ideal opportunity to test them. Weighing less than a fifth of batteries that produced the same amount of energy, the cells would generate electrical power without any moving parts by bringing together oxygen and hydrogen at very low temperatures. The gases would react to form water—a valuable by-product on extended flights—and energy would be released in the form of electricity.

The flight would also be the first to be equipped with radar, specifically one developed to aid future crews in orbital rendezvous and docking techniques. During their flight the crew would release a free-flying optical and electronic device with a flashing light and transmitting-receiving beacon known as the Rendezvous Evaluation Pod (REP). Weighing around sixty-six pounds, the suitcase-shaped pod (quickly renamed "The Little Rascal" by the astronauts) would be released from its stowage point in the spacecraft's adapter module, and then tracked as far as fifty miles away using the onboard radar. The astronauts would then try to close to within fifty feet of the pod.

There was also much interest in one particular experiment to be carried out by the crew from orbit. During his mission in May 1963, Cooper had reported seeing from space many man-made objects on the ground, including objects as small as one truck in a desert. Scientists wanted to further evaluate such observations in a controlled test, so as the astronauts prepared for their mission, designated areas in the deserts of Texas and Western Australia were also being readied with huge rectangular patterns. The Texas grouping, made from panels of plastic and gypsum, ranged in size from 25 by 100 feet to 100 by 400 feet. Sixteen similar checkerboard patterns were also being laid against the red-brown soil 130 miles southeast of the Carnarvon tracking station in Western Australia. These, however, were made from eighteen hundred tons of white seashells transported by bulldozers from a beach thirty-seven miles away and then evenly spread an inch thick to form the vast geometric shapes. It was hoped that the white-brown patterns would be used in visual tests by the crew to assess the effects of prolonged weightlessness on vision.

Interestingly, the first mission patch worn on an American spacesuit was designed by the crew of *Gemini 5*. It was essentially a very tongue-in-cheek effort, showing a stylized Conestoga wagon with the motto "8 Days or Bust" embroidered on its side. The motto was a reference to the legendary "California or bust" that was the rallying cry of early western pioneers. According to Cooper, the two men had earlier thought about naming their spacecraft *Ladybird*, in honor of the wife of then-president Lyndon B. Johnson. NASA, however, vetoed the idea after the *Gemini 3* crew had irreverently named their spacecraft *Molly Brown*. The two astronauts then decided to design a mission patch as a unique souvenir of their long-duration flight.

It was Cooper who hit upon the idea of "8 Days or Bust," a resolute term that aptly described their primary mission objective, which was to stay in space for the same amount of time it would take to get to the moon and back. He discussed the concept of a mission patch with NASA Administrator James Webb, but Cooper said that the administrator "wanted to designate the missions with terms like Mission No. 2 and Astronaut No. 7—no personalities, nothing personalized." Webb could hardly have found two more renowned rebels than Cooper and Conrad, and despite his rebuttal of the idea they went ahead anyway.

"As it turns out, two nights before launch he invited us back to Houston to have dinner with him and we confessed to him what we had done," Cooper told *Space Flight News* in 1987. They informed Webb that the patches were already sewn onto their spacesuits, and they planned on wearing them. "So he had me send him a copy of the patch, along with a little quick memo-wire telling him the details about it and how the covered wagon idea of the patch had arisen. When it was finally approved, he said, 'Okay . . . on one condition. You don't have a provision for the fact that you might not make the full duration.' " Although the two astronauts were supremely confident they would fly the whole eight days, Webb asked that they simply tape a small patch over the "8 Days or Bust" wording—until they had actually returned from the mission. This was satisfactory to Cooper, and he said, "Right—we'll do it!" The slogans were eventually covered by small squares of silk loosely sewn over the slogan, which could easily be ripped off after the flight.

Thereafter, James Webb gave permission for a mission patch to be designed by the crew commander or senior pilot for each successive mission, to be worn on the right breast beneath the astronaut's nameplate. This was subject, of course, to the patch design being approved by both the director of the Manned Spacecraft Center and the associate administrator for Manned Spaceflight at NASA headquarters. It is a tradition that continues to this day.

On 19 August 1965, Cooper and Conrad sat atop their Titan II booster, waiting to be launched. This first attempt would end in frustration, firstly due to a problem caused by a stopcock getting jammed while the fuel cells were topped up, and then by a sudden thunderstorm that swept across the Cape. A power surge resulting from a lightning strike caused the primary telemetry system between Mission Control, the Cape, and the spacecraft to fail. "The whole complex went flat-dead for an instantaneous period of time, until the backup generators came on," Conrad later told *Space Flight News* magazine. "All power dropped off the spacecraft—and the rest of the pad for that matter—when the lightning hit whatever it was, a transformer somewhere in the vicinity of the Cape. I can remember very well turning to Gordo—I think we were at about T-minus ten minutes in the count—saying 'Well, I think that's it for the day!' because every gauge in the thing went to zero and came back on again. There was no way, in my mind, NASA would ever launch us having a glitch like that go through the system."

Two days later, *Gemini 5* stood ready once again for launch, with the two astronauts listening in anticipation as the countdown neared zero. Careful checks had determined there had been no damage to the ship's onboard guidance system, and all communications had been successfully restored. Conrad was particularly interested in all the sights, sounds, and sensations of his first launch. "The thing about the Titan that I remember was somewhere in the count, like at T-minus thirty seconds, they open the oxidizer pre-valves. The oxidizer was all the way up in the tank, and that line runs all the way down through the fuel tank down to the base. So when they opened that thing up, you are sitting on top of this thing—you are not that far away—and you can hear it going '*glah, glah, glah.*' "

There would be no delay this time: the launch took place almost right on schedule—in fact just one second before 9:00 a.m. Cape time. It was a ride that Cooper told the authors felt a lot different from his Mercury-Atlas launch:

The Titan was a lot heavier . . . solidly made vehicle. I certainly had a smoother ride in some way. The Atlas, of course, you had to keep it pressurized to keep its shape, when it was on the ground. So you can imagine, the Atlas was like pushing against a balloon going up; it had a lot more give to it, a lot more oscillation. Although it was certainly a good ride, it was a different . . . solid feel.

The Titan, however, had a fuel pogo. These big fuel lines, they'd build up these passage semi-blocks, pressurization cycles within the lines, giving you a pogo effect. This pogo effect occurred in the low frequency range, and caused you to lose vision. The instruments would start oscillating at the point where they'd vibrate, and you'd get where you could hardly see them. Pete Conrad and I had previously been the ones to run the testing on the pogo, on a big shake table, to determine the levels we had were acceptable at worst, but very marginal. We fixed the problem, but when they recycled our launch, they had to re-purge the pneumatics, refill the air on one side. They did not do it properly. It was ironic that we were the ones to have fixed it, yet we got it! They fixed it properly for the next flight.

Cooper has described the Titan II rocket as "a Cadillac" in comparison to the Atlas booster, but Conrad told *Space Flight News* that it was more like a sports car, "because you lifted off at 1.18 G, first stage burn-out was 6½ Gs, and second burn-out was 7¼ Gs. You made orbit in about six minutes and forty seconds . . . so you knew you were going somewhere in a hurry. I mean, that hummer got up and got with it!"

Gemini 5 entered an initial orbit of 216 by 100 miles, and Cooper later used the maneuvering rockets to tweak their perigee up to 106 miles. Cooper, becoming the first person to fly twice into orbit, was obviously delighted to be back in space again. He reported: "It feels mighty good. It's been a long time getting back . . . everything is going fine." But trouble loomed just ahead for the crew. On the second orbit, after a successful radar lock-on with the free-flying REP, Cooper suddenly reported a steady pressure drop in the liquid oxygen tank that supplied oxygen to the power-generating fuel cell. At some point earlier in the flight the oxygen supply's heater element had failed, which seemed to be causing the new problem. It was a critical situation, and the rendezvous evaluation practice was swiftly abandoned. By the sixth orbit it seemed that the spacecraft would have to make an emergency return, and contingency plans were put in place for a Pacific splashdown. Navy airplanes were hurriedly ordered to take off from Hawaii to watch for the spacecraft as it descended by parachute.

From a high of 850 psi, the pressure dropped to an alarming 60 psi, but this fortunately stabilized after Cooper carried out an emergency power-down of the spacecraft's electrical systems. Meanwhile, Houston told the crew that a new flight program was being evaluated. After ground engineers and designers analyzed the problem and came up with solutions, they gave the crew instructions for a cautious powering-up procedure on the seventh orbit. Fortunately, the pressure began to build again, and power levels soon became sufficient for the mission to continue. The crew was informed they could "Keep going," while Flight Director Chris Kraft reported that "At the moment, I don't see anything to stop us going eight days." Cheered by the decision to let them complete the flight, the two astronauts reported that "Everything is *go* up here!"

Because the planned rendezvous practice with the REP had been canceled, the crew carried out an attempted rendezvous with an imaginary target: a nonexistent Agena rocket spinning in space near the spacecraft.

They closed to within a short distance of the phantom rocket after a two-hour chase, and the test was regarded as a vicarious success.

In one small milestone near the halfway point in their flight, Cooper and Conrad observed and photographed the fiery launch of a Minuteman intercontinental ballistic missile from Vandenberg Air Force Base in California, which had been launched to coincide with their pass over the U.S. Pacific coast. Conrad reported seeing the 58-foot rocket's vapor trail a few seconds after it emerged from the firing silo. "I see it, I see it!" he exclaimed with evident excitement. "There it goes. . . . He's over the water. . . . See it, Gordo? It's bigger'n heck . . . we can still see it quite clearly." A similar experiment was carried out the following day, and once again the crew was able to observe the launch. When talking about their observations, Conrad would later say: "As I remember it, Vandenberg was covered with ground fog. You could see them coming up out of the overcast, and of course being solids [solid-fuel rockets] they were leaving a super smoke trail."

The experiment in sighting huge geometric ground patterns from space was not altogether successful, however. Conrad would later say: "We never did—as I remember—get a good look at the visual acuity targets that they had in Australia and Texas. We had trouble finding both of them, but we did find the one in Texas on one or two passes." Other unusual sightings that the crew did make were a rocket-sled test taking place at Holloman Air Force Base in New Mexico, in which a vast cloud of water sprayed from a trough served to brake the sled, and their prime recovery vessel, the USS *Lake Champlain*, steaming across the Pacific.

At one point Cooper complained that they had too many experiments to do all at once. "*Bang, bang, bang*, right together," Cooper complained irritably. "We've got every piece of gear in the spacecraft floating around trying to keep up." Conrad initially remained far more upbeat. When told by Mission Control that he was talking too much, he promptly began singing instead.

As Cooper recalled, the Gemini spacecraft was a very small area for two people to live and work in for more than a week, particularly with the amount of equipment that was stored around them: "We designed it to give us more effective room. In reality . . . it had less room per person than Mercury had, but there was a little more usable room. You could scootch around a little better." Unpacking several items just to get one out, while keeping track of it all, was a nightmare under weightless conditions.

It was very small, but there was a lot of housekeeping to do. When you used equipment, you had to put it right back in its container, otherwise there wouldn't be room to do anything. You could leave a piece of equipment floating in front of you, and it would stay there, unless you used the thrusters, then the pieces of equipment would drift to one side. You had to keep watching. It was easy to misplace things—you'd have to hunt for them. You felt weightlessness most when you would bend around to get something out from behind you. You had storage lockers just aft of you, so you could get things out of the way, store them away. You could completely turn around in there.

Toward the end of their flight, the crew of *Gemini 5* began creating new space endurance records. They eclipsed the Soviet Union's total time in space of 507 hours and 16 minutes and beat the single flight record of cosmonaut Valery Bykovsky when they sped beyond the 119 hours and 6 minutes he had flown in June 1963. Conrad's Gemini commander may have become the first person to fly into orbit twice, but Conrad took great delight in the fact that he had become, for a while at least, the person to have spent the second-longest amount of time in space. As *Gemini 5* flew on, new problems cropped up, putting the crew's "8 Days or Bust" motto at risk. The water by-product from the oxygen-hydrogen mixture feeding the fuel cells was rising in the tank, threatening to drown the cells and cut off their power. Then, two of the spacecraft's sixteen maneuvering rockets gave out. To conserve power and maneuvering fuel, Cooper put the spacecraft into a slow-tumbling, drifting mode. It was going to be a tight finish, but Mission Control continued to give the crew a cautious okay to keep going.

Although they were now in a drifting flight, Pete Conrad loved the way the craft reacted when it was controlled. "Gemini was like flying a high-performance fighter," he observed after the flight. "You did everything manually. You flew it." However, as he later told *Air & Space* magazine, eight days was a long time to spend in space with very little to do:

I call it eight days in a garbage can. The fact is, you can't do anything. You can't go anywhere. You can't move and have no great desire to sleep because you're not doing anything to make you tired. You don't have anything to read and there isn't any music.

Near the end of the mission, most of our thrusters had crapped out, so all we could do was drift and, as the fickle finger of fate would have it, when we were over the ocean, that's when the spacecraft would be pointing down and all we'd see was the water. By the time it got to land, which is much more interesting to look at, that's when we would start pitching up and we'd be looking at black space.

I spent half my life opening stuff up and re-wrapping it real tight so we didn't have garbage all over the cockpit. If it wasn't for that I would have probably shot myself because there was nothing else to do.

After almost eight days in Earth orbit, Mission Control was piping music up to the crew, trying to keep them entertained. It was truly time to return, and it was almost a relief that the end of the flight had to be trimmed by one orbit due to a forecast of bad weather moving in on the Atlantic recovery range. Cooper later said that Conrad had some concerns about the reentry. "He was worried about the attitude control thrusters. The attitude control thrusters we were using during the flight were gradually going out, one by one. They were failing. So he was concerned—we all were concerned—that the re-entry thrusters would not work properly." Just as Cooper had encountered on his previous flight, *Gemini 5* presented him with reentry complications that required his piloting skills. Incorrect data had been fed into the computer, which meant that they were reentering at too steep an angle. Deceleration forces shot up to seven and a half g's.

"The computer didn't set us up very well for an accurate landing," he later told the authors.

We had to ignore the computer guidance—because the wrong data was loaded in, and it was guiding us in too steeply—and go to full lift to recover part of the distance. We had to roll about one hundred and eighty degrees; Gemini was certainly an incredible machine, and with the offset center of balance, you certainly could do a lot of longitudinal and lateral maneuvering, although we were still off a little bit on our landing area. I was on top of the situation; the g's were very acceptable. You learn with every flight, and with each flight you make progress, and you learn the right way to go on and take that into the next flight.

After the critical reentry blackout period, the jubilant voices of the two astronauts could be heard reporting in, saying "We feel okay." Then, as the 84-foot orange-and-white parachute billowed out to lower them toward the water, Conrad radioed: "Roger, main chute out!"

Gemini 5 came down to an off-course but perfectly safe landing in the choppy Atlantic southwest of Bermuda, about eighty miles from their waiting carrier, USS *Lake Champlain*. An HU-97 search airplane was circling over the bobbing spacecraft within minutes. When a helicopter from the *Champlain* reached *Gemini 5*, three navy frogmen jumped into the water, attaching and inflating a large flotation collar around the spacecraft. Both astronauts, now sporting scruffy whiskers and huge grins, climbed into a waiting life raft before being hauled up into the helicopter for transportation to the carrier. Their spacecraft was retrieved by the destroyer USS *Dupont*.

The two astronauts were later declared to be in good health after their history-making flight of seven days, twenty-two hours, and fifty-six minutes. With the successful conclusion of their mission, Cooper and Conrad could finally remove the small silk squares covering their flight patch, and there it was for all the world to see—"8 Days or Bust." But it had certainly been a close thing.

Although Conrad's astronaut career was now in its ascendancy, Cooper's was in serious decline. In fact, *Gemini 5* would be both his second and last spaceflight. Following the record-breaking mission, Deke Slayton assigned Cooper the job of backup commander for the 1966 *Gemini 12* mission. It was a clear dead-end assignment, as this would be the last Gemini flight, and training for it would keep Cooper from flying any of the earliest Apollo missions. Despite having commanded two very successful spaceflights, Slayton later said that Cooper was "basically marking time" by this point. His relationship with NASA managers had not improved, leading Flight Director Gene Kranz to describe Cooper as "the loner and rebel against the spaceflight bureaucracy."

"That's fairly accurate," was Cooper's response when reminded of Kranz's words. "The manned space program started with one hundred and fifty people involved in it. Ten years later we had sixty thousand . . . doing the same thing. Admittedly, the program had become a lot bigger, but *not* sixty thousand people. Everybody had built their little empires. Typical bureaucracy set in, in a big way. The way we caught the Russians was by not really having a bureaucracy to start with. Then, you could get all of the key people in . . . make a total change in fifteen minutes, because it was all first-name acquaintance. You scribbled out a little piece of paper, handed it around, and the change was made. By the time you have sixty thousand people in, that same change would take you a month!"

When his work on *Gemini 12* was over in late 1966, Cooper received yet another assignment—this time as backup commander for the *Apollo 10* flight. Under the crew rotation system then in place, Cooper, Donn Eisele, and Ed Mitchell could reasonably expect to be subsequently assigned as the prime crew for *Apollo 13*. But when Cooper found out that Eisele had been placed on his crew, he began to suspect that something was seriously wrong. Eisele was going through a divorce, and it was common knowledge at NASA that he was unlikely to fly again.

Following the *Apollo 1* fire, NASA Administrator James Webb made a very telling remark during the 1967 Senate hearings on the tragedy. Commenting on Wally Schirra's assignment to command *Apollo 7*, he said that, "He will become the only astronaut to fly three generations of U.S. spacecraft—Mercury, Gemini, and Apollo." Although it was probably just an off-the-cuff observation recognizing Schirra's many contributions to the space program, to some it sounded like Cooper's fate may have already been sealed.

Despite his rebellious reputation, which he knew did not sit well with NASA's management, a restless Cooper was still drawn to the heady competitiveness of sports car racing, and entered a race at Daytona while training for *Apollo 10*. As soon as Slayton was informed of this, he told Cooper in no uncertain terms to withdraw from the event. "He wasn't very keen on me racing at Daytona, and pulled me out of it," Cooper reflected. "The rules hadn't been spelled out loud and clear before then. But after that, it got spelled out loud and clear. I was very annoyed. We'd done a lot to get the car all qualified. I was having a lot of fun auto and boat racing. I got the bad end of it—but nothing ventured, nothing gained!" After being pulled from the Daytona race, Cooper did not endear himself to his NASA bosses either when he flippantly said to a news reporter, "I guess they want astronauts to be tiddlywinks players!" The widely published quote could hardly have helped his prospects for another flight. A legitimate concern for NASA, apart from Cooper's slight insubordinate streak, was that he might be named to a crew and then injured late in the game in a racing accident, forcing a last-minute replacement on a critical flight.

Spring 1969 saw the successful lunar flight of *Apollo 10*, but it also saw another event that was to change the Apollo assignments. Alan Shepard, America's first astronaut to fly, had been restored to full flight status following a lengthy medical grounding and was eager to command the first available flight. He had been chief of the astronaut office during his exile from flight status, and had the power and close connections with Slayton to be able to jump right back to the head of the line. The next flight to which a crew had not yet been assigned was *Apollo 13*, and Slayton decided to name Shepard as mission commander. As Cooper told it, Shepard and Slayton held a meeting with him, in which he was told he would also be training for *Apollo 13*—but as backup commander. He was devastated:

I felt that the higher-ups would really analyze everything carefully, with enough capability that whichever guy was selected might be the best guy for that particular flight. I wanted to get assigned to a prime crew. There were people in Washington who could have made that decision, but nobody would. I hated to see the typical government bureaucracy come crawling into spaceflight. Here was an opportunity to do the best job you could, with the best people involved, with the required funding to do it with. Actually, it ought to have operated ideally. Instead, a lot of politics got involved.

I had been planning on commanding a flight to the moon. But Al and Deke were trying hard to get back into space. Al looked after Al, and could be cutthroat when it came to what he wanted. He wasn't to be trusted when it came to his best interests. I was the youngest of the original seven, and the only one on active flight status at that time. I had been completely trusting, thinking they'd choose people on their strengths. But they wanted missions for themselves, and all bets were off.

I guess maybe I was a little naive. It was kind of unexpected that they did that, and very disillusioning. I was very disappointed that I did not get to fly Apollo. It was a terrible blow. It took me years to forgive them for that unfairness. In the end, however, we made up; we forgave each other and became good friends again.

Cooper said he turned down the backup assignment and then, believing he would never get another prime crew assignment, left NASA and the

The quiet nobility of Gordon Cooper, photographed in September 2004. Courtesy Francis French.

air force in early 1970. The *Apollo 13* and *14* crews were switched to give Shepard's crew more training time. By the time both missions had concluded and Shepard had walked on the moon, Cooper was already gone. Following his resignation he maintained an active schedule, working on numerous aerospace and electronic projects with a number of companies. His strong belief in the existence of other civilizations beyond our planet, and the possibility that they might be visiting us, influenced much of his post-NASA career. The post-NASA years also brought about reconciliation with his former Mercury astronaut colleagues. Massive public interest stirred up by the book and film *The Right Stuff* had them all talking to each other again. As Slayton said at the time: "Old disagreements didn't seem so important anymore." They set up a charitable foundation together.

Apollo astronaut Walt Cunningham once said that Gordon Cooper was "an adventurer, who was willing to pay the price, and did." When asked about this by the authors, Cooper's face showed mixed emotions as he pondered the statement for a moment. "I suppose I wouldn't disagree with it," he finally responded.

On Monday, 4 October 2004, the forty-seventh anniversary of the launch of the first Soviet Sputnik, a stubby experimental vehicle with the unpretentious name of *SpaceShipOne* shot into history, climbing to an altitude of 69.6 miles over California's Mojave Desert with solo pilot Brian Binnie at the controls. When the spacecraft reached an altitude exceeding 62 miles for the second time in a week (the first flight in the same craft had been under the command of Mike Melvill) Binnie earned his astronaut wings.

A non government consortium called Scaled Composites, founded in 1982 by famed aero designer Burt Rutan, claimed the $10 million Ansari X Prize as the first team to achieve this historic feat. Peter Diamandis, who headed the X Prize Foundation, spoke enthusiastically of the historic event later that day. "What we have here, after forty years of waiting, is the beginning of the personal spaceflight revolution," he stated. To many, this audacious event was almost like a ceremonial passing of the torch to a new generation of spacefarers. Just as Gordon Cooper had become the last NASA astronaut to journey into orbit alone, so Burt Rutan's pilots could be seen as precursors to what will eventually become a whole new generation of space pilots and commercial spacefarers.

But even as the space community celebrated the historic feat, some sad news was also sweeping around the world. Gordon Cooper, the famed solo space pilot, had passed away from a heart attack that morning at his home in Ventura, California, aged seventy-seven. Among the tributes were these from the three surviving Mercury astronauts. "Gordo was one of the most straightforward people I have ever known," reflected John Glenn. "What you saw was what you got. Pride in doing a great job, whatever his assignment, was his hallmark. You could always depend on Gordo."

Scott Carpenter said: "This is truly the passing of a beloved member of a unique fraternity. He had no equal anywhere. We celebrate his contribution, but at the same time we remind ourselves that nothing in the construct of man stands forever. I'm really going to miss that guy." Wally Schirra remembered Cooper in a droll but respectful way that his Mercury colleague would doubtless have appreciated: "not too bad of a water skier, not too bad of a pilot, but a heck of a good astronaut."

2. A Rendezvous in Space

Do not go where the path may lead;
go instead where there is no path and leave a trail.

Ralph Waldo Emerson

As a test pilot, Wally Schirra knows that almost every space mission from the mid-1960s until the present day has relied on one thing: the ability of two spacecraft to find each other in orbit. Without such a capability, there could have been no moon landings, no space stations—in short, no space program. Schirra, who piloted the key spacecraft in the first-ever orbital rendezvous, knows how vital that task, the centerpiece of the *Gemini 6* mission, was to any future plans, and he was thrilled to be assigned to it.

The mathematics needed for such meetings in orbit are complex, but were in fact the easier part of the attempted feat. Perfecting the engineering and piloting capabilities would prove much harder. Though the Soviet Union had orbited two spacecraft at the same time, they still lacked the ability to rendezvous in space. With the successful completion of *Gemini 5*, the next three American missions would not only be demanding test flights, during which NASA gained that vital experience, but they would also demonstrate orbital rendezvous techniques that the Soviets were unable to emulate at that time.

It would not be easy: rockets malfunctioned and exploded, and one Gemini crew would come perilously close to losing their lives. These setbacks could have delayed the program. Instead, individuals and the choices they made once again made the difference. Decision-makers on the ground devised bold mission changes that kept the program on track, while the astronauts made equally risky, life-saving decisions. Without this ability to adapt, NASA might have faced many frustrating years of delay.

Early in the Gemini program Wally Schirra was given command of the *Gemini 6* mission. It was a mission the veteran Mercury astronaut wanted because he regarded it as the piloting high point of the Gemini program. *Gemini 6*'s original mission was to rendezvous with a separately launched, unmanned Agena—the upper stage of an unmanned Atlas rocket—and then dock with the specially designed conical docking collar mounted on it. After conducting attitude control tests while docked, *Gemini 6* was to undock and return home. It would be a short, busy flight, and an interesting challenge for a couple of experienced test pilots.

Joining Schirra on the crew was Tom Stafford, a rookie from the second group of astronauts, and one of the more outstanding pilots in an already elite group. Mike Collins later wrote that Schirra and Stafford made a good team, with "Stafford's penchant for things electrical and mathematical acting as a strong right arm for the cool commander." The crew worked hard at becoming rendezvous experts, training in a variety of complex mathematical approaches to a target. As Schirra told the authors, not only did the crew have to learn the most practical rendezvous method, but they also felt obliged to ignore the advice of the one astronaut who had a doctoral thesis in orbital rendezvous:

Buzz Aldrin tried to get credit for our success, but he almost screwed us up. Buzz had this academic mind, and realized that there were two ways of doing a rendezvous. One way—his way—was mathematically pure, but if you messed up a little bit, you'd really mess it up. I told him we were not going to do it that way. He was talking about perfection, and perfection is when you're docked—not doing the maneuver. So we looked at a different way: the right way of doing it. We spent a lot of time in simulators. Dean Grimm, a NASA engineer, was one of the guys I worked with. I am indebted to those kinds of people. Of course, Young and Grissom were our backups and they worked on it too—about five or six of us really spent a lot of time on it. Stafford and I were going to do it right; it was a crucial step.

Yet for all the hard work, Schirra still had time to invent rendezvous-related quips, some of which he later shared with the authors: "Did I tell you exactly what a rendezvous is? When a man looks across a street and sees a pretty girl, and waves at her, that's not a rendezvous, that's a passing acquaintance. When he walks across the street through the traffic and nibbles on her ear, *that's* a rendezvous!"

Thomas Patten Stafford was born on 17 September 1930 in Weatherford, a place he describes as "the dust bowl of western Oklahoma." The son of a dentist and a teacher, he was born severely underweight, and would later write in his memoirs that he was "fortunate to survive infancy." When he was fourteen he took his first airplane ride, and it made him eager to learn to fly himself. He was also an accomplished scholar, largely because his mother encouraged him to read. He grew fast, shooting up to six feet while in high school, and as a result did very well in sports, particularly football. In fact, he was good enough at the sport to be considered for a university football scholarship. However, he was also enamored with entering the Naval Academy and becoming a naval aviator. By studying very hard, he eventually earned an appointment to Annapolis. Mike Collins says that Stafford was so accomplished in engineering at the academy that he "sopped up wiring diagrams with the same ease that some teenage girls memorize the words of popular songs." In 1952, he graduated from the academy with a bachelor of science degree, before entering the air force. He had decided to switch services because, at the time, the air force had the fastest jets.

The following year, Stafford had not only married Faye Shoemaker, a girl he had met in high school, but he also received his wings. He then served as a fighter pilot both in the United States and overseas, flying the hottest jets with delight and expertise. In 1958 he applied and was selected for the USAF Experimental Flight Test Pilot School at Edwards AFB, winning a trophy as the outstanding student of his 1959 graduation class. Staying on at the school as an instructor, he wrote comprehensive manuals on performance flight testing, taught students such as future astronaut Frank Borman, and became chief of the performance branch of the Aerospace Research Pilot School. He had not been considered for the first group of astronauts because he was not only one inch too tall, but had also not yet finished test pilot school. Following a relaxation of the height limit he was a natural to be selected as one of NASA's second group of astronauts in 1962.

The day he was chosen was in fact his third day at Harvard Business School's master's in business administration program. He'd chosen to attend because he believed that the air force "had good technical ability, but their big problem was management," as he told interviewer William Vantine. He also thought the experience would prove useful when he finally retired from the air force. By joining NASA instead, he earned what he calls "the notoriety of being the first dropout of the class." Stafford's business sense served him well as an astronaut. He quickly impressed the right people and became the first person in his group to score a flight assignment, although circumstances beyond his control caused him to be bumped from a prime crew role to backup pilot. "I was to be on the first Gemini crew with Al Shepard," he told the authors, "but Al got grounded due to medical problems." Stafford wound up being reassigned to the *Gemini 3* backup crew, under the command of Wally Schirra.

Stafford's quieter demeanor was a noticeable contrast to the more jovial, wisecracking Schirra, and his personality was hidden even more by a thick Oklahoma accent that some found hard to decipher. As cosmonaut Valeri Kubasov told the authors, "It is very difficult to understand Tom Stafford's language, because it's not the English language, it is Oklahoma language!" The sober, sometimes strict Stafford seemed more like a schoolteacher than a pilot to many of his colleagues. Yet his calm mannerisms masked a quietly efficient, determined, and extremely ambitious military officer.

Of all those in the second group of astronauts, said by many to be the most talented group ever selected by NASA, Stafford was the one who would go the furthest in the astronaut pecking order. He would even move into a position where he made crew selections, an almost sacred duty previously reserved for the most powerful of the Mercury astronauts. Stafford would later command a mission with a Mercury astronaut as his backup, and have another even take a subordinate position to him on a prime crew. He seemed to ease into the right circles without any of the trepidation felt by some of his colleagues. "Some people were intimidated by Al Shepard, who could turn icy at the snap of a finger," he would write in his autobiography, "but I got along well with him." For *Gemini 6*, his first mission, he was off to an impressive start as a crew member on one of the program's keystone missions.

Stafford still fondly recalls that exciting time in the space program. "It was a great time to be there," he explains,

because every mission that we would fly would be something new, something that had never been done before. Each one depended on the previous mission.

At the time, it was obvious that we were in a tremendous race. It was a race against time that President Kennedy had set, and also a race with the Soviet Union. They had already had two spacecraft flying at the same time, although not rendezvousing. Our response was very slow. We finally started flying the Gemini program, and we did not know it at the time but it was Gemini that put us well ahead of the Soviets. This was a classic example of how you take a sword and beat it into a ploughshare. We had fifty-four of those Titans on alert, and normally they would have carried a nine-megaton weapon on top. We substituted a nine-megaton weapon with an eight-thousand-pound Gemini.

On 25 October 1965, Schirra and Stafford prepared to launch aboard *Gemini 6*. "There was a lot of pressure on Wally Schirra and myself for this rendezvous," Stafford recalls. "It had been decided that the best way to approach a landing on the moon would be to do a rendezvous around the moon. We were sizing the giant Saturn V, and the Lunar Module and Command Module, all based on the premise that we would do a rendezvous around the moon. The problem was, nobody had ever done a rendezvous—so it was up to Schirra and me to prove the theory. If we could not learn to rendezvous and dock, we would be stalled on the moon trip."

Before their launch, their unmanned target—an Atlas Agena—was launched ahead of them. Although it had never been used before as part of a manned mission, the Agena was considered very reliable, having flown frequently and successfully since 1959. Schirra and Stafford were already strapped into their Gemini spacecraft when the Agena lifted off, just a few launch pads away from theirs. The Agena countdown had been flawless, and all looked good—for the first six minutes of flight. "We could hear our target vehicle roar off down on Pad 14, where the Mercury Atlas people were launched from," Stafford describes. "Unfortunately, the Agena exploded over the Atlantic Ocean. Now we had no target vehicle, so we scrubbed our launch. What do we do now?"

When the Agena had separated from its booster the vehicle began to wobble, and then exploded into five large pieces. Mission controllers tried to work out what had gone wrong with the usually reliable rocket. There was little telemetry that could give a clue as to what had gone awry, and they could only speculate on how long it would be until they were ready to try again. While some were dwelling on what had caused the explosion, others were already thinking about ways to rebound from this upset. An obvious thought was to replace *Gemini 6* with the next Gemini spacecraft and fly that mission instead: there was no rendezvous planned for the *Gemini 7* mission. However, it was soon realized that this was not an option; the second spacecraft was much heavier, in fact, too heavy for the booster now sitting on the pad. Literally minutes after the Agena explosion, senior officials were already discussing a bolder new plan. Instead of having *Gemini 6* rendezvous with an unmanned target, why not have it rendezvous with *Gemini 7*?

The main objective of *Gemini 7* had already been set: send two astronauts into orbit and keep them there. The crew would try to remain in space for two weeks, undergoing a range of medical experiments to see how well humans adjusted to long-term weightlessness, and whether missions of the length needed to explore the moon could be endured. A number of NASA doctors worried what the results might be. As the Gemini flights steadily increased in length, the astronauts had returned with increasingly higher heart rates and lower blood pressure. If the rate of deterioration was the same as in previous flights, the *Gemini 7* astronauts might find that their hearts were not able to effectively pump blood when they returned to Earth. Instead, it would pool in their legs, and they would black out. The doctors even put plans in place to remove the astronauts from the spacecraft and resuscitate them in a life raft if necessary. It is arguable whether the crew was thought of as chiefly pilots, or medical specimens.

Gemini 7 was hardly the best mission of the program, and perhaps this is why it was avoided by the Mercury astronauts. Instead, they allowed two rookies from the second astronaut group to be assigned. Deke Slayton, when choosing a commander, believed that Frank Borman's tenacious nature would be a good fit for a long mission, where a grim sense of duty would be the only way of enduring the long weeks in space. Borman summed it up as "not much of a pilot's mission," but it was an opportunity to impress. "I figured when they gave me Gemini 7, we'd do the best we could with it." Joining him on the flight would be Jim Lovell.

Frank Borman and Jim Lovell were both born in March 1928 within eleven days of each other, and both joined NASA at the same time. Both were only children, born into poor families who struggled hard to make ends meet, and the duo even had the same hair and eye color. They have both enjoyed two of the most enduring marriages of any astronauts. Although one had served in the air force and one in the navy, their flying careers were very similar. However, their personalities were quite distinct.

Frank Frederick Borman was a graduate of West Point, and to many who worked with him it seemed like he had never really left the place. Born in Gary, Indiana, on 14 March 1928 into a strict Lutheran family, he suffered in his youth from persistent breathing disorders. When he was five, his parents decided that the only way for his health to improve was to move to the desert conditions of Tucson, Arizona, where in fact he did greatly improve. He tried to read science fiction, but found that it bored him. Yet, like many other future spacefarers, he grew fascinated with aviation and saved up money for flying lessons by working part-time jobs. He was still a teenager when he earned his flying license. "I was rather small in stature," he would write in his memoir, *Countdown*, "and like many undersized kids I tried to compensate by being feisty and maybe a bit too overaggressive; . . . the consensus amongst my peers was that I was too bossy and aggressive, a complaint that carried over to my adult years."

His father, who ran an auto repair shop, was not a wealthy man. There was no money to pay for his son's flying lessons, and so the only chance to become an aviator was to join the military. In 1946 Borman was admitted to West Point as a cadet. "The boy was about to become a man, molded and forged," he would write. "I was not only indoctrinated but inoculated." He was lucky to get in—technically, he had waited too late to apply—but he had been helping the son of a local judge build model aircraft, and the judge helped him get on the list. Also on his side was the fact that he was a straight-A student: in his memoirs, Borman writes that he "was one of those lucky kids who never had to study hard." He graduated from the prestigious military academy in 1950 with both a bachelor of science degree and a strong idea of how he should run his life: for duty, honor, and country. When he graduated he also got his wish to be commissioned into the U.S. Air Force. "I was into airplanes," he told interviewer Catherine Harwood. "I'd never even thought about rockets or space. . . . I had no idea, no concept, nor any desire to participate in that part of the business."

The same year that Borman graduated, he also married Susan Bugbee, a girl he had met in high school when he was just fifteen. "When we got married in the 1950s, it was like a team," Borman told Harwood. "She had her job and I had my job. And one of her jobs was to not in any way show any kind of fear or trepidation over what I was doing." By 1957 he had also obtained a master's degree in aeronautical engineering from the California Institute of Technology, and in the meantime had served in the air force as a fighter pilot. After obtaining his master's degree he returned to West Point for three years to teach science as an assistant professor, and in 1960 attended test pilot school at Edwards, ranking first in his class. By this time, he was not just interested in flying within the atmosphere. He had grown curious about the possibilities the space age could offer and wanted to best position himself for selection as an astronaut. By the time NASA chose him in 1962, he was an instructor at the school in a new kind of aerospace program. "They were starting in what they called the graduate program, an advanced test pilot program, that would . . . hopefully, prepare people to fly in space," he told Harwood. As he made the move to Houston, Borman had little idea what NASA would be like. In fact, he would treat the assignment very much as a continuation of his military career, not bothering much with the world beyond it. In 1969, for example, he had still never heard of movie icon James Dean, who had died a full fourteen years before. He later described his NASA years as "Eight years of total dedication. . . . I was with NASA with the objective of being part of the team that beat the Russians."

His personal ambition was probably the equal of Tom Stafford's; Mike Collins described Borman as "attracted to money and power." Yet although Stafford's career progressed smoothly, Borman never quite reached the same heights. He would command the first-ever mission to the moon. But after only two flights he chose to leave active astronaut training to gain some administrative experience and apparently because he believed his astronaut status would transfer well into other areas of service for his country. At first, it did. He served as a special ambassador for President Nixon on the thorny issue of Vietnam, and was even offered a job on the White House staff. But he was not a Washington insider, and it was never made clear to Borman what his actual assignments would be. Instead, he carved out a career as an executive in the world of commercial aviation.

It seemed that although the decision-makers always respected Borman's personality and flying skills, he was perhaps too much of a straight arrow to ever be considered one of the insiders in the top fraternity where Shepard, Slayton, and, eventually, Stafford resided. "He wasn't really one of the guys," Gene Cernan would later write in his memoirs, "and somewhat holier-than-thou." While some of his peers enjoying living up to their dashing public images, the ever-practical Borman once told a panel of senators that "I am afraid that sometimes the newspapers and the magazines attest a great deal more of the 'silk scarf' attitude to the astronauts than actually exists." In place of a silk-scarf attitude, Borman said that "If you ever want to get anything done you have to cut through the mass of things. You can generate an awful lot of enemies along the line. I've got a few." He also had no time for the pranks that other astronauts played, and admits "I was something of a loner; . . . the most impatient and outspoken of them all. . . . I had started being that way when I was at West Point and I never changed." Several years later, when interviewer Nancy Conrad asked him if he had a humorous anecdote about his astronaut days, Borman's prim response was "Actually, I didn't think much of it was funny. . . . To me it was hard work and there wasn't any frivolity to it." It was an honest attitude, but one that was not going to make him many close friends or move him into the astronaut inner circle. To his credit, Borman can joke somewhat about his hard-nosed personality today, admitting "I'm stuck with it!"

Borman had been under strong consideration to join Gus Grissom on the original backup crew for the first manned Gemini mission, but there was something about him that Grissom did not like. "I went over to his house to talk to him about it, and we had a long talk," Borman related to Harwood. "After that I was scrubbed from the flight. So I guess that I didn't pass the test with Grissom. Don't know why. Nobody ever said anything to me about it. I never asked. Could have cared less." Instead of training with Grissom he found himself serving as backup commander for *Gemini 4*, paired with Jim Lovell. "I wasn't the most popular astronaut," Borman wrote in his memoirs, "but I didn't give a damn."

James Arthur Lovell was born on 25 March 1928 in Cleveland, Ohio, and grew up in Wisconsin. His father, a coal furnace salesman, died when he was twelve, and his mother worked as a secretary to pay the bills. They shared a one-room apartment, and struggled as best they could to make ends meet. Just like Frank Borman, he met a girl in high school whom he ended up marrying—in Lovell's case, Marilyn Gerlach. "She has supported me throughout all these years in what I was trying to do," Lovell told interviewer Glen Swanson. On many of their dates they would sit on the roof together while Lovell pointed out the star constellations.

A model rocket buff as a teenager, Lovell had teased information out of his chemistry teacher about rocket fuel chemicals and built three-foot-long rockets with the help of friends. At age sixteen he was lucky to escape serious injury when one of his gunpowder-filled creations exploded. Just like Borman, he was interested in a career that required a college education his family could not afford; Lovell dearly wanted to be a rocket engineer. "I was interested in rockets and astronomy long before the Glenns and Shepards of the world could spell *rocket*," he told interviewer Ron Stone. Like several other astronauts, the military would be the only option for achieving his dream. Lovell did not make the selection for the Naval Academy, instead joining the U.S. Navy under a program that allowed him to attend college for two years and study engineering, then receive naval flight training. By then, he had grown as interested in aviation as he had been in rocketry.

While still in this program, Lovell was finally admitted to the academy at Annapolis, graduating in 1952 with a degree in science and a senior thesis on the subject of liquid fuel rockets. His lengthy schooling meant that he missed out on combat flying in Korea, but in its place he had received a valuable six years of college training from the navy. After graduation he achieved his goal of being assigned to flight school. Lovell would spend the next six years as a naval aviator and flight instructor, which included the ever-risky assignment of landing on an aircraft carrier at night. "In those days," he told Stone, "jet training was the epitome of everything, and not too many people were in it. I was very, very lucky . . . and I learned an awful lot. Night flying off carriers separates the men from the boys."

The work was dangerous and challenging, but Lovell wanted to try something even more cutting edge. In 1958 he applied for and obtained a transfer to the Naval Test Pilot School in Maryland, where he was in the same class as future astronauts Wally Schirra and Pete Conrad. Lovell finished first in his class, and all three of them were considered for Mercury astronaut selection, as Schirra told the authors:

I knew Pete Conrad and Jim Lovell when they came to NASA as they had been classmates of mine at test pilot school—Class 20. Pete and I water-skied a lot back then, and he and I tied for second in our test pilot class. Lovell studied, so he became first! All three of us went through those early Mercury tests together, to see which one of us would survive, and somehow I did. Lovell had some kind of anomaly with his liver at the time of those Mercury tests, so didn't get picked.

Interestingly, Lovell believes that the navy initially thought that he had been selected as a Mercury astronaut. As he told Ron Stone, when he returned to his squadron after being rejected in the final round of medical tests, he immediately received orders to report to Wright-Patterson Air Force Base and join the Mercury selectees:

I went out to Wright-Patterson, and they were sort of expecting me. The people . . . said "Oh yes, we got seven people, and you're number seven." I said, "Oh, I'm great!" The next morning we all went down to breakfast to get ready to start our physical, and a fellow walked in and said, "I'm sorry I'm late. I just came in from Edwards. My name is Gus Grissom." And he was number seven, and, of course, I wasn't supposed to be there. . . . I was packing my bags, going back to the squadron! Very, very disappointed.

At the time, it was not known whether there would ever be future astronaut groups, so Lovell was upset that he had missed out. But, "before I knew it," there was a call for a second group, and Lovell was asked by the navy if he wanted to try again. Naturally, he agreed. When selected by NASA for its second astronaut intake, Lovell was the safety engineer for a fighter squadron in Virginia. It wasn't his choice—he would have preferred to be in flight testing—but, like Gordon Cooper, he had turned the assignment into a positive experience. Lovell was now in the right place to put his rocketry and aviation enthusiasm to work.

In contrast to Borman's spit-and-polish bearing, Lovell was, and still is, a far more outgoing, relaxed, and jovial character who combines hard work with a smooth, pleasant attitude. After decades of relative anonymity, he ironically wound up being one of the best-known astronauts in the mid-1990s, a full quarter of a century after his fourth and last spaceflight. The reason was a movie made of his autobiography, *Lost Moon*, chronicling his *Apollo 13* mission. Lovell would in fact have one of the quirkier spaceflight careers, twice flying missions that were originally intended for others to fly, losing a place on the *Apollo 11* crew in the process, only to finally end up commanding a moon-landing mission that never got to land. First, however, he'd be piloting *Gemini 7*.

The *Gemini 7* crew would have to be able to get along well given such a long, intense mission. Luckily, they bonded from the start. Decades later, Lovell could joke, as he did with the authors, about the pairing and the unexciting mission. "*Gemini 7*, of course, was a two-week mission. I had a very small spacecraft, so it was quite a challenge. Of course, two weeks with Frank Borman anyplace was a challenge! You work for so long to get a spaceflight, I'd have gone up with just about anybody. But we were guinea pigs, absolute guinea pigs." It was perhaps the most unrewarding mission in the program, but any first spaceflight had to be considered a good one. With the sudden addition of a rendezvous, it had just become a lot more interesting.

To accomplish a rendezvous between *Gemini 6* and *7*, some major problems would have to be overcome. First, fewer than two weeks could separate the spacecraft launches. Though the Russians had been able to accomplish this with relative ease, it was a risky proposition to NASA planners—especially as they would have to use the same launch pad for both missions. NASA's tracking network had also never been used to track two manned spacecraft before. Doing so would be difficult, but a method was soon formulated to achieve it.

Just over a day after the Agena explosion, all of the vital decision-makers were on board with the idea, and it was adopted. It was remarkably fast action for a federal agency, and almost unimaginable today. Flight Controller Gene Kranz still describes it as a "gutsy management decision," saying the positive attitude of those on the ground toward making it happen was even more impressive than what happened in space. Schirra and Stafford were in favor of pushing the new plans even further, and having a *Gemini 6* and *7* crew member change places during an EVA. It was wisely decided that this would be too ambitious—a rendezvous would be enough. "We decided it was time to get the theory proved," Stafford jokes.

Once the plan was set, events moved fast. *Gemini 6* and its booster were carefully dismantled from the pad and stored in pieces, ready for a fast reassembly and launch. Plans were formulated for a speedy pad refurbishment after *Gemini 7* lifted off, and the spacecraft's planned orbital path was changed to make it an easier target. Borman and Lovell maintained a grueling schedule of running and weightlifting, getting themselves in the best possible shape for the upcoming flight.

Borman wanted to keep the *Gemini 7* mission as simple as possible. He rebuffed attempts to add EVAs to the flight and to try a docking between the two spacecraft using an inflatable docking collar. "Maybe it's the military background," he explained to Harwood. "You always say, 'Okay, what's the main objective?' Let's make sure the main objective gets done, and put all of your resources into your main push." When training for the mission, he was concerned less with how he and Lovell would carry out the flight plan than with how they would physically and mentally endure so long a flight in such a confined space. The mission was significantly longer than any astronaut had ever undertaken, and the previous flights had shown that many problems arise as a mission stretches on. For example, it had been proved that it was almost impossible for crew members to sleep in alternate shifts. As the crew of *Gemini 5* had also found, trash had to be stowed very carefully or it soon became impossible to work effectively in the cabin. For *Gemini 7* another change was attempted: a lesson that NASA would have to relearn many times in its lifetime. Instead of assigning tasks for a precise time, the crew was given more leeway to fit different assignments in where they could.

Staying inside a standard Gemini spacesuit for two weeks would also have been extremely uncomfortable, and so a lightweight suit was created that had a soft hood and could be stored far more easily in the spacecraft if taken off. Some astronauts had hoped that on Gemini missions without EVAs the suits could be dispensed with altogether; but no NASA official was willing to take such a bold step. Instead, it was decided that only one crew member would be allowed to remove his spacesuit at any given time.

It had only been forty frantic days since *Gemini 6* and its crew had sat on the launch pad ready to go. Now, on 4 December, after a breakfast with Schirra and Stafford, Frank Borman and Jim Lovell were inside the *Gemini 7* spacecraft, ready to fly ahead of *Gemini 6* on a completely redesigned dual mission. From the beginning, all went well. *Gemini 7* launched smoothly and entered its intended orbit. Borman quickly turned the spacecraft around for the first rendezvous procedure of the mission: to stay abreast of their spent Titan booster's second stage. It was not as easy as planned; frozen fuel flakes poured out of a broken line on the booster, causing it to tumble and shake. This meant Borman was using more fuel than he intended just to stay next to it, and the erratic tumbling was a little too close for comfort. Soon they left the booster behind and began the major purpose of the mission—the medical experiments.

These experiments were key to understanding the effects of long-term spaceflight, but they were hard to conduct in such a confined cockpit. The astronauts' brain waves were studied using scalp sensors, and Lovell and Borman also had to keep meticulous track of how much food they consumed and excreted. They'd even had to bathe in distilled water before the flight so calcium loss through their sweat could be accurately measured postflight. Other experiments involving active piloting were kept to a bare minimum so fuel could be conserved for the planned rendezvous.

At first, while performing their medical tasks, both men had to wear their spacesuits. The suits were hot, sweaty, and uncomfortable, and Borman made frequent requests to the ground that they be allowed to take them off. After two days they began to take turns removing them, and eventually they only wore them during the riskier phases of the mission. It made the long flight more comfortable for the astronauts, although the biosensors they wore were still irritating. Conditions in the cramped cabin were worsened early in the mission when a used urine collection bag came apart, and it proved impossible to completely clean up the mess. Lovell later wryly commented that the flight was like being locked in a restroom for two weeks.

Meanwhile, on the ground, preparations were continuing at an even more furious pace to ready the pad for a second Gemini launch. Fortunately, the pad had sustained very little damage from the *Gemini 7* liftoff. *Gemini 6* would now be a quick, stripped-down mission with the primary objective of rendezvousing with *Gemini 7*; there was little else in the flight plan. On December 12, a mere eight days after *Gemini 7* launched, *Gemini 6* was back on the pad, with Schirra and Stafford ready for a second try at launching. Once again, they would not lift off—and this time, the risks to them would be far higher.

As the countdown reached zero and the Titan's engines ignited, *Gemini 6* seemed on its way at last. But moments later, with only one second on the clock, the engines shut down. Had the rocket lifted even a fraction of an inch off the pad, it should now topple and explode. The planned procedure for this kind of shutdown called for the astronauts to pull the abort handle, which would have blasted them out of the spacecraft on ejection seats. The emergency procedure was designed to save their lives, but the astronauts stayed put. As Schirra explained to the authors, he could feel by the seat of his pants that the rocket had not moved. "I would have ejected if I'd had to," he said.

But I had heard the booster liftoff in Atlas, and this Titan didn't work exactly the same way, so I knew in milliseconds that something had gone wrong, that we had not lifted off. Having had the experience of lifting off in Mercury, I knew the difference. I had my butt working for me. The rule was to eject, to punch us both out. That was kind of a mission rule, and another of those rules that was kind of a what-if. And the what-ifs were not all necessarily in a row, or in the proper sequence.

Tom Stafford also described the tense moment, disclosing to the authors one other reason they decided to stay put:

We came real close; we got down to three seconds, two seconds, the engines lit off, you feel the booster start to shake, rattle, and roll, build up—and suddenly, at T-0, the engines shut down. We had a little light, which was kind of a liftoff light; there was a computer start and also a timer. They'd told us that when you have that light, the plug has pulled out the bottom of the booster, and that takes fifty pounds of pull force, so you know you've lifted off. We had the liftoff signals and everything, but we knew from the seat of our pants that we hadn't lifted off. We didn't have that push on our back. Wally's responsibility—and I was backing him up—was to pull the ejection D-rings. We didn't do that, which was a good thing. We could feel we hadn't lifted off, but there was also one other last cue, a very valuable one. We never heard a liftoff call: that's the final thing you should hear. We always had a fellow astronaut watching the TV camera monitor screen, and he's the only one who can call liftoff. And that was Al Bean. Al was looking at the booster, and it didn't move. He didn't say a thing, and that's exactly what he should have done—say nothing. It was kind of a close call, and I was quoted as saying, "Aw . . . shucks!" And a few other words!

It was later determined that an electrical plug had disconnected from the rocket a fraction too early, shutting down the engines before the rocket was able to move. This anomaly activated a clock that was not supposed to start until the vehicle was actually moving. The clock display was visible to the astronauts in the cockpit, and if Schirra had believed it he would have assumed the vehicle was rising and quickly ejected. Instead, Schirra trusted instincts over instruments, and he was correct. Stafford, trusting his commander implicitly, followed his example. Had Schirra and Stafford ejected from the spacecraft, it would have been damaged beyond repair by the ejection-seat rockets, and the rendezvous with *Gemini 7* would never have happened. Instead, the duo sat through the tense moment, remaining seated until the rocket was made safe and they could unstrap and exit the spacecraft, over an hour and a half later. "We climbed down, very disappointed," Stafford recalls. It was a moment that forever impressed their astronaut colleagues. "Schirra had balls," Gene Cernan would later write. "He was a cold-nerved pilot, by God."

Fixing the plug issue was easy. After a review of the rocket data indicated another anomaly, NASA engineers carefully examined the rocket and found a plastic dust cover in an engine inlet. It was a lucky find, as the cover would very likely have caused other problems during the launch. Most probably, it would have led to a shutdown after two seconds, by which time the rocket would certainly have been moving upward and the use of ejection seats would have been mandatory. "It turned out we had two failures that morning," recalls Stafford. "Normally, you are not supposed to have a double failure when you fly in space, just like when you fly an aircraft. We got hit with two of them simultaneously. That really saved our lives." Because of Schirra's risky decision, the mission was saved. The spacecraft was intact, and NASA was ready to try and launch again. They shaved a day off the usual four-day preparation schedule, and successfully launched on 15 December. It was the third time that the crew had tried to launch, and they rocketed from the pad with a deep sense of relief.

Schirra and Stafford, by now highly attuned to the slightest movement of the booster, sensed some shimmying as the rocket lifted off. Schirra talked aloud to the booster, urging "For the third time, *go*!" And the rocket burn was a good one. Borman and Lovell, orbiting far overhead, had a view of Cape Kennedy through their windows but did not spot the moment of launch. They did, however, see the booster's contrail and knew that they would soon have company in orbit. "On the Gemini-Titan," according to Stafford, "we were in orbit in five minutes and thirty-five seconds. It was a real hot-rod, riding that! The first launch, you never forget. You had about 6 g's first stage, then bang, you stage in a fireball; then the second stage, we were out there pushing nearly 8 g's. You're pushed back, and you go from 8 g's to zero-g in a tenth of a second. That was a real ride!"

When piloting the spacecraft, Schirra did not have to rely on information from the ground to make a rendezvous. If he needed it, he had all the guidance and navigation equipment on board to fly his way to a precise meeting in orbit unassisted. Gemini became Schirra's favorite flying vehicle, bringing him the closest of all the spacecraft he flew to that ultimate for a test pilot—the harmony of man and machine:

Without a doubt, Gemini was my favorite flight; I appreciated it the most. In Mercury, you couldn't translate. You could just change attitude. But you were actually flying it like a flying machine in Gemini. Apollo was just too big. I did *Apollo, but I* enjoyed *Gemini. Gemini was just about the right size—it was not much larger than Mercury, really, and it was optimized for both my right and left hand. I did translation with my left hand, which is a very delicate maneuver—a bit like when the shuttle docks with a space station. My left hand was better than my right hand for that. I made tiny, tiny thrusts, and developed that technique to perfection.*

As he told the authors, Stafford was also impressed by the piloting pleasure of the Gemini craft in comparison to the later Apollo spacecraft: "Gemini was a lot more of a sporty ride than Apollo on launch, and in terms of spacecraft, the real flying was Gemini, too. This was in part because the Gemini was real small, of course, like an F-104 fighter plane."

After performing four maneuvers soon after reaching orbit, Schirra turned on the radar and locked on to his target, still hundreds of miles away. "It was as if we were two horses on a racetrack," Stafford would later state. Five hours into the mission, *Gemini 7* was only a hundred miles away, and to Schirra the other spacecraft looked like a bright star. Stafford gave his commander precise navigational guidance, verifying it with radar data from the ground, and the duo's months of training paid off perfectly. Less than six hours into the mission, Schirra was already braking as he sidled up to within 120 feet of *Gemini 7*, with plenty of maneuvering fuel still in his tanks. The rendezvous had been even easier to carry out than in the simulators. The Gemini program had just achieved its most important objective.

Borman and Lovell had watched them slowly approach, surprised by the bright flames of *Gemini 6*'s thruster firings. "I can still recall them coming up from below at nighttime, their thrusters firing as we came together," Lovell recounts. The *Gemini 7* crew needed some excitement at this point in the mission, and the rendezvous gave it to them. Borman had remarked a few days before that "the days seem to be lengthening a little . . . we're getting a little crummy." Considering how the spacecraft smelled after eleven days, Borman's comment was a tactful understatement.

There was another surprise for both crews when they drew even closer and took a good look at each others' spacecraft. Neither had disconnected cleanly from their boosters, and both were trailing the separation cords behind them; some of the cords were sixteen feet long. "It was quite a shock when we saw all the dangling particles," Schirra recalled.

With so much fuel still in the tanks, *Gemini 6* remained the active partner in the rendezvous as a variety of station-keeping and fly-around maneuvers were practiced. "We were the target craft," Lovell explained, "while *Gemini 6* rendezvoused with us." The spacecraft remained together for more than three orbits, which included remaining very close to each other in the darkness of the night side of the planet. For Schirra, it was the high point of his NASA career—the moment when he could feel the spacecraft respond crisply to his every command, as he flew it precisely around *Gemini 7*. "The ground was talking in terms of feet," Schirra explained, "but we were closer—six to eight inches. That's how deftly the spacecraft could be controlled." Stafford was also very impressed with the handling of their spacecraft as he and Schirra took turns flying. As they drifted nose to nose with *Gemini 7* Schirra quipped that the only downside was the view; he had to look at the tired-looking Borman and Lovell through the windows.

In an orbital meeting of three naval aviators and a pragmatic West Pointer who disliked practical jokes, it was time, Stafford told the authors, to have some fun at Borman's expense:

It was right around the first of December when we started planning this. We were sitting around the crew quarters, we were super trained; there was Al Shepard, a Naval Academy grad, Wally Schirra, Naval Academy, and myself. I don't know which one of us came up with it—I think it was Al Shepard—he said "How about, if everything works out right, let's put up a sign, Tom, in your window, a blue sign with a gold Beat Army.*" So in orbit I said, "Jim, the doctors are concerned about your condition: you have been up here longer than anybody has before in space, and they want to check your visual acuity. Take a picture of this and say what you see!" So with that I put up the* Beat Army *sign, and he shot a picture of it. It ended up on the front page of the* Washington Post *a few days later. You can see that plaque in the Naval Academy museum back east. It's the highest Beat Army rally we've ever had!*

As Borman later related, "Wally was always one to inject some levity into the program. . . . He has that little quirk of being able to include some fun with things. I never had that. I didn't think much about the 'Beat Army' sign. . . . I guess it is just the way I am built. . . . I wanted to do the mission and I didn't care about the other crap. . . . All the funny fun and games, it was a bunch of crap as far as I'm concerned."

Though there had been time for joking, it had been a busy day for the *Gemini 6* crew, and all too soon Schirra and Stafford thrusted away from *Gemini 7* to the safe distance of ten miles, where they enjoyed a well-earned sleep. They had made rendezvous look easy, but in reality only a huge amount of preparation had enabled them to pull it off so masterfully. *Gemini 6* was Schirra's kind of mission—it involved a lot of piloting, and it was relatively short. After little more than a day in space, with the rendezvous objective a stunning success, they readied themselves for the return to Earth. Schirra was also feeling the effects of a head cold, which made him want to return even more. On his next spaceflight, he would not have the luxury of a fast return when suffering from a stuffy head.

It would not be a Wally Schirra space mission, however, without one last quip. Tom Stafford radioed *Gemini 7* and the ground and excitedly reported, "We have an object: looks like a satellite going from north to

Postflight, the *Gemini 6* crew of Wally Schirra and Tom Stafford on the recovery carrier uss *Wasp*. Courtesy NASA.

south, probably in polar orbit. . . . Looks like he might be going to reenter soon." Astounded mission controllers were mystified at this report of a potential UFO, and were even more astonished when Stafford next reported that he was picking up communications from the unknown craft, which he would relay to the ground. As Stafford explains, it was a perfect prank for the time of year:

That was just before Christmas time, and about one round before retrofire I called ground control and told them we had something coming towards us, which looked like it was coming from the north to the south. The CapCom was getting excited, and I told them the angular rate was nearly zero, and this thing was heading straight towards us. "It's getting nearer," I said. "We can hear some noise: listen." And about that time, Wally played "Jingle Bells" on a little harmonica, and I jingled some little bells. You can see them in the National Air and Space Museum, in the gallery upstairs.

With a final comment to *Gemini 7* from Schirra—"We'll see you on the beach"—*Gemini 6* began its reentry. Schirra flew manually for much of the journey down, banking the spacecraft to reduce lift and range. "In Gemini, you could change your flight path by rotating," Schirra told Roy Neal. "You were actually controlling it through the atmosphere on reentry." On 16 December they splashed down in the western Atlantic Ocean, where they were recovered by helicopters from the aircraft carrier USS *Wasp*. "We were really a happy bunch of fellows," Stafford recalls, "although it had taken us three launch attempts to get off." Three days later, the carrier would collect another Gemini spacecraft.

For Borman and Lovell in *Gemini 7* an already tiresome mission grew steadily worse. With almost three days left to go, they now had nothing left to look forward to except homecoming. As Tom Stafford would describe it, "two weeks in a Gemini . . . was like spending two weeks eating, sleeping, working, and going to the bathroom stuffed into the front seat of a sports car wearing an overcoat." The crew ate food that was covered in wax to help preserve it, hating the taste and the unpleasant feeling of wax coating their mouths. Adding to the discomfort, their skin was flaking in the dry, pressurized atmosphere, and their hair was full of dead skin particles. They had also lost a toothbrush, and were forced to share the remaining one. "We were a pretty odorous pair," Borman would later recall.

With the rendezvous over, at least the spacesuits could be removed again. "My grandkids say that I spent two weeks around the Earth in my underwear," Lovell jokes. Both crewmen, now with noticeable beards, occupied themselves reading novels. Borman later admitted that they were bored. "We were waiting for time to pass . . . checking off days . . . We did not have enough attitude control fuel to maintain an attitude so we just drifted and tumbled for days at a time. And that is rather frustrating." He found those last three days to be endlessly long and tiring, and says his "sense of mission" was the only way he could stand the boredom and fly for the full duration.

As Jim Lovell told the authors, "Frank says that his *Gemini 7* flight was worse than our *Apollo 8* flight, in terms of his two flights. *Apollo 8*, of course, was such an historic mission, so many things were happening, whereas *Gemini 7* was just really, hang in there. It was tedious work. Frank really wanted to come down after about twelve days, because our fuel cells were really dying, and we were losing a lot of our reaction control thrusters. I argued with him: told him not to worry." As they circled the Earth

Glad to be home: Jim Lovell and Frank Borman grin with relief after their long-duration *Gemini 7* mission. Courtesy NASA.

together, all alone, they sang a Jim Reeves country-and-western song to each other, over and over again. Borman later recounted, "We were too tired even to take each other's pulse." Both of them had already decided that, immediately after splashdown, they wanted to get out of the cramped cockpit instead of waiting to be lifted to the deck of the recovery carrier.

At last, the two weeks were up. Relying mostly on instruments and making minor adjustments to the course the computer was flying, Borman and Lovell flew a smooth reentry, working together as a team so one person could look at the instrumentation while the other checked the horizon through the window. As they did so, an experience that Borman compares to "flying in a neon tube," they fought off the deadening effect of returning gravity. After two weeks in space, the weight was uncomfortable, but bearable. Deploying their parachutes, they hit the ocean with a satisfying thump. Fighting off a dizzy feeling, Borman tried to take off the annoying spacesuit, but found it too difficult a task after two weeks of weightlessness. They soon began to feel better, however, and by the time they were aboard the recovery carrier USS *Wasp* they were able to walk unaided. The doctors who examined them were delighted. They were in much better shape than had been feared; there now seemed to be no physical obstacle to flying long missions to the moon. "The only thing I really felt after two weeks like that was that our leg muscles were shot," Borman recalled.

NASA rounded out 1965 with five successful Gemini missions completed and a huge number of important objectives achieved, all vital to future space plans. Borman and Lovell, who had endured the misery of long-duration flight to prove it could be done, would get their reward. They would end up being on the very first flight out to the moon together. Despite the launch setbacks and last-minute change of plans, or perhaps because of them and the initiative they inspired, the rendezvous of *Gemini 6* and *7* was an incredible achievement for the program. "The mission will be there in history forever," Stafford stated decades later. And as Jim Lovell told the authors, the rendezvous success was the culmination of a fast and effective process: "On *Gemini 4*, when McDivitt and Ed White tried to do the rendezvous, they kept circling around, because they didn't realize the aspects of orbital mechanics, where you retrograde to get up there, and you posigrade to go down—all that sort of stuff. It took a little while for that type of rendezvous to be developed. *Gemini 6* and *7* was a real rendezvous—by that time we understood orbital mechanics. We completed the first rendezvous in space."

As for the medical experiments, they had also been very important, but the astronauts were far less likely to enthuse about them. For Borman, it seems, *Gemini 7* was enough for one lifetime. In his memoirs, former scientist-astronaut Brian O'Leary vividly recalled a meeting the astronaut corps had with NASA Administrator Jim Webb two years after the *Gemini 7* flight. According to O'Leary, when one of the other scientist-astronauts tried to ask Webb a question about science in future manned space plans, Borman interrupted loudly and without warning: "To *hell* with the scientific community!"

NASA looked hopefully toward a year of similar success in 1966. Instead, the first flight of the new year saw their first major mission failure, and the most dangerous moment astronauts had ever faced in space.

Now that two Gemini spacecraft had rendezvoused in orbit, it was time for NASA to try to accomplish the original objective of the *Gemini 6* mission—a docking with an unmanned Agena target. The *Gemini 8* mission was scheduled to last for three days, and in addition to the docking plans included an ambitious EVA. The spacewalk was to last for two hours, with the pilot strapping on an Extravehicular Support Package (ESP). Then, using a maneuvering gun, he would float over to the Agena, while remaining tethered to the Gemini spacecraft. Once there, he would remove a package from the side of the spacecraft that had been designed to detect micrometeorites and test a special power tool. Having accomplished these tasks, the spacewalker would then float to the end of his long tether and be pulled for a distance by the thrusting Gemini spacecraft. It was an ambitious leap from the relatively easy floating EVA conducted on the *Gemini 4* mission. Dave Scott, chosen as pilot and spacewalker for the mission, put in a huge number of training hours preparing for the difficult assignment.

David Randolph Scott was born on 6 June 1932 on an air force base in San Antonio, Texas, the son of an air force general. His family life was one of strict discipline and numerous rules, but Scott seemed to happily accept these facts and was keen to follow his father's career path. He attended a rigidly strict military school, where his father pushed him hard to be assertive and successful. Scott was accepted to the University of Michigan but only stayed a year because he earned a chance to achieve his ambition to attend West Point. His intensely competitive streak enabled him to graduate fifth in his class in 1954. Joining the air force and receiving his wings the following year, he served in a tactical fighter squadron in the Netherlands, marrying the daughter of another air force general in 1959. He then resumed his academic studies in 1960.

Scott had been told that obtaining a degree was the best way to get the best test-piloting jobs, so he earned a master's degree in aeronautics and astronautics at the Massachusetts Institute of Technology, specializing in interplanetary navigation. In 1962 he achieved his wish and returned to piloting, attending the air force's test pilot school at Edwards. Graduating in 1963 as the best pilot in his class, he moved up into the Aerospace Research Pilot School, also at Edwards, where future spacefarers were groomed. Possessing the perfect background for the job, he was selected to join NASA in 1963 as part of the third group of astronauts. He saw the assignment as less of a career move than an "evolution"—a natural move for a promising test pilot. He planned to fly in space a couple of times and then return to a more mainstream air force career.

At NASA, Scott was quickly recognized as being one of the best pilots in his astronaut group. Very highly ranked by his peers, they described him as "All-American," and also joked that Scott came closest to the iconic persona of the tall, rugged, patriotic Cold War hero and was the one who looked most like the movie version of a daring astronaut. Over thirty years later, Scott still has that air about him; although time has altered his features, he now bears a striking resemblance to another American icon, the late actor Lee Marvin.

Given Scott's test-piloting credentials, it was little surprise to his peers that he was the first of that group to receive a flight assignment. He didn't even have to serve on a backup crew first. Scott had concentrated on spacecraft guidance and navigation during his training, which fit the *Gemini 8* mission objectives perfectly. Indeed, Mike Collins later wrote that seeing Scott picked first reassured him that NASA knew what it was doing.

Commanding the mission would be another space rookie—Group 2 astronaut Neil Armstrong. Armstrong was the last of his selection group to be given a prime crew flight assignment. He had in fact been waiting so long that he was now commanding a member of the next group. Although Scott found Armstrong "a taskmaster" at times, the two had a great mutual respect and worked well together. Armstrong has become one of the best-known personalities of spaceflight, primarily because he happened to be in the right place at the right time, picking up command of the most famous Apollo flight of them all. In the context of his astronaut peers at the time, however, there was nothing especially remarkable about him. His relative blandness could almost be said to be his most distinguishing feature. Armstrong has spent his life following his own interests and instincts, not allowing fame to change him at all. Four decades after he first flew in space, he comes across like an amiable college professor who loves his subject and would be happy to discuss it all day with his students. As he said to reporters in 1989, "I am still hemming and hawing in front of the cameras just like I did twenty years ago." His mild manners and modest demeanor color Armstrong's attitude toward himself: he loves aerospace engineering and sees his astronaut career only in the wider context of advancing that technical field. He's a keen enthusiast of the airplanes and cars of the 1930s, and one almost has a feeling that he would have preferred to have been an adult in that era, when one small workshop could lovingly design and craft a cutting-edge vehicle, in contrast to the tens of thousands needed for Gemini and Apollo.

The other, sterner and warier side of Armstrong's personality sometimes comes to the fore when unwanted demands of fame invade his relatively private life. Like John Glenn, Armstrong has a profoundly deep sense of right and wrong, but an occasionally overdefensive sense of propriety has led him in the past to rebuke and frustrate some well-intentioned people. "Neil doesn't waste words," Mike Collins once revealed, "but it doesn't mean he can't use them." As pad leader Guenter Wendt also noted, Armstrong has never liked other people telling him what to do. There are few who don't respect him, but for some his refusal to do anything except on his own terms can seem overly stubborn.

Neil Alden Armstrong was born on a farm near Wapakoneta, Ohio, on 5 August 1930. His father was an auditor for the state's county records, and so the family moved all the time—six times in his first six years. A bookish child, he showed an unusually early academic intelligence, so much so that he was able to skip second grade completely. He learned to fly before he could drive, making his first flight at the age of six and obtaining his pilot's license on his sixteenth birthday. He became an aviator primarily because he wanted to learn how to improve aircraft design based on experience; he'd decided he wanted to be an aircraft designer while still in elementary school. When other teenagers were out having fun, Armstrong was earnestly building a wind tunnel in the family basement, studying calculus outside of school time, and collecting vintage air magazines, which he kept well into adulthood. "I filled notebooks with scraps of information about makes of aircraft, specifications, and performances," he would later admit. Not only was he shy; he also looked strikingly younger than he was. Armstrong once said that when he was sixteen he looked like a twelve-year-old. He was, unsurprisingly, never particularly athletic in school, and instead played the horn in the school band. His brother later told reporters that Neil rarely dated and that when he left for college he was "immature and withdrawn." He may sound like the ultimate nerdy teenager, and in many ways he was, but he also had a firm wish to do something record-breaking. By the time Armstrong became a pilot, however, he felt he had missed his chance to do anything dramatic in the aviation field. It seemed that all of the great record-setting flights around the globe had been completed; he was, he believed, a generation too late.

History would soon prove Armstrong wrong. While studying aeronautical engineering in college just after World War II, he benefited from the huge technical strides the conflict had spurred—in particular, the jet engine. He was only able to go to college because the navy was paying for his tuition, and in 1949, after just a year and a half in college, Armstrong was pulled out for active duty as a fighter pilot. The nerdy young man would now have to grow up whether he liked it or not. "I was very young, very green," Armstrong would later say of that time. He was assigned to a squadron fighting in the Korean War, and quickly became a seasoned combat aviator, flying seventy-eight combat missions from aircraft carriers.

Resigning from the navy and returning to college in 1952, he graduated three years later with a degree in his beloved aeronautical engineering. He then went to work at the National Advisory Committee for Aeronautics (NACA), the aviation precursor to NASA. Transferred to the NACA station at Edwards Air Force Base, he began work there as a civilian test pilot. His duties included flying simulated launch and landing approaches for the X-20, a proposed air force winged spacecraft. In 1960, he was assigned to the program as a pilot, although there was as yet no X-20 to fly. In the meantime, he also flew the X-15, the rocket plane that was capable of brief suborbital spaceflights. Armstrong, however, was never permitted to take it that high. "Neil had been considered one of the weaker stick-and-rudder men," Mike Collins would write, "but the very best when it came to understanding the machine's design and how it operated."

Now married to his first wife, Janet, Armstrong characteristically chose not to live in any of the towns near the base. Instead, the family lived in a former forest ranger's cabin, in the foothills of the nearby mountains. "It was one of nature's quiet hideaways," he would later describe, adding that it was "an outpost of serenity." Living away from everyone else and flying solo in the most cutting-edge aircraft, Armstrong was in engineering heaven; he described these experiences as "the most fascinating time of my life." By this time he could see that becoming an astronaut would allow him to further explore his love of engineering by going beyond the "atmospheric fringes and into deep space work; . . . throughout my seventeen years at NACA and NASA, a very large percentage of my time was involved in real engineering work."

When he joined NASA's astronaut second group in 1962, much press coverage was made of him being one of two "civilian astronauts." In fact, of course, he'd had an active military career, including combat duty. The other so-called "civilian" in his group, Elliot See, had also been a naval aviator. The two men were assigned together as the backup crew for the *Gemini 5* mission.

Armstrong continued his somewhat monastic lifestyle when he moved to Houston. "Probably not the most charismatic individual; . . . most would agree that he did not make friends easily," Guenter Wendt would later write. Reporters noted that Armstrong "sought utter privacy" and not just from media intrusions. One writer noted that he was so reticent that even his parents did not know what his philosophical and religious beliefs were. Another, a little less tactfully, wrote that Armstrong "surrendered words about as happily as a hound allowed meat to be pulled out of his teeth." Socializing or sharing thoughts about anything but the job at hand was never going to be something he cared about. "I am not very articulate," he told reporters at the beginning of the new millennium. "I am, and ever will be, a white-socks, pocket-protector, nerdy engineer."

As *Gemini 8*'s launch day approached, there was concern within NASA over the Agena rocket's reliability. One had blown up in a test only a few weeks before. Mission Control instructed the astronauts to "get out fast" if the Agena showed any signs of malfunction during their flight. But all began well on 16 March when the Atlas-Agena target vehicle launched successfully after a one-day delay. Its trajectory was a little low at first, but the engines on both Atlas and Agena worked efficiently, with none of the problems that had disrupted the prior attempt. A suitable orbit was attained.

Given a definite target, Armstrong quipped, "Beautiful, we will take that one." *Gemini 8* followed the Agena into orbit an hour and forty minutes later, exactly on time. It was the first time that two space vehicles had simultaneously counted down and launched at the precise second planned, and all looked good for another successful rendezvous. Schirra and Stafford had demonstrated how to catch a target in space; now the *Gemini 8* crew followed much the same procedure when chasing their Agena, making nine maneuvers to synchronize their orbit. During the first orbit, on the dark side of the Earth, the crew observed their thrusters firing, compensating for an unwanted movement. It resembled the situation encountered on *Gemini 3*: the radiator was venting and turning the spacecraft. It was not a serious problem, however, and the mission controllers paid it little attention. In between making the maneuvers, Armstrong and Scott even took time to eat a meal and to look out of the windows at the Earth below. Armstrong looked down with particular interest at Edwards Air Force Base, where he had spent so much time as an aviator.

As they drew to within eighty-seven miles of the Agena, the two astronauts spotted it shining in the sunlight. Drawing ever closer, Armstrong made slight braking maneuvers, trusting his eyes while Scott read the radar readings to him. Eventually, the two spacecraft were moving at the same precise speed and in the same orbit. Less than six hours after launch, *Gemini 8* was flying in formation with the Agena, with only 150 feet separating them.

For half an hour, as Armstrong kept *Gemini 8* in precise position, Scott carefully scrutinized the Agena, making sure that it looked stable and intact. With everything looking perfect, Armstrong gently coaxed their spacecraft to within three feet of the target, and then slowly, carefully, he eased into the docking collar. "Flight, we are docked!" he reported, adding happily that the docking was "really a smoothie—no noticeable oscillations at all." For the first time, two spacecraft had docked in orbit, and one of the major objectives of the Gemini program had been achieved.

With a communications station in Zanzibar closed because of political troubles, however, *Gemini 8* was out of contact with ground controllers during the dramatic, life-threatening events that followed. The Agena had been instructed to fire its control thrusters in order to keep the two spacecraft at the same angle relative to the direction of orbital travel. Agena rockets had malfunctioned before, and the flight controllers had not been able to verify this Agena's program commands before losing contact. So as *Gemini 8* was passing out of communications range, the astronauts were instructed to turn the Agena thrusters off and take control of the spacecraft if such a malfunction occurred.

Dave Scott commanded the Agena to turn both spacecraft to the right, a maneuver that initially seemed perfect. The docked spacecraft were in darkness, so it was only by glancing at his cockpit instruments that Scott noticed they had yawed thirty degrees from their planned position. Armstrong also took a look, and confirmed the yaw with his own instrument indicator. Unless both indicators were malfunctioning, the yaw was real. Armstrong carefully moved the two spacecraft back into the correct position using Gemini's thrusters. However, as soon as he released the hand control, the joined spacecraft resumed a slow, unwanted yaw. With no telemetry information from Gemini or Agena available via Mission Control to explain the motion, the astronauts had to analyze the problem on their own.

Both astronauts felt the problem probably lay with the Agena's control system thrusters, so Scott sent and continued to send shutdown commands to Agena's attitude control. This, however, only made the situation worse; as well as yawing, the connected vehicles now began to spin together, faster and faster. Worried that the docking adapter connecting them might break under the strain, they used their thrusters to slow the spin enough to disengage from the Agena. Half an hour after they had first docked, Scott hit the emergency release, while Armstrong fired the thrusters, pushing the spacecraft backward and away from the Agena.

The astronauts did not know it at the time, but the thruster problem was in their spacecraft, not the Agena. Now separated from the other vehicle, things only got worse aboard *Gemini 8*. With its total mass now suddenly reduced almost in half with the Agena detached, it began to roll to the left and tumble even faster. Soon it was rotating once a second. Items that had previously floated in the cabin now stuck to the spacecraft walls due to the centrifugal force. The cause, Armstrong and Scott surmised, was one of *Gemini 8*'s thrusters firing on and off intermittently. They had little time to identify which one. Armstrong had not initially believed their Gemini craft was the culprit because its thrusters normally made a loud cracking sound when they fired. In this emergency, he hadn't heard anything.

Armstrong and Scott's vision began to blur, and their heads banged together. Scott later recalled that both of them were getting close to the onset of tunnel vision: a dangerous development. Without their eyesight, they would have been helpless. He also felt he was close to blacking out, after which he knew it would have been all over for them. But there was no time to think about such things—there was only time to try and find a solution. "A test pilot's job is identifying problems and getting the answers," Armstrong would later relate. "We never doubted that we would find an answer—but we had to find it fast."

The spacecraft was once again within range of ground stations, but as the radio antenna was spinning with the spacecraft, communication with the ground was almost impossible. Chris Kraft in Mission Control later admitted to feeling frightened and helpless, having two astronauts "alone up there with an out-of-control spacecraft." Armstrong and Scott, however, were both highly experienced test pilots and were used to solving problems independently. The ground could not have helped them anyway. As Scott later compared, if a test pilot gets into a spin in an airplane only the pilot can work his or her way out of it; the ground cannot help. On previous missions, the astronauts and mission controllers had worked well together as a team. This time, all the ground could do was hope the crew knew their spacecraft well enough.

Armstrong was admittedly not an intuitive pilot. "It has never been a strong point," he once told reporters. His engineerlike credo was instead to "interpret the problem properly, then attack it." The astronauts were now also facing the grim possibilities of colliding with the Agena while rotating wildly or their spacecraft disintegrating as it spun beyond its structural strength. "We were concerned that the stresses might be getting dangerously high," Armstrong would explain. "The spacecraft might break apart." With blurred vision making it difficult to see individual thruster switches, they had no choice but to shut down all sixteen of the OAMS thrusters, which were used to change orbit and attitude. Now, at least, they would not roll any faster. However, with no atmosphere or gravitational effect to slow the relentless tumbling, simply stopping the thrust did not slow them at all.

With no time left to work out which OAMS thruster had caused the problem, Armstrong had to use the spacecraft's other set of thrusters, the Reaction Control System (RCS), to halt the rotation before they both lost consciousness. Finally, the spacecraft's roll slowed, then stopped. When Armstrong began to individually test the OAMS thrusters, the rolling began once more, so he shut them off again. Something was short-circuiting within the system that controlled them, causing a twenty-three-pound thruster to fire at first intermittently, then continuously.

The mission rules stated that once the RCS reentry thruster control system had been activated, Gemini had no choice but to return to Earth immediately. "I made that decision reluctantly," Armstrong reported, "because once the decision was made, the mission had to be terminated. That excluded Dave's EVA . . . and that hurt." Having trained for so long, especially for the EVA, both Scott and Armstrong had hoped to continue the mission. But much of the fuel reserved for reentry had already been used saving their lives, and only one set of thrusters could now be relied upon. It was wise to return as soon as possible.

Armstrong and Scott hurriedly packed away a floating pile of items, including bulky equipment that had been designed for the now-cancelled EVA and was never intended for return to Earth. A fleet of recovery ships rushed into position as the two astronauts quickly prepared for an early return during their seventh orbit. Armstrong, concerned that he might make a navigational error that would lead them to return far off course, asked Scott to double-check everything he did. "Being an old navy guy, I much preferred coming down in the water to coming down in Red China at the time," he'd later explain.

After being in space for almost eleven hours, the *Gemini 8* spacecraft successfully splashed down in an emergency recovery zone in the western Pacific Ocean, 500 miles east of Okinawa. At first, the recovery forces were too distant for them to make contact, but soon the ships grew close enough. Over three hours after the troubled spacecraft had hit the ocean, the USS *Leonard Mason* picked up the disappointed astronaut duo. After their long wait in the hot spacecraft, both men were nauseous from the ocean swells.

It had been a frustratingly brief mission, and yet it was hard for NASA to be completely disappointed. After all, another historic first—the first docking of two spacecraft—had been accomplished. "The whole program learned something," Scott told reporters after the flight. "It proved that the systems we have are good systems and that the backup capabilities are there. It showed that if things go wrong and we have major problems that we can take care of the problems." Flight Controller Gene Kranz felt there was even more to learn. "We were just lucky," he stated. "And luck's got no business in spaceflight. So we had to eliminate that luck component."

Skill had been even more important than luck, and the flight had publicly demonstrated the test-piloting skills of both Armstrong and Scott. When the spacecraft had begun its wild tumbling, Julian Scheer, NASA's

Neil Armstrong (*left*) and Dave Scott, crew of the troubled *Gemini 8*. Courtesy NASA.

first public affairs director, had been concerned that he was about to hear Armstrong and Scott's last words. Listening to the tapes of the dangerous moments shortly afterward, he was amazed to hear how calm and in control the duo sounded at all times. As Gus Grissom wrote, "Gemini 8 reminded the public, briefly, that spaceflight was not yet as simple as a ride on a trolley car." Frank Borman put it even more starkly when he said, "they damned near became the first U.S. casualties in space."

Scott, in particular, had shown incredible presence of mind during the unexpected events of the *Gemini 8* mission. Even in the middle of an emergency, out of contact with Mission Control, he had thought to reenable ground control command of the Agena before the two vehicles separated. This meant that for the following two days, the ground could thoroughly test the Agena's thrusters and ensure that it had not been to blame. The Agena was in good enough shape that it would remain in orbit for another visit by a second Gemini crew. The skill Scott had shown was reflected in the fact that, only five days after returning from *Gemini 8*, he was assigned to an Apollo crew.

And yet, not everyone was full of praise for the duo. Just as Gus Grissom and Scott Carpenter had endured a whispering campaign after their Mercury missions, there were those who believed the mission failure might be due to something the pilots had or had not done. "There was some talk behind our backs," Scott wrote in his memoirs, "by those who thought we had screwed up . . . unnecessarily turned on the re-entry control system, which aborted the mission." If this was the case, it certainly didn't affect Scott's assignments at all. Armstrong initially had a far less promising assignment: a dead-end job backing up *Gemini 11*. This meant that he would neither fly Gemini again nor be immediately available for an early Apollo crew. If it had not been for a lengthy delay in the Apollo program due to the death of some of his colleagues, it is possible that he would have been waiting for a second flight for quite some time. As it was, Neil Armstrong would ultimately do very well with his next, and last, prime crew flight assignment.

3. The Ballet of Weightlessness

No man's knowledge here can go beyond his experience.

John Locke

It had taken eight Gemini missions to achieve, but it seemed that rendezvous and docking was now, if not perfected, at least understood. Rendezvous and docking still had a critical role to play in the final four Gemini missions, but now there was another function that the astronauts had to master: spacewalking.

Unfortunately, NASA had no real understanding of the EVA difficulties they would soon face. On *Gemini 4*, a super-fit Ed White had made it look relatively easy, while the Soviets were keeping tight-lipped about the extreme EVA difficulties experienced by Alexei Leonov on *Voskhod 2*. Dave Scott had planned to conduct an ambitious EVA on the curtailed *Gemini 8* mission, but it had not been possible. *Gemini 9* had some catching up to do.

Yet it seemed that nothing was ever going to go as planned for this flight. Originally, four astronauts were considered for *Gemini 9*'s prime and backup crews; only one of them would be alive on the day the mission launched. On the last day of October 1964, Group 3 astronaut Ted Freeman was undertaking a routine flight out over the Gulf of Mexico in order to maintain his proficiency in one of NASA's T-38 jet airplanes. He was asked to abort his first landing back at Houston's Ellington Field, as there was other airborne traffic in the area. On his second approach a flock of snow geese intersected his flight path, and one bird smashed violently into his airplane. The Plexiglas windscreen shattered and was ingested into the engine, which promptly choked and flamed out. Freeman managed to dead-stick the T-38 away from a populated area, but he ejected too late to save himself. His parachute did not have time to fully deploy, and he hit the ground at high speed. Had he lived, Freeman was said to have been in line for assignment to the backup crew for the *Gemini 9* mission, together with Tom Stafford. With Freeman gone, astronaut Gene Cernan was assigned to the backup crew, supporting prime crew astronaut rookies Elliot See and Charlie Bassett.

On 28 February 1966, both the prime and backup crews for *Gemini 9* were flying into St. Louis, Missouri, to inspect their spacecraft, then under final construction at the McDonnell plant. Bassett and See were flying in one airplane, with Stafford and Cernan at their wing in a second T-38. Conditions at the field were atrocious, with rain, heavy cloud cover, and blinding snow flurries. Both airplanes missed their first landing attempt and aborted. Stafford and Cernan flew back up into the clouds, but See decided to attempt a visual approach. Unfortunately, he was traveling too low and slow, and one of the T-38's wings clipped the top of the three-story building in which their spacecraft was undergoing completion. The airplane slammed into the roof, bounced, and exploded, before tumbling into a courtyard parking area. See and Bassett were killed instantly. In the aftermath of the tragedy, the backup crew of Stafford and Cernan became the prime crew, and the follow-on effect of this changed crew assignment would later directly influence the selection of crewmembers for the Apollo missions.

It had thus taken the deaths of three men in line ahead of him for Gene Cernan to end up as a prime crew member. In some ways, it is characteristic of Cernan to end up somewhere he never expected to be. Even today his enormous energy and enthusiasm is underpinned by a bashful sense of modesty. "If you're not careful you feel like you are some kind of hero," he told the authors. "I don't want to feel that way."

Gene Cernan looks and sounds like the kind of presidential politician you only ever see in Hollywood movies. He is sharp, sincere, and can charm anyone, from a small gathering to a crowd of thousands. He talks from the heart and is skilled in working an audience up to a standing ovation. Yet, unlike many real-life politicians, he is never too slick. He is genuinely at ease in the highest social circles and believes in the words he uses so convincingly during public speeches. He never quite seems to accept the mixture of skill and circumstance that allowed him to have a successful astronaut career. "I had ambitions and dreams," he confessed, "but I also made a lot of mistakes. Even today, I sometimes wonder, why me?"

Eugene Andrew Cernan was born on 14 March 1934 in Chicago, and says he "came from an average blue-collar family." He was an "unplanned Depression baby," with a father who lost his job soon after his wife gave birth. It was not a luxurious upbringing. During World War II, Cernan began to make plans for his future while watching newsreels at his local movie theater; the idea of being a fighter pilot greatly appealed to him.

Cernan's father encouraged him to become an engineer, but there was not enough money in the family to send him to the best school, MIT. Instead, he attended Purdue University, in the year behind Neil Armstrong, from which he received a degree in electrical engineering in 1956. He was only able to attend because the navy gave him a partial scholarship. Cernan still dreamed of being a naval aviator flying to and from aircraft carriers, yet he had still never flown an airplane, not even as a passenger. His first experience would be as a passenger on a flight back to college when he was already a student.

Luckily, his career instincts were correct: when he was finally able to fly naval aircraft, he loved it as much as he expected to. Going on active navy duty after graduation, Cernan quickly entered naval flight school, becoming a naval aviator in December 1957. He then served as a fighter pilot with attack squadrons in San Diego. "San Diego is sort of a second home for me," he told the authors. "It is where I spent most of my Navy life." It was while he was in San Diego that he really began thinking seriously of becoming an astronaut:

I remember that day well: it was 5 May 1961. I was stationed at Miramar, just back from my second cruise. I was getting married the next day, but I was not sure what I was going to do with my life at that point in time. That day, Alan Shepard, first American in space, flew little more than fifteen or sixteen minutes. And Al became my hero. The whole world was watching America's response to the Soviets then, and I remember after Al went up someone said, "Gene, how would you like to do that?" And I said, "I'd love it—boy, just give me a chance!" I was not ready to do something like that, but I'd love to do it, I said. But by the time I'd be ready, I thought, by the time I met all of the re-

quirements and was qualified, was good enough—there won't be anything left to do in space exploration. But that was not true. Because it was only ten years, ten years to the month that I was backup commander for Apollo 14, backing up that same Alan Shepard.

By 1961 his obligation to the U.S. Navy was at an end, but that same year Cernan received an enticing offer—the navy would allow him to return to college, and he could continue to fly jets. Cernan agreed to the deal, and in 1963 he earned a master's degree in aeronautical engineering from the Naval Postgraduate School, while also interning at a rocket propulsion facility. He was hoping that the degree would make him more likely to gain entry to the navy's test pilot school and was about to begin a third year of college. Then, to his immense surprise, he discovered that the navy wanted to put his name forward for the next round of NASA astronaut selections. After an exhausting series of interviews, he was told that he had made it; NASA had included him in its third group of astronauts.

It took Cernan quite a while to really believe that he was there. At first, he felt like an "imposter astronaut," sharing the same offices with his Mercury idols. He even asked them for their autographs. He was delighted to be assigned to the *Gemini 9* backup crew, as the rotation system then in place meant he should fly the twelfth and last Gemini mission. "Tom Stafford and I, as different as night and day, became like brothers because we spent so much time together training," he would later write. Cernan felt it was quite an honor to be backing up Charlie Bassett. He considered him one of the best test pilots in the astronaut corps. As well as admiring his professional skills, Cernan also liked Bassett as a close personal friend, and their families spent a lot of off-duty time together. To get into shape for what might be a demanding spacewalk, the two of them lifted weights together in the gym almost every day. Working as a very close team, the duo did everything they could to prepare for a complex and ambitious mission.

Within hours of Bassett's death, Cernan received official word that he was to replace him. He attended Bassett's funeral and then threw himself back into the training with renewed vigor. Tom Stafford was inheriting a mission under difficult circumstances, but there was not really time to think about that. The mission was going to push Gemini's rendezvous and spacewalking capabilities into new areas. "It was a very ambitious flight," Stafford told the authors.

We would do three different types of rendezvous. One would be a faster rendezvous than Schirra and I had done before; it would be standard to what we would use around the moon. We would do an overhead rendezvous—always before we had come up from below, with the light of sunrise or the stars in the background, so it was like flying an instrument landing system approach: it was a very easy course of reference. Here, we'd come from above, with all the Earth going past us, to simulate a command module coming down to pick up a lunar module that had had some major problems down in low lunar orbit. Then we would do an optical rendezvous, using optics only.

As if that were not enough for one mission, the plans for the EVA sounded like something out of science fiction. Stafford concentrated on his rendezvous duties and left the spacewalking details to Cernan, viewing the addition of a rocket pack as a "Mickey Mouse idea." As Stafford recounts,

Cernan would fly a rocket pack on a 125-foot tether—an air force experiment, using hydrogen peroxide for maneuvering fuel. We knew it was a risky-type mission flying that rocket pack. What would have happened if one of the thrusters had stuck on, and how would we get it off? Remember, the only thing that we had had experience with on our side was Ed White. There was a lot we didn't know about working and walking in space. We didn't have any good simulation. All we'd used to simulate maneuvering around was a polished steel table on an air bearing. Cernan would get on that and we'd scoot him around—we didn't even have three-dimensional simulation, anything like that. In fact, we were so naïve, we didn't even think about putting defog on our visors, like you would to go snorkeling or scuba diving. But it was a start; every step was a new step.

Gene Cernan agrees that it was a new step, but perhaps a step too far. "We were to build on the spacewalking experience of *Gemini 4*. *Gemini 8* was supposed to do a very extensive EVA, which might have taught us a lot prior to *Gemini 9*. Of course, they had to abort their mission. So we'd lost something in between. The bottom line is we had to jump from *Gemini 4*'s experience of only twenty minutes of EVA. Ed White was smaller than me, and had trouble getting back into the spacecraft, but apparently we didn't pay too much attention to the problem."

Many of the flight control team, mission trainers, and astronauts thought that the rocket pack, called the Astronaut Maneuvering Unit or AMU, was too dangerous to use in space. Its use was really only being attempted at the insistence of the air force. NASA tried to reduce the risks by insisting on the safety tether and also making the spacesuit even thicker and thus harder to move in, as Cernan explains:

I trained to fly the AMU at the end of that tether, all around the sky, like you have seen them do on the shuttle. One of the unique things about the AMU, unlike the cold gas nitrogen systems they use in these backpacks they fly around in the shuttle: we had hydrogen peroxide. We had real rockets. I remember when we first were designing this unit, they had one of those hydrogen peroxide rockets between our legs—we changed that design pretty quickly! But on top of that, I had to wear steel pants. Full metal-woven steel pants, to keep this thing from burning a hole in my suit. I should have got the message!

Tom Stafford must have had a horrible sense of déjà vu. Sitting in a Gemini spacecraft on the pad on 17 May, ready to launch, he and Gene Cernan awaited the liftoff of their Agena docking target. "We got ready to launch," Stafford recalls. "We heard the Atlas Agena target vehicle lift off from the pad just like we'd heard before on *Gemini 6*, and this time it was the Atlas that blew up. As I'd lost two Agenas, an Air Force major, one of the support crew members, wrote me a poem. He wrote: 'I think that I shall never see, an Agena out in front of me.' And I never did!" As with his previous mission, the launch was scrubbed, and a substitute target was hurriedly devised. Having learned not to rely on the Agenas, Stafford says that NASA was already prepared with a backup option: "McDonnell Aircraft had this idea of a target vehicle where you would take the Agena docking collar, put some batteries in it, some rate gyros, and recycle a used part from Gemini, the attitude control, so that it could hold attitude once that was put into space on top of an Atlas. As a matter of fact, it turned out that it was the reentry control section from my *Gemini 6* spacecraft." The backup vehicle was called an Augmented Target Docking Adapter—or ATDA, by the acronym-loving NASA engineers—and it was quickly readied for launch: as it turned out, perhaps too quickly.

Gene Cernan, in the meantime, had just sat through his first-ever launch attempt, and he later characterized the loss of the Atlas Agena as slightly unnerving. "This was my first time around; I had never been on a rocket before. You could hear that big Atlas go up—and you could almost hear it blow up. We were next in line, and all of a sudden I wasn't sure that this was the right thing to do. But frankly, I wanted to fly so badly I would have gone on that flight if they had strapped me to the outside!"

On 1 June 1966, ATDA was launched and reached the perfect orbit. Telemetry, however, suggested that the launch shroud covering the docking collar had not fully opened. ATDA was also tumbling, and a suspicion that Stafford had held before was being confirmed:

The telemetry on the ground showed that they couldn't stop the rotation and that they had already got through one of the two rings of attitude fuel. Now, I had questioned the program manager at McDonnell Aircraft at the time—he said that that attitude fuel would last for three to four months. I asked, how could it, you only had twenty-five pounds of fuel in each ring, and just the gravity gradient alone would use that. He says, "Tom, trust me, it will last three to four months." And here in about half an orbit one ring was gone, and it had started working the other one. And this is one thing: if something just doesn't feel right to you, particularly if you work in engineering, you ought to go and question it. Everything was measured in specific impulse. When they went to change from seconds to units, and divide by sixty, whoever was working as a sub engineer multiplied it by sixty. So it was off by a factor of 3600! I never let the guy forget it!

Despite the problems, Stafford and Cernan were ready to launch and join ATDA in orbit, but problems with launch control computer equipment meant that the attempt was postponed for forty-eight hours. Stafford had now sat through five launch attempts and had still only flown in space once. His patience, however, was about to be rewarded.

Stafford's sixth launch-day visit to the pad, on 3 June, was a successful one, and as he and Cernan sped into orbit the commander monitored their launch trajectory closely. Stafford was able to use the spacecraft thrusters and data from Mission Control to fly a precise course to their target. He soon had a radar lock on ATDA. As he grew nearer, he could see the target shining in the strong moonlight. On their third orbit of the Earth, as Stafford drew closer to the tumbling ATDA, he could see that the telemetry information had been correct: the shroud on their target had not completely detached:

We finally broke out into daytime. The ground control center wanted us to describe what we saw. I said, "The nose cone of the shroud is still attached, it has opened up and is slowly rotating. It looks like an angry alligator." I also called it a few other things to Cernan! It turns out that, through a whole series of human mistakes, instead of securing the lanyards by hooking into the socket support, they wrapped them around with another piece of wire and put duct tape over them!

Stafford considered using the nose of his spacecraft to try and dislodge the half-opened shroud, but the ground quickly rejected this idea as too risky. Pulling within a few inches of the rotating ATDA, Stafford looked at it closely. "Gemini was such a maneuverable spacecraft," he remembers. "I could put my window up where the shroud was, match the spacecraft roll rate, and just stay right there taking pictures of it." Though the explosive bolts that should have freed the shroud had fired correctly, Stafford could see that the straps holding the cover in place were still fully secure. The ground sent up signals to tighten and relax the docking cone, hoping that this might nudge the shroud free. Nothing worked.

Unless there was some way to remove the nose cone, a docking would be impossible. Stafford told the authors that brief consideration was given to removing it by hand: "There were two big 300-pound springs trying to push it apart, and they wanted Cernan to go out, try and take a pair of surgical scissors and cut the two straps. I saw the sharp, jagged edges where the pyrotechnics had fired, and said, no, we'll just wait and do the EVA later." If *Gemini 9* could not dock with the target, at least it could be used to perform rendezvous exercises. First, Stafford and Cernan tried a simple exercise. Speeding the spacecraft up, they watched as ATDA seemed to drop away and under them. Then, slowing themselves back down, they gradually rendezvoused with the target once again.

It had been a long, mostly disappointing day, and it was time for the crew to get some rest. Moving away from ATDA, they tried to sleep, but succeeded in getting only a little rest. It would have to do for the busy day that lay ahead. Their first task the next day was to rendezvous a third and last time with ATDA: this time, a rendezvous from above attempting to use only visual cues. It was not easy to track a moving object by eye with the Earth as a backdrop, but they pulled it off and were soon alongside their target once more.

The faulty docking target had now outlived its usefulness, and it was time to leave it behind. "It had served its purpose as a target vehicle," Stafford relates. "We hadn't really docked with it, but the big thing was, we'd done the rendezvous." It was hoped they could do the mission EVA next; but the crew was still very tired, and it was decided to leave it for the third day of the mission. Stafford and Cernan instead carried out a variety of science experiments. With little else to do that day, they gave the experiments more attention than any Gemini crew had ever been able to do before.

On 5 June the crew depressurized the spacecraft, and just before sunrise the hatch was opened. Cernan stood up in his seat, ready to push away into space and begin his first task. From the beginning, he could tell that the EVA was not going to go to plan. The first part of the spacewalk was supposed to allow Cernan to acclimatize himself to working in space and see if he could move around easily by pulling on his tether. Instead, he recalled later, it quickly exhausted him: "Ed White went out there with a gun to stabilize his body. The protocol was for me to evaluate the umbilical dynamics. The first thing I knew was this just wasn't going to work, because I could never pull on the umbilical through my center of gravity. I would go tumbling through space—the most helpless feeling a human being can have, particularly if you are the kind of person who likes to be in control of what is going on."

Cernan's lack of control meant that Stafford had to work hard, firing thrusters to keep the spacecraft oriented and hoping they were not firing too close to his colleague. Every movement Cernan made pulsed down the tether like a ripple through a whip, shaking the spacecraft and then rebounding back to him. "I was flying with the hatch open," Stafford recalls. "I have just about one picture of Cernan, he was right outside, and he was really huffing and puffing. There was no doubt he was having one heck of a time; he was torquing the spacecraft all over the place."

After a short rest, Cernan tried to work his way to the back of Gemini's adapter section, but found the provided handholds were completely inadequate for an easy translation. In addition, just as Leonov had found, Cernan constantly had to fight the restrictive movements of his pressurized suit. "I compare that Gemini suit to hardened plaster of Paris when I was pressurized. It did not want to bend at all, anywhere. Moving my arms and legs was very, very difficult."

It had been an endurance test just getting to the back of the spacecraft, but now Cernan planned to try something even more ambitious:

I had to basically assemble the AMU: turn a valve, push the arms of this maneuvering unit back into the seat, twist them and turn them. I had a couple of lights back there, one of which didn't work, and I had a couple of little handlebars and a footrest. And you think we'd have been a little bit smarter than that, because a footrest doesn't quite hack it in zero gravity; you have to have some place to anchor your body. I would put one foot under that footrest, and another foot on top, and try and hold myself there, wrap around the handlebar and try and twist. It would twist me back! I would try and telescope an arm on the AMU to get it assembled, and instead I would just go floating somewhere out there in space and then had to work my way back. It got pretty tiring.

With the insufficient support, assembling the AMU was taking far longer than expected. Adding to these problems, Cernan's exertions were causing his spacesuit to overheat. As Stafford explains, the spacesuit design was simply not up to the task: "We didn't have any liquid cooling garments at that time, just a simple evaporator with an air blower that was trying to keep him cool. At one point, he told me that his whole back was burning up; he couldn't figure it out. Turns out, the seven inner layers of insulation that were in his suit had torn, right through by the zipper on his back, and when we finally landed a couple of days later he still had about a second-degree burn on his back where the sun had scorched him." According to Cernan, the overheating soon took a dangerous new turn: when the sun set, his faceplate fogged over: "I overpowered my environmental control system, cooling system, and pressurization system in the spacesuit. It was literally all fogged up, like the windshield on your car on a cold winter morning. I was going over Australia at night one time; I could rub a little hole with my nose and see the lights of Sydney down there. I wondered, why the hell am I up here and not down there?"

The AMU was almost ready to fly, and Cernan was hoping to press on with the task. He sat in the maneuvering unit, hoping he could cool down and that his visor would clear. Then, to make matters worse, he had communications problems and could not hear Stafford clearly. In addition to being effectively blind, he was also deaf. Stafford says it was a difficult time for his crew member:

The moment that the sun went down, zap, he completely fogged over. He could not see, absolutely could not see. Then when he tried to hook up into the rocket pack and use that, we lost one way of two-way communications. We were going over the South Pacific at night, watching the Southern Cross come up, and I thought "Darn, it's lonely up here. We are unpressurized, no ground stations around, my buddy is twenty feet back, he can't see, and we have lost one way of two-way comm." A hell of a lonely place. He could hear me, so we worked out kind of a simple signal. I could hear kind of some static, and I could get a yes or no binary type of system worked out with him. So I said, "Look, Gene, when we get into sunrise, in ten minutes, if you don't defog, we are calling this quits, get out of there and come back in."

It was not what Cernan wanted to hear. Despite his risky predicament, he had no intention of wasting those long months of training. He desperately wanted to finish the job he was so close to fulfilling:

I had by that time connected the oxygen and a radio cable from the backpack, because I was going to be an independent satellite: tethered, yes, but independent. I was breathing the backpack oxygen; all Tom had to do was flip the switch and I was gone, I was out there. But the doctors were getting pretty concerned. My heart rate was going, I guess, 160, 170. Bottom line, it was ready to go, but Tom and I couldn't talk to each other, I was pretty tired, and so we decided to abort the next part of the mission.

It was frustrating for me, because I couldn't get the job done that I was sent to do. I was extremely disappointed, because I know that if we had had the time, we could have flown the next part of the mission successfully. I was also disappointed because I felt I'd let these guys down. I didn't do what I was sent up there to do. In retrospect, looking back, had Tom cut me loose, and I had flown around out there, our plan to get back out of the AMU was to hold on to the docking bar with one hand, unbuckle and unstrap and put the new umbilicals in. That all required a tremendous amount of force; in retrospect, I am not sure that getting out would have been any easier than getting back in.

Reconnecting to the spacecraft umbilical while still blinded by a fogged visor, Cernan groped his way back to the front of the spacecraft and the cockpit as Stafford tried to guide him over the radio. Stafford says he kept reeling in the huge tether and stuffing it into the cramped cockpit: "As he floated up he was still completely blind, fogged over, and I kept pulling on the umbilical, hooked into the spacecraft there on the center console. I gave him like a ground control approach, told him where to put his hands. He got ahold of some handholds and started working his way up to the hatch. Finally, I reached over and grabbed hold of his foot. Then I could get him stood up in the ejection seat, and turn him towards the sun and hope that he could defog. It still didn't work."

Cernan, back in his seat after over two hours outside, believed that he was relatively safe. But the most painful and life-threatening moment of the EVA had just arrived. "Getting back into the spacecraft itself," Cernan recalled,

was like putting a champagne cork back into a champagne bottle. It was next to impossible. But let me tell you, if it is you out there—you are not going to not get in! The Gemini spacecraft was extremely small, and I was a little bit bigger than Ed White. We could do it in training every time, but we never were fully aware of the problems that we might encounter in zero gravity. To get into the spacecraft in the pressurized suit with the steel pants, I had to bend my legs under myself, like a limbo dancer: it was very painful. My inflated suit still would not bend. Tom Stafford reached over to fully close the hatch. That hatch came down on my head, and I was literally like a pretzel with my legs doubled up. I felt ready to die. I have never known such pain, and I could hardly breathe or stay conscious. I told Tom, "Get this spacecraft pressurized!" When he did, of course, my suit softened, and I was able to unbend and get my helmet and gloves off.

The effects of the strenuous spacewalk were evident the first time Stafford saw Cernan's face again: "When he finally opened his helmet, he looked like he had been in a sauna too long; his face was absolutely pink. Gene lost thirteen and half pounds in two hours and five minutes outside. That is not a good way to lose weight, believe me!" Cernan felt swollen, weary, and as if his skin was on fire. What Stafford did next was an incredibly welcome relief for him, as Cernan recalls: "At that point in time, Tom broke a cardinal rule. Water can be the worst possible problem you can have in a spacecraft. But Tom looked over at me and said, 'You look like a boiled turnip.' And he took that water gun and squirted it right in my face—just blew it all over me. We had water all over the spacecraft."

Cernan still had a long night ahead of him. As he cooled off, the huge amount of perspiration floating in his suit cooled with him. "He nearly froze that night because of all the water in his suit," Stafford recalled.

On 6 June, an exhausted *Gemini 9* crew finally came home, splashing down hard in the Atlantic Ocean. Landing just over six hundred yards from their target, very close to the aircraft carrier USS *Wasp*, they were on board within the hour. As Stafford explains, he was able to truly pilot his way down from orbit, thus reaching the splashdown zone accurately. His targeted landing was even closer to the designated spot than the first shuttle landing in 1981 was able to achieve.

It was the most precise touchdown. You have to remember with Gemini, retrofire attitude, we all did it by hand, although we did have an automatic sequencer for the retros. But we now had a new computer program. Through work in the simulator, we could read out the longitude and latitude during reentry as we were approaching, particularly on those trajectories where would be on about the same latitude, so the main thing we were changing was the longitude. And we determined how fast or how slow we were approaching in longitude, and somebody had written a computer program to lift vector up, so we could reduce the g-load. The g-load wasn't that much. I called the skipper of the Wasp, *the same carrier I'd landed on with Schirra six months before, and said, "Captain, I think we have a good way to really do a precise touchdown. If you can get the* Wasp *right on the aim point, I'll try for the best you'll get in Gemini."*

So we went into the reentry, went into the fireball, and coming right near the end, Gene was pushing out the longitude and says, "Tom, we are getting there too fast, split S." So I rolled the spacecraft to full lift vector. Of course, the skipper had a difficult time because back then there was no inertial navigation system on board, no global positioning satellites. He was still taking sunrise at high noon with a sextant. But he did a great job there with the Wasp *and his aim, and that was the closest touchdown of any of the Gemini or Apollo missions. We set the record for the closest splashdown to the carrier.*

Despite the incredible landing, the mission had been a deeply frustrating and disappointing experience for Cernan. He later wrote that "the horrendous mission had been filled with glitches." Wally Schirra teases

Cernan to this day that he did not make a good spacewalk until 1972. "After we landed," Stafford recalls, "they flew Cernan's suit right away back to Houston. The technicians poured a pound and a half of water out of each boot: all perspiration. Cernan called it the spacewalk from hell; it was unbelievable. We said, this is something we have got to understand better than this!"

In the end, it was all worth it. Jack Schmitt, an astronaut who later walked on the moon with Cernan, thinks that the EVA lessons of *Gemini 9* were vital to the successful exploration of the moon. "If there was one enabling technology for what we did scientifically and in terms of exploration on the moon, it was water-cooled underwear. We could not have done that job! We could not have spent more than an hour, and maybe not that much, outside. Gene's *Gemini 9* mission testified—if you just have air cooling, it gets so hot, so sweaty, that you just can't do it, it just won't happen. But as a result of developing that enabling technology, we were able to spend seven, eight hours outside the spacecraft."

In hindsight, Cernan now thinks that his *Gemini 9* spacewalk was a necessary failure if the next three EVAs in the Gemini program were going to have any chance of success. "You know, there were a lot of cues," he emphasizes, "including the cues that we learned so loud and clear in fourth and fifth grade: that for every action there is an equal and opposite reaction. I'd like to look back and think that we eventually learned something, and that Mike Collins, Dick Gordon, and Buzz Aldrin picked up from the mistakes I made. Maybe this is self-serving, but I like to think that maybe *Gemini 9*, like *Apollo 13*, was not a failure—it was a success."

It was easy to forget that Stafford and Cernan had once again pushed the envelope when it came to advanced rendezvous techniques. Some of their trials would be directly applicable to rendezvous around the moon. Interestingly, Stafford and Cernan would be the first two astronauts to experience that in person.

By the time NASA was ready to launch the *Gemini 10* flight, mission planners were aware that the program was rapidly coming to an end. It was hoped that the first Apollo mission could be launched as early as the same year, and there were some techniques that the planners sorely wished to practice before then. Both *Gemini 8* and *9* had slipped a little; less dock-

Dee O'Hara is reunited with Gene Cernan in Burbank, California, in 2004. Courtesy Francis French.

ing and EVA experience had been gained than hoped. It was time to try something ambitious yet straightforward. It was also time for the Gemini program to enjoy a run of good luck. For this mission, Mike Collins wrote, they would try "a little bit of everything."

John Young was back, this time commanding the mission. Now a confident and experienced commander in his own right, he was going to do things his own way. Concerned about how he would be able to handle certain connections and plugs with his spacesuit gloves on, he asked if he could take a pair of pliers with him. When his request was refused, he smuggled them on board anyway. Joining him was Collins, a rookie from the third group of astronauts.

Like Dave Scott, Michael Collins was a military man from a military family. He was born on 31 October 1930 in Rome, Italy, the son of an American career army officer who was constantly moving his family from posting to posting. Collins's brother would later become a colonel, and Michael could also boast of two uncles in high-ranking positions. He was a graduate of West Point, receiving a bachelor of science degree from the academy in 1952. Like Scott, he'd also chosen the U.S. Air Force as a career.

Winning his wings in 1953, he served in the United States and Europe before entering the test pilot school at Edwards in 1960. Three years later, he was selected for the prestigious Aerospace Research Pilot School and the most coveted assignment of all—fighter ops. He was another natural for NASA's third group of astronauts, and was selected in 1963, having made it to the final thirty-two in the selection process for the second group the year before.

Unlike Scott, Borman, and some of the other astronauts who'd grown up in the military, Collins had a droll, cultivated personality that was far removed from his straight-arrow colleagues. He'd later characterize many of the military leaders he'd served with as "lemming-like. . . . I guess I am more of a civilian at heart." Collins was always going to have his own opinion: he joined the air force instead of the army primarily to do something different than the rest of the men in his family. "I'm stable, diversified, and I am not narrowly focused," he once said of himself. "I'm not extraordinarily good in one area and extraordinarily bad in the other. I'm kind of even. . . . I'm balanced."

His casual, graceful, and slightly mischievous exterior meant Collins sometimes seemed a little out of place in a group of unashamedly ambitious astronauts. Unlike most of the others, he arrived with a well-cultivated taste for the best food and wine, and even impishly graded the astronaut food samples as if he was writing a gastronomical guidebook. Yet his subtle joking was really just so he would never be seen as taking himself too seriously. Underneath, he was as determined as any of them.

With probably the driest, most self-deprecating wit of any astronaut, he managed to combine what many of his colleagues never could: immense piloting and engineering capability with the perception of an outsider. He'd later say that he became an astronaut through "one part shrewd logic and nine parts blind luck; . . . it's a peculiar twist of circumstances that got me here." Quiet and unassuming while in the astronaut corps, he'd prove to be the most talented writer of the astronauts after he left. Novelist Norman Mailer would sum him up in three words: "Collins was cool." His coolheadedness would be an advantage for the mission ahead.

For the first time, an American manned mission was not scheduled for an early morning launch. Instead, the crew would lift off just before dusk. As a

Mike Collins (*left*) and John Young at the Cape beach house prior to their *Gemini 10* mission. Courtesy NASA.

result, Young and Collins found themselves under instructions from Deke Slayton to "keep partying" until at least two o'clock on their launch day. Slayton told reporters that he wanted the crew to be tired enough to sleep late into the day, hopefully until about 1:00 p.m., when they would be awakened for their flight. In fact, the two astronauts—dubbed the "Night Owl Patrol"—had stayed up until after midnight the previous ten days in an attempt to adjust their sleeping patterns to the later launch. A smiling Slayton added: "We've told them to try to keep awake . . . if we can find a party going late enough for them!"

It had been planned many months before to launch *Gemini 10* on 18 July, and for once there were no launch delays. The spacecraft lifted off exactly on time in perfect weather, streaking into space in pursuit of the Agena 10 target rocket, launched exactly 100 minutes before them. Earlier, to his delight, John Young had been handed a four-foot pair of pliers by a laughing close-out crew as he prepared to enter the spacecraft, in response to his denied request to carry a real pair on board.

To the crew's relief, Agena 10 reached orbit, 183 miles above the Earth, perfectly. Performing a night rendezvous would prove a little trickier, they knew. If all went well, they would start up Agena 10's 16,000-pound-thrust main engine and be propelled to a high point of 468 miles above the Earth. Should this succeed they would later attempt yet another rendezvous—this time with the Agena 8 booster that Armstrong and Scott had docked with on their earlier mission.

Trying to navigate by using star sightings, Young and Collins found that their numbers were not matching up with those given by Mission Control. Relying on the more trustworthy ground numbers, Young began to change orbit. Unfortunately, the spacecraft was slightly out of alignment when he made his thruster burn, and he had to use a large amount of fuel—three times more than planned—to get back on course to find the Agena. Nevertheless, less than six hours after launch, Young reported to the tracking station in Hawaii, "I can see it out the window." He closed in on the Agena and flew formation with it for a while. He then nosed the *Gemini 10* spacecraft into Agena 10's docking collar and reported a successful docking. "John made it look easy," Collins would say after the mission. "On the first try we glided right into the docking cone."

It had been planned to undock and then practice docking again, but Young had wasted too much fuel, a mistake that made him gloomy and morose for the rest of the day. Instead of redocking, the crew pressed on with a second rendezvous. This time, they would be using one Agena to find another. With *Gemini 10* still secured nose to nose with the docking target, the Agena's engine was fired up. This was not without its risks. "A serious malfunction during this critical operation," Gus Grissom would later write, "could blow both spacecraft to smithereens." For a second, Collins thought that the engine had not fired, but then "the whole sky turned orange-white" as the engine thumped into life, and he shouted in excitement, "That was pretty wild!" Sitting facing the rocket, Young and Collins were pushed against their straps rather than back into their seats for eighty seconds, with a force equivalent to Earth's gravity. "When that baby lights, there's no question about it," Young reported. He later compared the bizarre feeling to hanging from a wall: "There was a pop, then a big explosion, and a clang. We were thrown forward in the seats. We had our shoulder harnesses fastened. Fire and sparks started coming out of the back end of that rascal. The light was something fierce, and the acceleration was pretty good. The shutdown . . . was just unbelievable. It was a quick jolt . . . and the tailoff . . . I never saw anything like that before: sparks and fire and smoke and lights." With the engine shut down again, *Gemini 10* was propelled into a looping orbit 474 miles high—further away from the planet than humans had ever been before. Unfortunately, the Agena blocked much of the view from their window, and they could not fully appreciate the new, loftier vantage point.

Still docked to the Agena, it was time for the two astronauts to sleep. When they awoke, it was to good news: they were to fire the Agena and change their orbit again. Once more, the push against their straps got the astronauts' attention, with Young commenting "It may be only 1-g, but it's the biggest 1-g we ever saw! That thing really lights into you!"

Now it was time for the first EVA of the mission. The mission planners were going to keep events deliberately slow and simple this time around, breaking tasks up into more frequent, shorter spacewalks. For his first spacewalk, Mike Collins did not even leave his seat. Instead, his hatch was opened and Collins stood up, photographing constellations while the docked spacecraft sped across the night side of the Earth.

"Spacewalks are when you really notice the differences and similarities between airplanes and spacecraft," Collins told the authors. "In an airplane, you try as hard as you can to keep the pointy end going forward. In a spacecraft it doesn't really matter. You don't have any rudder pedals: instead, you have a three-axis hand controller, and if you turn your hand you can go sideways—which is kind of cool! On Gemini there you are, going sideways around the world, the door is open, you are standing up on the ejection seat—you hope the pins are in! That was just a very unusual feeling."

Even given such a simple EVA, something began to go wrong. Both Collins and Young felt their eyes begin to water with irritation, and soon Young could not see at all. They suspected that an antifogging chemical on the inside of their face masks, introduced after Cernan's experiences, was causing the problem. They cut the EVA short by a few minutes and quickly closed the hatch. When they turned off one of the suit compressor fans, the irritation thankfully cleared. "I was crying a little all night," Young would joke, "but I didn't say anything about it; I figured I'd be called a sissy!"

A third Agena burn sped the orbit up slightly and, more importantly, circularized it. Because of Gemini's fuel shortage, the Agena had been used longer than planned: without it, they would never have been able to accomplish the second rendezvous now ahead. They were growing closer to the orbit of another target Agena—the one left in orbit by the *Gemini 8* astronauts. It had been a long day, and the astronauts were now tired. They slept long and deeply.

The next day, Collins and Young undocked from their Agena booster. It was a welcome moment for Young, who had grown bored of having the large rocket blocking their windows and view. Nevertheless, as it drifted away, he reflected, "it was a mighty good train." Leaving it behind, they maneuvered over to the orbit of the *Gemini 8* Agena target, which had been launched four months earlier. Essentially, Agena 8 was now a "ghost ship" drifting in space—a lifeless target. The battery for its radar had expired. They soon found it, and confirmed that it was not tumbling. The Agena would now be the target not for a rendezvous, but for a spacewalk.

Collins prepared for the spacewalk on the dark side of the planet, while Young stayed in tight formation with the Agena, keeping it in sight with Gemini's searchlight. Uncoiling the fifty-foot umbilical cord for the EVA meant that the cockpit was soon full of snaking hose. Then, as dawn broke, it was time to go out. Moving slowly and carefully, Collins first retrieved an experiment from the outside of the Gemini spacecraft, while Young kept *Gemini 10* stable, being careful not to fire any thrusters that would hit his colleague.

Next, Collins had to connect a nitrogen line to the side of the spacecraft to power his maneuvering gun. Once again, it quickly became evident that the EVA had not been sufficiently well planned. Connecting the line was a job that needed both hands, and there was nowhere for Collins to place his feet and hold himself in position. Collins let go of the spacecraft for a second, jammed in the connector, then quickly grabbed onto Gemini again. His quick grab caused him to slam his legs against the spacecraft, however, which automatically fired thrusters to maintain its position. It was an inelegant start, but at least Collins was not overheating. It was a good thing that his visor did not fog up; Collins suffered from claustrophobia and had suppressed feelings of panic at times during his spacesuit training.

Now it was time to try something new and daring—to jump the eight feet over to the Agena. Collins pushed off carefully, trying to aim himself precisely for the docking adapter where Armstrong and Scott had made the first-ever docking. But the adapter was smooth, designed to softly cushion a spacecraft; there was little for a spacesuit glove to grab onto. Collins somehow managed to keep a tenuous hold on the adapter lip and worked his way hand over hand toward a micrometeorite experiment package on the opposite side that he planned to retrieve. He had been strengthening his hands by squeezing tennis balls during the months of training, and that extra edge was probably all that was allowing him to hold on to the Agena at this moment. Working his way around the circular lip, he knocked loose a two-foot-wide metal ring that was part of the docking apparatus. It was a sharp-looking protrusion that he had to avoid so it did not cut his suit or line.

Reaching the experiment package, Collins found that his slippery grip on the spacecraft was not enough to pull himself to a stop. Instead, his hands slipped off the docking ring, and he cartwheeled away into space—as Collins puts it, "ass over teakettle." As he spun away, Collins grew concerned that his umbilical was still close to the Agena and its metal protrusions. If it snagged, the combined movements of the Agena, *Gemini 10*, and Collins could entangle him. He also worried that he would now loop around on his line and squarely hit the Gemini spacecraft. To right himself, he used his maneuvering gun, similar to the one that Ed White had carried. It did not have enough power to stop Collins's loop back toward the spacecraft, but it was enough to slow him. As he swooped in toward the back of the spacecraft, he urgently signaled to Young not to fire the thrusters. Slamming into the hatch, Collins managed to grab hold of it before he spun off into space again.

Despite the dangers, it had been a fair first try. Collins was determined to get it right and decided to have another go. This time, he used the gun to move himself out of the Gemini cockpit. But as he rose upward his left boot snagged on the spacecraft, sending him into a face-down spin. Almost missing the Agena altogether in his efforts to correct, he stretched out his left arm and just managed to grasp the docking collar. It was not enough to hold him there, so risking a puncture to his gloves he thrust his right arm into a recess in the Agena and grabbed a handful of wiring.

By now the Agena spacecraft was beginning to tumble, knocked out of balance by Collins's repeated clawing on its end. With little time to spare on the increasingly unstable spacecraft, Collins managed to remove the experiment package and pull himself back along his umbilical to the safety of the Gemini cockpit. It had been planned that he would install a replacement package on the Agena, but this idea was wisely abandoned. Back in the cockpit, Young spent a few moments untangling the umbilical that had wrapped itself around his colleague. At some point in the chaos, Collins's camera had come loose from his spacesuit and floated away; the photos of the EVA were forever lost.

The maneuvering fuel, always in short supply from the start of the mission, was now very low. A plan to further test what Collins calls "that dorky little gun" was abandoned, and the hatches were closed. Once again, the crew had to contend with the huge umbilical taking up all of the available cabin room. Because of it, the astronauts could not see each other or their instruments, and accidentally hit the switch that turned the radio off. When they realized this and turned it on again, they told Mission Control that the cramped cockpit "makes the snake house at the zoo look like a Sunday school picnic."

Fortunately, a third and final EVA was undertaken soon afterward. The hatch was opened, and the umbilical and maneuvering gun, together with the assorted trash of a couple of days' life in space, were unceremoniously thrown out. It was all done within three minutes, and was the easiest and most welcome EVA of them all.

Struggling with the umbilical had put the astronauts behind schedule, and they hastily prepared for their next important task: changing their orbit to put them in a better position for reentry. With this done, and some more experiments completed, the crew spent their final night in space. Reentry was smooth the next day and on target. "Hey, John, you're on television," Gemini control told Young as the spacecraft drifted down beneath its parachute within sight of the aircraft carrier USS *Guadalcanal.* Television cameras aboard the carrier relayed the landing live to the mainland, and the assembled crew cheered as they watched the craft splash down in the western Atlantic just 3.4 miles away from them, after a mission lasting seventy hours and forty-seven minutes. In the meantime, the Agena 10 rocket they had used to boost their orbit was fired up again and eventually placed into an orbit that would allow it to serve as a future docking target.

The mission not only demonstrated techniques that would be used directly in Apollo; it also showed what else was possible once Apollo was over. Launching a manned spacecraft to dock with an unmanned "space tug," then changing orbit to rendezvous with yet another, showed NASA's increasing confidence when working in Earth orbit. Unfortunately, many of these promising developments have still to be built upon in human spaceflight programs.

Reflecting on his spacewalking experiences years later with the authors, Collins readily admitted that he had been underprepared:

We were just stupid. Ed White wasn't really out long enough, and didn't do enough to start getting into those problems. Cernan did, and to some extent I did and did not, depending on how you look at it. We were uninformed, and we hadn't really thought through what would happen in weightlessness if we tried to exert a force on something: where the equal and opposite force went. When we had trouble during training in the zero-g airplane, the KC-135, we attributed that to imperfections in the flight path of the airplane rather than the zero-g environment. We had not really thought it through; the fundamental thing was stupidity.

Yet, despite the difficulties, Collins also found his spacewalk experience to be the most interesting thing he ever did in his astronaut career—more so even than his participation in the first moon-landing mission. His favorite souvenir of his NASA years is a painting of himself making a Gemini EVA.

For all of its dangers, the spacewalk had been a relative success compared to Cernan's. Yet the lessons were still not being learned fast enough to employ on the very next mission. *Gemini 11* would almost be a step backward in the learning process.

The *Gemini 11* crew was comprised of two of NASA's most gung-ho and enthusiastic pilots. Pete Conrad now had his own command, and after the boredom of the long-duration *Gemini 5* mission was looking to make his mark on the spaceflight record books in a far more dynamic way. One idea that had been floating around NASA headquarters for some time was to send a Gemini spacecraft on a looping flight around the moon's far side, and Conrad was eager to give it a try. But with attention shifting fast to the first Apollo missions and with no sign that the Soviets were about to launch a manned lunar craft, the idea was rejected. Still, Conrad wanted to do something spectacular. If he couldn't go all the way to the moon, at least he could journey further away from Earth than anyone had done before.

Conrad wasn't going to be able to persuade NASA planners to let him fly high just to break a record: he would need to come up with a practical reason. He found one in the newly emerging field of weather satellites. If he could journey almost as high as the orbits where they transmitted black-and-white television images, he could take color photos, and the two could be compared. Such a comparison might help determine whether a color-image satellite system was needed. It wasn't much of a reason, but it was enough: Conrad would get to try for his record.

The other astronaut on this flight was just as eager to make his mark. Born in Seattle in 1929, by the dawn of the space age Richard Francis Gordon was a naval aviator starting a tour of duty as a test pilot at the navy's test pilot school. He'd been a naval aviator since leaving college and served as a fighter pilot for many years. For a while, his test-pilot instructor had been Alan Shepard, later selected as one of America's first astronauts, and he also knew John Glenn via the Bureau of Aeronautics. Dick Gordon became the first project test pilot for the F-4H Phantom II, and helped to introduce that aircraft to the Pacific and Atlantic fleets. In the same month that Alan Shepard became America's first spacefarer, Gordon was also making history, winning the prestigious Bendix Trophy for a flight from Los Angeles to New York, establishing a transcontinental speed record of two hours and forty-seven minutes. He had almost been selected for NASA's second astronaut group, but just missed the cut. An almost automatic choice for the next group the following year, he looked like he was on track for a stellar astronaut career. Others in Gordon's group, such as Gene Cernan, regarded him as one of the top astronaut pilots and likely to get some of the best mission assignments—perhaps even command of a lunar landing mission.

Gordon had other advantages over his new astronaut colleagues. He was the highest-ranking military member of the group and had been chosen by Deke Slayton as their spokesperson. In an astronaut office where it paid to have powerful friends, Gordon had more than most. In addition to Shepard, Gordon had also known Slayton and Wally Schirra personally before his selection. Gordon's easygoing personality and impressive piloting assured he would not lack friends at NASA. Walt Cunningham would write that he had "a touch of devil-may-care; . . . he knew half the guys in the space program by their first names before he was selected, was sociable, well-liked, and appealing to women." He is, even today, a gregarious, outgoing, and roguish type with a fun-loving, devilish charm. Ruggedly built with a dark, brooding look, he was athletic enough to have considered becoming a professional baseball player. In his usual mischievous way, Conrad began calling him "the Animal," a nickname that Gordon only grudgingly accepted.

Perhaps most importantly, when obtaining his first flight assignment, Pete Conrad was Gordon's good friend. In fact, he had been his roommate on the carrier USS *Ranger*. Although delighted for his friend, Gordon had been very upset when he missed out on also being named to the second intake of astronauts. "He was selected," Gordon said of Conrad to interviewer Catherine Harwood, "and I wasn't. I had some very strong emotions about that. I was bitterly disappointed, and almost resigned from the navy." After thinking hard about his career, Gordon decided to stay on, and fortunately was selected for the very next astronaut group. Back with Conrad again, the duo worked incredibly well together during astronaut training, as Gordon later recalled: "Pete and I could communicate without talking. We thought alike, and we trusted each other completely. We reacted the same." Guenter Wendt characterized the two as "swashbuckling wisecrackers; . . . that pair was nearly inseparable." The duo were legendary for being able to work hard in the simulators all day, and yet party all night, blowing off any stress with laughter and funny stories. "Were we better than anybody else?" Gordon asked Harwood rhetorically. "Sure we were! We didn't say it; we didn't have to say it."

Now the two test-pilot friends would get to fly together in space and carry out some of the space program's most ambitious piloting before or since. Conrad described their teamwork and camaraderie as "sometimes startling; . . . it allows you to think and act alike, which in certain situations can be very comforting." As Gordon also told the authors, he could not wait to fly with his commander: "We were friends. We were both naval aviators. We had known each other for many, many years, and appreciated each others' talents and capabilities. We were very, very compatible, and had a lot of fun. We relied on each other. Life is full of squares—you go through life filling as many of those squares as you can. This was a great opportunity for us to experience together one of the greatest opportunities that there was."

In addition to trying for an altitude record, the mission would attempt something planned for Apollo operations around the moon: a rendezvous with a target on the very first orbit after launch. After the experience of *Gemini 10*, mission planners were concerned that this rendezvous method might consume a lot of fuel. A careful plan had to be devised in which ground telemetry and onboard information were quickly and accurately combined. Any mistakes, and not enough fuel would remain to complete the entire mission.

Showing its growing confidence, NASA planned another experiment for this mission that had no direct bearing on the Apollo program. Instead, it might have practical uses in the far future; *Gemini 11* would attempt to create artificial gravity. To do this, it was planned to tether *Gemini 11* to an Agena booster on a 120-foot Dacron line and then rotate the two spacecraft. To attach a tether from Gemini to Agena would require a spacewalk. But the lessons learned from *Gemini 9* were only just making their way into mission training, and most were too late for *Gemini 11*. As Gordon later explained, he had no idea what he was in for:

I don't think we did a very good job of sharing previous astronaut experiences with EVA. I didn't learn about some of their problems until well after books were written. It wasn't until I read Gene Cernan's book that I realized the problems he had. Or if I knew it at the time, it didn't sink in. We didn't have enough time to assimilate the information from previous flights; or we didn't grasp the significance of the information that was being brought back. We were on a two-month launch schedule, and it was almost like passing each other to and from the pad. We didn't have any discussion about handholds and footholds, and I didn't anticipate any problems with the EVA.

If mission planners had thought it through, they would have delayed *Gemini 11*'s launch until newly devised spacewalk training methods had been implemented. As it was, because every other element was ready a launch date was set for 9 September 1966. There was a lot of confidence that things would go right this time. "Those guys were two of the best we had in the program," Cernan would write. "If anyone could slay that EVA dragon, it would be Dick Gordon."

One launch attempt was scrubbed due to a leaking oxidizer tank, and another when the Agena target's autopilot sent out faulty readings. But finally, on 12 September, the Agena was sent into orbit. If Conrad and Gordon were going to catch their target, *Gemini 11* had to launch less than two hours behind it and within two seconds of the designated launch time. They made it with half a second to spare, and the Titan booster shot them into orbit on a near-perfect trajectory. As soon as they were in orbit, the crew speedily made corrections to their orbital plane, rapidly closing the hundreds of miles between themselves and their target. Locking on with radar, Conrad and Gordon chased down the Agena until they could see it in the distance. "It's bright, fantastic," Conrad reported. "The prettiest thing I ever saw." Soon, they were station-keeping within feet of its docking cone. It had only taken eighty-five minutes since launch, but the two test pilots had caught their target, and with fuel to spare.

This time there was enough propellant left to accomplish what *Gemini 10* had failed to do: undock and then redock. Dick Gordon was allowed to pilot this maneuver, an unusual honor for a copilot. "If Pete hadn't let me do that," Gordon later joked, "I'd have thrown him out of the spacecraft!" Docking with the Agena reminded Gordon very much of aerial refueling, which he had often carried out as a naval aviator. "It was all feel," he commented on the technique. Once the docking maneuvers had been completed, Conrad fired up the Agena's engine to test its efficiency and change their orbit slightly. Then, after a meal, it was time for their first sleep period in space. It had been a perfect mission day.

Their first EVA on the following day was supposed to be relatively easy. However, as they began their preparations, four hours before the set time to open the hatches, Conrad and Gordon began to realize that some things are *too* easy. They were suited and ready to make the spacewalk with just under an hour of preparation time remaining. So the two of them sat there, fully suited, waiting for the appointed moment. As the time for the EVA grew closer, things suddenly began to go awry. When Gordon hooked into the heat exchanger that would be used on the EVA and began to test the oxygen flow, the temperature within his suit quickly rose to an uncomfortable level. The system was designed to be used in the vacuum of space, and in the still-pressurized cabin it was having unintended side effects. Gordon quickly switched back to the cabin systems.

With the moment to open the hatch approaching, Gordon realized that it had been a mistake to prepare ahead of schedule. He was supposed to fix the sun visor on his spacesuit helmet earlier in the checklist, and found that doing it later with all the EVA equipment on became a real struggle. Conrad could help with the side of the helmet facing him, but as Gordon pushed on the other side to try and lock it in, the visor suddenly cracked. "I couldn't get the damn thing on," Gordon told interviewer Michelle Kelly. "I struggled and struggled with that thing; . . . it got messed up." Making matters worse, the exertion had already exhausted Gordon.

Only now did the real spacewalk begin. The hatch was opened, and Gordon floated upward and out. "I had totally exhausted myself," Gordon later explained. "I was behind the curve. . . . I had really worked up a real heat load in there. That damn visor! I wanted to throw that thing as far as I could." With Conrad playing out the umbilical line, Gordon tried to install a camera on the outside of the spacecraft to record his EVA. It would not fit on the bracket, and the frustrated spacewalker had to pound on it with his gloved fist to finally fix it in place. "I've got to rest here a minute," Gordon gasped after the hammering. "I'm pooped!"

It was time to try and tether the two space vehicles together. Gordon pushed off toward the Agena's docking adapter but missed and floated off above it. Conrad tugged on Gordon's umbilical to reel him back in, instead sending the spacewalker spinning and jerking around at the end of his line. "Easy!" Gordon shouted as he was whipped around. Conrad was finally able to pull him back to the hatch, where he could try to leap across to the Agena once again. This time, Gordon's aim was good. He sailed over to the adapter and then, grabbing a handrail, tried to wedge his legs around the Gemini spacecraft's nose where it met the docking adapter.

It was a primitive and impractical solution to a problem that NASA was only slowly understanding: that for any work to be done, a spacewalking astronaut had to be securely anchored to something that would not move with him. In this case, it did not work. Gordon was hoping that he could clamp his legs around the spacecraft tight enough to stay in place, while using his gloved hands to secure the tether. Instead, the pressurized suit pushed back, and he felt himself beginning to slip off. "In training, I had always been able to always wedge my legs between the spacecraft," Gordon told Kelly. "I couldn't keep myself in that position. I kept floating away." He was forced to hold on with one hand, trying to secure the tether with the other. Watching his colleague struggling to stay on the spacecraft like some rodeo rider, Conrad shouted "Ride 'em, cowboy!" as encouragement. Despite his exuberance, his worry over Gordon was growing. "That was probably the only time that I really got concerned on any of the flights," Conrad told interviewer Don Pealer. "He was obviously in trouble."

It took Gordon six long minutes to secure the tether in space, and by the end of it he was facing some of the same troubles that Gene Cernan had encountered. He could feel his body overheating, perspiration was pooling on his face and stinging his eyes, and he could no longer see clearly. "Dick's breathing awfully hard," an anxious Conrad reported to the ground. Gordon was hoping to make his way to the back of the Gemini spacecraft where a maneuvering gun was stored and also to test a special power tool. But as he blindly groped his way back toward the hatch, Conrad decided that enough was enough. He canceled the rest of the EVA and helped Gordon back into the spacecraft. "I'm very tired," Gordon panted. Fortunately, the hatch closed without incident. The spacewalk had lasted just over half an hour, instead of the planned hour and three-quarters. Gordon was bitterly disappointed.

"Gene Cernan warned me about this, and I took it to heart," Gordon would later remark. "I knew it was going to be harder, but I had no idea of the magnitude. I got myself in a real bind. Using large muscles in my legs to hold myself in position, I became oxygen deprived. My heart rate went very high as well. It was painful for me. When I go back now and listen to the mission tapes, it scares me just to hear them!" An hour after the spacewalk had come to a premature end, the duo reopened the hatch and threw out the unneeded EVA equipment. Gordon now had some time to rest and help Conrad with the less strenuous task of putting equipment away in the spacecraft cabin before enjoying a well-earned sleep. But as he slept, many questions were being asked on the ground about the future of astronaut EVAs. With only one Gemini flight left in the program, NASA seemed to be back where it had started.

When asked by the authors to discuss the Gemini EVA, Gordon at first replied with a laugh, "Why do you ask frustrating questions?" It is clear that, for an accomplished test pilot, it is never enjoyable to reflect upon failures:

Alexei Leonov may have been the first to walk in space, but it was more of an accomplishment that he saved his own life out there! Ed White started this country on our EVA adventures, but my experiences, and the experiences of the others who followed him, were a bit different. With Gene, Mike, and me, we hadn't learned how to utilize that environment very well. Mine was a frustrating experience, not all that successful. I kind of wish I had the opportunity to do it again. The tasks I had were very difficult ones to perform. I equate the challenge I had to trying to tie your shoelace with one hand! We did learn from it, but it was a very difficult lesson, and certainly a frustrating one for me.

Although the EVA had only been a partial success, the two aviators could at least accomplish some record-setting test piloting. The next day, Conrad was given the go-ahead to try for the altitude record. At first, it didn't look like he was going to get the chance, as the Agena only began accepting commands after a puzzling delay. Once the glitch was cleared, however, Conrad was given the okay to fire the Agena engine for twenty-six seconds. "Whoop-de-doo!" he shouted as the astronauts were pushed into their seat straps with a sudden surge of fiery acceleration. "The biggest thrill of my life." Gradually, the planet beneath them fell further away, until its spherical shape became distinctly clear. Conrad and Gordon could not view Earth as a distant globe, as they later would during the Apollo program, but they were still seeing their home planet from a vantage point no one had ever been able to attain before. "The world's round," Conrad exulted from their high point of 850 miles. "You can't believe it. . . . I can see all the way from the end, around the top; . . . the curvature of the Earth stands out a lot."

One concern of NASA planners, that the spacecraft would be flying into the Van Allen radiation belts that surround the Earth, was quickly dispelled. Conrad reported that their dosimeter was reading a radiation level only half that experienced by *Gemini 10*. Firing the Agena again, *Gemini 11* plunged back toward a lower orbit, four times closer to the Earth. Soon afterward, it was time to carry out another, simpler EVA.

This time Gordon stood on his seat, tethered to the spacecraft so he could stay in place and take photos of the stars. As this could only be done on the night side of Earth, the crew had to pause as they passed over the sunlit side before they could resume work. As they waited, with nothing to do, they both fell asleep. "He was asleep hanging out the hatch on his tether, and I was asleep sitting inside the spacecraft," Conrad told the ground in surprise when he woke up. As Gordon explained, there was "nothing between me and oblivion except a pressure suit . . . but it was nice and warm and cuddly. Back in the womb, you know . . . nice and snuggly. You get that warm, fuzzy feeling, and I dozed off." Remarkably, in stark contrast to the struggles of the first EVA, the two of them had fallen asleep in raw space, zooming around the globe at seventeen thousand miles an hour. Now awake at last, they could finish the EVA tasks. Once the photography session was completed, they closed the hatch and had a second rest in safer conditions.

Now it was time for Gordon's EVA efforts to pay off. The tether that he had manually attached would be used. The Agena was turned so that its engine faced the Earth, after which Conrad carefully undocked and backed away from the docking collar, trying to move in a precisely straight line so the tether and the vehicles could be lined up precisely. It did not go as planned. The tether wanted to retain a certain shape, which pushed Gemini slightly off course. The tether then stuck a number of times when coming out of its stowage container, and Conrad had to move out of alignment to keep pulling it out.

With the lineup plans disturbed, it was decided to push on to the second tether experiment. For this, Conrad was to fire the spacecraft thrusters to keep the tether straight and the two spacecraft cartwheeling around the tether's central point. Centrifugal force should then have kept the tether taut, but this proved difficult. At first, the tether curved and rotated "like a skip rope . . . making a big loop," as Conrad described it. Every time he pulled on the tether with Gemini's thrusters, the spacecraft would bounce back when the tether reached full extension, and the line would go slack. Eventually, the tether pulled and stayed taut, and the joined spacecraft began to rotate around each other as they passed onto the night side of Earth. But when the crew tried to accelerate the spin, the line grew slack once again, and Gemini began to whiplash in response. Once the craft

Astronauts Pete Conrad and Dick Gordon (*right*) demonstrate the tether procedure used on their *Gemini 11* mission at a postflight press conference. To Conrad's right is George Low, deputy director of the Manned Spacecraft Center in Houston. Courtesy NASA.

settled again, the astronauts noted that a slight amount of artificial gravity was being created. They placed a camera on their instrument panel, then watched as it slowly drifted to the back of the cockpit. "We could feel the gravity," Gordon later recalled. "It was very, very low. But you could feel it, you could sense it, and you could observe it."

After three hours, it was decided that they had learned everything they could from the experiment. The tether was jettisoned along with the spacecraft's docking bar, and the crew carried out some more rendezvous maneuvers with their Agena. Finally, they placed their spacecraft in a station-keeping orbit, and went to sleep. When Gordon and Conrad awoke, they closed the gap between the two space vehicles and rendezvoused one last time. Then it was time to leave the Agena for good. "We were sorry to see that Agena go," Gordon said wistfully. "It was very kind to us."

Conrad then had to do something very counterintuitive for a test pilot: he had to disengage his hand controller and allow the computer to fly the spacecraft automatically through reentry. Although he watched the readings very closely, ready to intervene if necessary, the test of the system went perfectly. The spacecraft splashed down close to the recovery carrier, and the mission was over.

The rendezvousing, docking, altitude record, and tether experiment made *Gemini 11* one of NASA's most successful flights to date. The primary spacewalk, however, had been a relative failure.

The last space flight of 1966 and last ever Gemini mission suddenly had a new urgency. Preparations for *Gemini 12* had been suffering as attention shifted to Apollo. It had even been hard to find booster replacement parts during prelaunch testing, for example. But now *Gemini 12* was considered a vital mission. It would be the last chance to solve the puzzle of EVA before the Apollo program began. Weightless spacewalks would not be overly important to the first Apollo missions: the crucial EVAS would be on the moon. However, spacewalks were an integral part of emergency plans if anything went wrong, and it was hard to imagine human spaceflight progressing much further if the techniques could not be understood and applied.

If this mission's EVAS failed, no possibility existed for another Gemini flight to explore further. On 11 November, as the Agena target vehicle thundered into orbit, followed soon after by the *Gemini 12* spacecraft, demolition crews stood by ready to cut the Gemini launch stand into scrap metal. As *Gemini 12* hurtled toward space, the Gemini spacecraft preparation team had already worked its last day together. Even as the mission started, some aspects of the Gemini program were over for good.

As Jim Lovell made his first orbit of the Earth as *Gemini 12*'s spacecraft commander, he might well have wondered if the Gemini Mission Control team had also decided to pack up and go home early. For almost half an hour, he could raise no signal from them. It turned out to be a tracking station error, and when it was fixed he could begin talking with the ground about his most urgent task—a rendezvous with the Agena. But there was another problem: Gemini's radar was not getting a good lock-on signal to the target. A rendezvous would have to be made using only the crew's navigational expertise.

It was a moment that Lovell's copilot, Buzz Aldrin, might almost have wished for. He already had his sextant in hand to look for the target and soon began computing the angle of the Agena against some rendezvous charts he had in the cabin. To Aldrin, this wasn't some under-rehearsed and barely understood backup plan, however. He had played a large role in writing the charts he was now using and had pushed for the inclusion of such techniques in the Gemini program. He had even written his doctoral thesis on the subject of orbital rendezvous and was nicknamed, though not always affectionately, "Doctor Rendezvous" by many in the astronaut office. As Aldrin took the place of the computer, working on the calculations that made possible a rendezvous in an incredibly fast three hours and forty-five minutes from liftoff—and one that used hardly any precious fuel—he was confirming what his colleagues already knew about him. He was a very different kind of astronaut than most of those who had flown before.

Edwin Eugene "Buzz" Aldrin Jr. was born on 30 January 1930 on the border of Glen Ridge and Montclair, New Jersey. His father, an aviation manager in the oil industry, had previously served in the Army Air Corps as a pilot and was one of commercial aviation's pioneers. Edwin Aldrin Sr. had worked with aerospace pioneers such as Orville Wright, Robert Goddard, Howard Hughes, and Charles Lindbergh, and took his son for his first airplane ride when Buzz was two years old. Aldrin remembers both of his parents as strong willed, his father as being away most of the time, and, when he was home, as being "rather remote . . . an extremely intense person."

By the age of seventeen Aldrin was at West Point, from which he went on to graduate with a bachelor of science degree in 1951, placing third in a class of 435. He had previously been uninterested in academics and far more eager to play football. But his father pushed him hard to excel academically, and he soon gave up the sport for good. "My dad and I didn't always agree on what direction I should take in my career," Aldrin would remark about his hard-driving parent. West Point was a formative experience for Aldrin, who liked knowing what was expected of him. "You knew exactly where you were at all times," he would later write. "You could measure your progress. . . . I fit in well." Yet his father wasn't impressed by his hard-earned graduation, Buzz remembered. "Third place doesn't quite hold the appeal for him that first place does." Now that Buzz had a degree, Aldrin Sr. believed his son should serve in the military first, before pushing on into a successful business career, just as he had done. "He planted his own goals and aspirations in me," Aldrin would later write in his memoirs. "Many fathers live vicariously through their sons."

Commissioned into the U.S. Air Force, Aldrin soon earned his wings and a prized assignment as a fighter pilot. Posted to a fighter wing in Korea, he flew sixty-six combat missions in two years, then spent three years at the Air Force Academy as a general's aide and a flight instructor, before returning to fighter pilot duties with a squadron in Germany. In 1959 he attended MIT, the same school his father had attended, earning a doctorate in aeronautics and astronautics in 1963. His thesis on rendezvous techniques for manned orbiting vehicles gave him, he later wrote, "an understanding of the complex orbital mechanics of spaceflight, a background few other astronauts shared."

Aldrin had been interested in becoming an astronaut for quite some time, but was rejected for the 1962 astronaut selection because he was not a test pilot. He had realized that earning a doctorate meant he would be too old to attend test pilot school by the time he graduated, but academic work held the greater appeal for him. His work at MIT was doing a lot to enhance his air force prospects, however, and Aldrin had deliberately chosen a study area that he believed would be useful to both the air force and NASA. There were enough American engineers studying reentry, he believed, but few were looking at orbital rendezvous. It was an obscure subject area at the time, so much so that his supervising professors had trouble understanding what Aldrin was trying to put across. "I'm not sure I ever did succeed in communicating successfully what I had in mind in certain aspects of rendezvous and docking," Aldrin would later say. "I guess I was sensitive about that for a long time."

His doctorate did spur the air force to reassign him to their Space Systems Division in Los Angeles, then to its detachment at the Manned Spacecraft Center, where he worked on military experiments that could be incorporated into the Gemini program. "Gemini was the realization of all the obscure astronautical theory I'd absorbed at MIT," he would later write. He was keen to get even more involved. He applied again to NASA for the third astronaut group selection, and received some valuable advice on how to make it through the selection process from his friend Ed White, who had been chosen for the second group. This time, it worked: he was in.

Aldrin hit the ground running as an astronaut. Understandably, given his background, he wanted to become involved in Gemini rendezvous techniques as quickly as possible, including flying a mission himself. "I got into the space program with high hopes for an early rendezvous mission," he told the authors. "That didn't happen." Aldrin soon found that his lack of a test pilot's background was counting against him more than his technical expertise was working for him. "I soon realized that there was a definite pecking order," he explained. "I was really an odd man out in the astronaut corps . . . alienated from some of the other astronauts." To his surprise and disappointment, astronaut status was apparently not based on skill or the results of any training tests. Instead, it seemed based on the personal perceptions of his superiors. To obtain the flight assignment he craved, he needed to keep Shepard and Slayton happy. As he told the authors, he knew that this was going to be tough because of his background: "I didn't start out to be a test pilot; I got into the business by being an egghead from MIT. Other people went to test-pilot training—I concentrated on education. I got fascinated with orbital mechanics, and I had some interesting things in my background, but they didn't really want somebody who hadn't been in the test pilot business."

Aldrin was assigned to mission planning, and he saw it as an opportunity to advance his ideas in rendezvous theory and put them into practice. As he wrote in his memoirs, it was not always an easy experience for him "translating these complex orbital mechanics into relatively simple flight plans for my colleagues. . . . I saw that most of these guys weren't really interested." Aldrin felt that most of them were "hard-core stick-and-rudder fighter jocks" who did not wish to know anything more than which way to point the spacecraft. They had nicknamed him "Doctor Rendezvous," and Aldrin found himself giving them "a hard time for being so intellectually lazy."

It was an entirely different world from West Point. "It was NASA's desire not to be competitive," Aldrin told interviewers Douglas MacKinnon and Joseph Baldanza, "and that means we were never given tests to see how well you could do something; . . . the people designing our training had no idea whether we observed the training that they gave us. . . . I don't think that's the way the military would do it."

Aldrin's zeal for pure theory and his wish to see measurable results represented a striking change for the astronaut corps, and they were unlikely to win him many friends or a coveted early rendezvous mission. He vividly remembers one occasion when a fellow astronaut bluntly snapped at him that he had a "reputation for trying to screw up guys' missions." It was an outburst that caught him by surprise. "I wasn't a good office politician," he would later admit. "My entire adult life had always been so structured. . . . I am by nature a very competitive, direct person, and when I want something I am not afraid to make my wishes known."

He waited three long years for his first flight assignment, and when it came it was a profound disappointment to him. He was to be the backup pilot for *Gemini 10*. This meant that he would normally rotate into a prime crew slot three missions later—except that there was no *Gemini 13* planned. Aldrin would not only be training to back up a flight he would never fly, but this dead-end training would also eliminate him from consideration for the early Apollo missions. It was about the worst assignment he could have received. "I just wasn't an organization man," Aldrin would later admit. "I now realized team playing was how you got ahead as an astronaut; . . . by being direct and honest rather than political, I'd shafted myself."

Then, a month later, Bassett and See died. In the shuffle of assignments that followed, Lovell and Aldrin moved up a place to the *Gemini 9* backup role. Now the duo was squarely in line to fly the last Gemini flight. "That was how I came to have a mission assignment," Aldrin would write, "but it was a hell of a way to get one." When asked bluntly by the authors how it felt to step into dead men's shoes, Aldrin paused for a long time before replying. "It was a very, very . . . enabling turn of events for me," he responded in a show of mixed emotions:

I am probably one of the most fortunate ones. Jim and I were scheduled to be on the backup crew for Gemini 10. I'd gotten involved in rendezvous at MIT, *and was hoping to be involved in an early rendezvous flight, and here my future was going to be backing up Gemini 10, and not even flying in the program. A lot of things changed; we lost Elliot See and Charlie Bassett in St. Louis. We lost a good number of people in aircraft accidents. A lot of things happened just because they happened, and we ended up benefiting, those of us fortunate enough to get the nod to fly on a significant mission. It was being in the right place at the right time—all of us certainly were there. They needed a backup crew, so Gene and Tom moved in. They became the prime crew, and Jim and I were backup. It took a tragedy to alter the course of that.*

Aldrin's focus would now change. "My hopes for an early rendezvous mission didn't happen," he told the authors, "so instead I got a chance to

One of the last photographs ever taken of astronaut Ted Freeman (*right*), shown here at the Manned Spacecraft Center with Buzz Aldrin the night before he was killed during a routine training flight. Courtesy NASA.

pioneer EVA." He would be the one to demonstrate beyond doubt that the puzzle of EVA had been solved. The first piece of the puzzle was underwater training. Like most puzzles at NASA, the solution was a team effort. Underwater training was something that Mercury astronaut Scott Carpenter had been thinking about ever since his leave of absence from the astronaut office, where he worked on the ocean floor as part of the navy's Sealab experiment. "What the route was that brought it to NASA, I am not sure,"

Carpenter told the authors. "I don't claim responsibility for it. But it was a concept that came to me in the water, and it finally got applied at NASA."

As Carpenter explains, zero gravity and underwater work have a lot in common:

Without weight, you don't have traction. And traction is important wherever you are if you want to do work: you have got to be stabilized. And the buoyancy of the water or the absence of sensible gravity in spaceflight makes that very difficult because you have no traction. And it occurred to me when we were doing work in the water, where there is no traction because of the buoyancy, you need to have foot restraints. That is something that is immediately apparent to a diver, because there is no way to react against the forces you apply. So you can intuit that, but nobody seemed to really do it. Somewhere in the aftermath of the difficulty that Gene Cernan had, it became apparent that we needed foot restraints and hand restraints in order to give the stability that would be provided by traction at 1 g.

Carpenter would soon be involved in the design and construction of NASA's first immersion facilities, "to work on the problems of zero-g handiwork," as he puts it. But there was no time to create something like that in a hurry, with Gemini missions flying every couple of months. Instead, following the *Gemini 9* mission in which Aldrin backed up Gene Cernan, both astronauts journeyed to an existing facility. As Aldrin told the authors, "we trained in a swimming pool in Baltimore, the McDonough Boys' School swimming pool!"

It is not clear who first came up with the concept of training underwater. "Not mine," Aldrin answered when the authors asked him if it was his idea. Possibilities include a group from NASA's Langley Research Center and two private aerospace businessmen from Baltimore. But as Jim Lovell told the authors, it didn't really matter; what mattered is that it worked:

I really don't know who originally talked about the swimming aspect of it, in the pool. I think it was one of our engineers who said, "Look, we are looking at this all wrong. We ought to build a crude spacecraft, at least the rear end of the spacecraft, get it down, and put Buzz underwater." I sat on the edge of the pool; we had a communications link. I went through the checklist, and he went down and tried various things, to see what would work and wouldn't work.

We spent a whole day in that pool up in Baltimore—and that, I think, was the breakthrough. That then allowed the engineers to go back and redesign the footholds and do everything like that.

The spacesuit works as well underwater as it does in space. All you have to do is put weights around to keep it neutrally buoyant, and tell the person, "Do not swim." You are defeating the purpose if you swim. And so therefore you have to learn to float neutrally, and then look at the kind of handholds and footholds. One thing that the early EVA people didn't realize was that for every action there is an equal and opposite reaction. On the Earth, gravity is such a strong influence on everyone that we don't realize when we are walking down the street and we put an action against the Earth, it reacts against us. But in space, when we neutralize gravity by centrifugal force and then we go to a body the size of a spacecraft and touch it, it repels us. And so unless we had the proper handholds, and knew how to use zero gravity to our advantage, we lost that touch.

The work of the astronauts in the pool was progressing well and providing good solutions, but the lessons it taught were not implemented in time to help the *Gemini 10* and *11* missions. Collins and Gordon did not have the opportunity to train underwater before their missions, and as they encountered problems in space they returned with largely the same conclusions that were being discovered in the pool. As Gordon told the authors, the primary lesson of his spacewalk in his view was a need "to create handholds and restraint systems to allow us to position ourselves to do useful work." Mike Collins gave the authors a similar response: "In two words . . . in fact, in one word—restraints. Specifically, foot restraints. Following my flight, when Dick Gordon had a lot of problems too, people said, at that stage of the game—stop. Instead of sending people out to do tasks, let the task be discovering what EVA is all about. So that is when Buzz came along on 12."

Just as Aldrin had backed up Cernan on *Gemini 9*, now Cernan was backing up Aldrin on *Gemini 12*. Based on Cernan's experiences and the reports that also came back from Collins and Gordon, NASA devised an entirely new training method and set of EVA tools. Although Aldrin had been afraid of water as a child he later became an accomplished scuba diver, so the techniques came quite naturally to him. He found that one important aspect of working underwater was that he could practice EVA according to the mission time line he would use in space rather than during brief spurts of activity in zero-g aircraft and on air-bearing tables. As he told the authors, through teamwork he felt they had finally cracked the mystery:

In retrospect, neutral buoyancy was the breakthrough. We improved, in an evolutionary way, foot restraints that were absolutely critical; feet have to be anchored, not temporarily attached. We looked at Velcro, little loops, and we canted the loops, and we finally said, anchor the feet. Well, that came along at the same time as neutral buoyancy. I think that's one of the major things that we got out of the whole Gemini program: the neutral buoyancy training that gave us the opportunity to exercise three-dimensional freedom, similar to scuba diving, in a neutral buoyancy tank. It was outstanding training for spacewalking.

Gemini 12 was therefore the first flight that was equipped with a reassuring number of effective handholds and footholds. Aldrin would also have special tethers, including some that attached to his waist, to help secure himself in place. "They were little different from the straps window washers use when working on high-rise buildings," Aldrin later wrote. A plan to fly the AMU backpack that Cernan tried to test on *Gemini 9* was canceled as overambitious, and the maneuvering guns were also finally abandoned as a concept. It was a bittersweet moment for Aldrin. Just before his selection as an astronaut, he had been assigned to integrate air force experiments like the Astronaut Maneuvering Unit into the Gemini program. "NASA chickened out and canceled the AMU—the Air Force's number one experiment," Aldrin told the authors. "Not good for Air Force–NASA relations!"

By the time Aldrin was sent into orbit with Jim Lovell, the *Gemini 12* mission had been drastically simplified. The EVA experiments were not designed to help achieve any other mission objective. Instead, they were designed as an end in themselves, providing a simple, step-by-step analysis of spacewalking techniques to see if basic tasks could be performed in zero gravity. It was an approach that should have been tried on earlier Gemini missions; only now had that lesson sunk in. "We had to take a couple of steps backward and start reviewing what we had learned from previous EVAs," Aldrin would explain. "On the last flight of the Gemini program we just couldn't afford *not* to have a success."

Before any spacewalk trials, however, Aldrin showed his prowess at manually calculating a rendezvous trajectory. It was a maneuver that Lovell is still very proud of, as he told the authors: "We did that rendezvous without radar, which was a little bit touch and go, because we didn't know our range rate—how fast we were going. We had practiced the backup procedures in training, but never really expected to use them. We could line up directly coming up towards it because we could line up with the background stars, but we didn't want to hit it, and we didn't want to go by it and go too far away. That worked out very well." Just as Conrad and Gordon had done on *Gemini 11*, Lovell and Aldrin also practiced undocking and then redocking with their Agena. However, the procedure did not work as well on this flight. When undocking, the Gemini caught on the docking latches and made a disconcerting grinding noise. Lovell was only able to shake free by firing the thrusters. Unlike *Gemini 11*, Lovell and Aldrin were unable to use their Agena to rocket to a higher altitude because the Agena's main engine had lost thrust pressure on its way to the rendezvous, and Mission Control decided that it would not be safe to fire it again with a Gemini spacecraft attached. Instead, the crew used Agena's secondary propulsion system to place them in a path directly under the shadow of a lunar eclipse. Perfectly aligned with the Earth, moon, and sun, Lovell and Aldrin were able to wake up for their second day in space and take some dramatic photos of the astronomical event.

The Agena failure had been a disappointment for the two astronauts, but they had the consolation of knowing it was not the primary focus of this mission. That was to carry out successful spacewalks, and Aldrin was ready for the challenge. He had learned a lot from his underwater training, and so even before Aldrin opened the hatch he was trying out his new methods. He worked slowly and methodically, analyzing his movements, resting for short periods, and allowing himself to adjust to the new working environment. For the first EVA he remained standing on his seat, photographing star fields with an ultraviolet camera, with the exception of a brief moment when he installed an outside handrail and retrieved a micrometeorite experiment.

Aldrin felt so at home during his spacewalk, in fact, that he was not even surprised when he noticed blue sparks jumping between the fingertips of his gloves. He attributes his calm to his more scientific way of looking at things. "Observational sensitivity," he called it when recounting the episode to the authors. "Being available to note the unusual. Rubbing fingers together and seeing a spark didn't seem unusual to me, but it sure did to the people who heard about it." As he brushed his gloved fingers together, he created a steady flow of current as the spacecraft passed through the electrons of the ionosphere. He was discovering new aspects of the space experience at every moment.

After two hours and twenty minutes, Aldrin floated back into his seat and closed the hatch again. So far, all had gone well with the EVA trials, and it was time for another sleep before preparing for a longer, more complex spacewalk.

The next day, Aldrin began the true test of the EVA training. Using the handrail he had installed the day before, he removed and replaced a movie camera, proving that the handrail could keep him in place without having to exert himself. Then he worked his way to the Agena adapter, carefully using handholds to make the journey. Once there, he positioned himself securely with a tether. He then found it was easy for him to accomplish the task that had given Dick Gordon so much trouble: tethering the Agena and Gemini together. "I thought about every move I made while I was out there," Aldrin would later write. "Even flexing my fingers or shifting my hips."

Next, Aldrin made his way along a handrail to the back of the Gemini spacecraft. Securing his feet inside special restraints that slipped over his boots, he rocked himself from side to side and backward, seeing how well they held him in place. He found he could lean backward to a horizontal position, and the restraints still gripped his feet securely. It seemed that the problem of providing adequate restraints was at last solved. This would not be certain, however, until Aldrin tried performing some basic tasks. He severed test cables and turned bolts at the back of the Gemini with ease, then once again made his way to the Agena, where he practiced unhooking and reattaching electrical connectors. This served no real purpose other than to prove it could be done—and Aldrin proved it. With his tasks accomplished, Aldrin returned to the cabin. He had been outside for over two hours and felt fine. "The EVA was just how I expected," he would happily recount after the flight. The last major objective of the Gemini program had been completed.

During the final two days of the last Gemini mission, there was still time to try and perfect some other spaceflight techniques. The biggest challenge was an attempt to line up the Agena and Gemini and point them straight at the Earth in a tethered configuration that was supposed to remain stable. The *Gemini 11* crew had not been able to accomplish this task, but Lovell was able to—for a while. Once again, it was frustratingly difficult to make the tether taut and then to keep it that way. But for four hours the joined spacecraft orbited the Earth at either end of the tether, and some useful lessons were learned for the future.

The third day of the flight saw the very last EVA of the Gemini program. It was a simple one: throwing out trash and taking more star photos. "The EVAS on this flight," Gus Grissom would later write, "three periods in all, went off far better than anybody had dared to hope; proof that the underwater training was a valuable tool." By now, however, the Gemini spacecraft was not holding up too well. Some of the thrusters were not performing, and the fuel cells were misbehaving as the water was not draining from them correctly. The astronauts drank as much of the water as they possibly could to help alleviate the problem—an almost comical solution—but it was not enough, and eventually the spacecraft had to rely on battery power alone. Trouble continued as the spacecraft reentered on the fourth and last day. A storage pouch pulled free of the cabin wall in the increasing g-forces and hit Lovell, right at the point where the commander was holding the ejection seat ring. If he had tried to catch the pouch with his hands, he might have accidentally ejected. Luckily, he resisted the impulse. Aldrin had a similar experience: trying to film the reentry out of his window with a sixteen-millimeter camera, the device became difficult to hold onto as the g's built up. "I let go, and it slammed into my chest," Aldrin recalled. It was an inglorious end to a mission that had had more than its share of failures, including an unreliable Agena and a failed Gemini radar. But all that now seemed forgotten due to the overwhelming success of the EVA. "It was a magnificent conclusion to Project Gemini," Grissom would conclude.

Jim Lovell also felt it was the perfect end to the Gemini program, as he told the authors: "*Gemini 12* is not too well highlighted in the history of our spaceflights. However, there were several things on *Gemini 12* that wound up the program on a very positive note. One, of course, was the EVA. When you look back on it, we determined how to train for EVAS, and that was using a water tank. Consequently, when we actually did the flight, Aldrin did an excellent job with that. *Gemini 12* really capped off the Gemini series and led into the initial Apollo flights quite nicely." Aldrin had also redeemed himself in the eyes of the decision-makers—at least for a while. "Buzz Aldrin, once and for all, banished the gremlins of spacewalking," Shepard and Slayton would later write. The deaths of two colleagues had allowed him to get one Gemini flight under his belt. Without it, Aldrin might not have been considered experienced enough to be assigned to Neil Armstrong's *Apollo 11* crew. "A lot of fate determines where you fit into the puzzle," he later reflected.

There would only be one other American spacewalk mission flown before Aldrin was in space again and once more making an EVA. His next one would be on the surface of the moon.

After their *Gemini 11* flight, Dick Gordon and Pete Conrad served as the backup crew for *Apollo 9*, along with another old naval aviator friend, Alan Bean. As was customary, the backup crew flew three flights later, as the *Apollo 12* moon landing mission. Since there had been no guarantee that *Apollo 11* would be the first landing mission and as *Apollo 9*'s mission was originally planned as the second manned Apollo flight instead of the third, Conrad and Bean could easily have been the first people to land on the moon. Dick Gordon, as the mission's command module pilot, is just as proud of the achievement as if it had been the first landing, as he told the authors: "I had the great distinction and privilege of flying the next flight, *Apollo 12*, in November 1969, where we accomplished the second lunar landing. Not only did we accomplish the mission that President Kennedy said we were going to do once—we did it *twice* before the decade was over." Yet Gordon dearly wanted to land on the moon with his two *Apollo 12* colleagues and found it hard to watch them journey to the surface without him. With the *Apollo 12* mission over, Pete Conrad planned to move on to the Skylab program, realizing that was where the next flight opportunities were. He urged Bean and Gordon to do the same. Bean agreed, and the two of them later commanded Skylab missions.

Gordon, however, had set his sights on commanding an Apollo lunar mission of his own, crossing the last few miles from lunar orbit to the moon's surface. He was soon assigned as the backup commander for *Apollo 15*, with a crew of Vance Brand and geologist Jack Schmitt. "In the normal rotation at that time," he told the authors, "I was hopefully preparing to command *Apollo 18* and go back to the moon." Budget cuts led to the cancellation of the *Apollo 18* lunar mission. Gordon and his crew were very disappointed, but continued to work hard at training for *Apollo 15*. The *Apollo 17* crew had not been formally announced, and Gordon was hoping to get that mission instead. In the normal rotation, that assignment would have gone to the *Apollo 14* backup crew of Gene Cernan, Ron Evans, and Joe Engle. However, this would be the last Apollo flight to the moon, and no one knew if the normal rules would apply. Meanwhile, the scientific community was putting great pressure on NASA to assign scientist-astronaut Schmitt to a lunar landing mission. Gordon felt that, with Schmitt as part of his crew, he had a good chance of getting the *Apollo 17* assignment. He asked Shepard and Slayton to keep his crew together and to fly them.

It was also time to call in the favors from his powerful friends, and Gordon had some impressive astronaut names on his side. Pete Conrad, Jim McDivitt, and Dave Scott all thought Gordon's crew should fly the *Apollo 17* mission. Cernan, however, had two of the most politically powerful astronauts—Tom Stafford and Alan Shepard—suggesting that he be the one to fly the mission. Gordon thought he had it for certain when, before any decision had been made, Gene Cernan crashed a training helicopter in a clear case of pilot error, raising questions in many minds about his command skills. Even Cernan thought at that point that he had lost his chance to Gordon. In the end, however, Slayton stayed with his rotation system and decided that it should be Cernan, Evans, and Engle flying *Apollo 17*. He submitted their names to NASA headquarters, but he was overruled; his bosses insisted that Schmitt fly. Slayton could have easily relented and nominated Gordon's entire crew. However, he chose to compromise instead: he broke up the crews. Schmitt would fly to the moon, but with Cernan and Evans. "He stole my lunar module pilot," Gordon says today, only half-jokingly. Gordon would not be returning to the moon after all. Schmitt pleaded with Slayton not to separate the crew and let him fly with Gordon—but the decision had been made. Instead of becoming the last man to stand on the moon, Gordon would never fly in space again.

"Unfortunately for me, *Apollo 17* was the last," Gordon told the authors; "*Apollo 18* never flew. If you can think of a way I can get those last sixty miles under my belt, please let me know: I'd be glad to go!"

For a hotshot test pilot who had come to NASA with such promise, it was a surprise to many that Gordon never did command a mission. The disappointed astronaut retired from NASA and the navy in January 1972. "I still miss the flying," he says. Jack Schmitt and *Apollo 17* flew that December, and humans have not returned to the moon since. Gordon is still most unhappy that the Apollo lunar program came to a premature end, denying him a well-deserved opportunity, as he told the authors:

Several things stopped it—economics, desire, and leadership. The reward-to-risk ratio went down. Put them all together. There's always this controversy too, over why spend all this money in space, and all that kind of thing. Not a damn nickel has been spent in space! It's spent right here: right here on Earth. I think of our advances in technology, and I think the space program has given them all to us. Our standard of living and the advances in technology have been accelerated because of our space program. The only other event that accelerates technology is war. You know which one I would choose; I think you would, too. You always hear about so many social ills that this country has to take care of. I propose to you that if our social ills had been a priority back in the 1700s and 1800s, the western boundary of the United States would be Virginia's Allegheny Mountains.

On leaving NASA, Gordon became executive vice president of the New Orleans Saints football team, before going on to work in the oil industry and serving as a director for Dave Scott's science and technology company. No longer an astronaut, he watched his *Apollo 12* crewmates fly missions to America's first space station—Skylab.

Pete Conrad commanded his country's first space station mission in May 1973. The space station had suffered a major problem on launch, when its micrometeorite shielding and a solar panel were ripped loose. In a hurried but determined effort, Conrad and his crew trained to launch just eleven days later and repair the station in a series of dramatic spacewalks. Without the EVA experience gained during the Gemini program, it is unlikely that such a risky mission would have been attempted. The effort was successful, and Conrad subsequently spent twenty-eight days on board the station. "Pete and his crew saved the Skylab," Tom Stafford would later remark. "He was one hell of a guy." For Conrad, it was the proudest moment of his space career, far more important to him than walking on the moon. "If people say 'space' to me," Conrad would explain, "I don't think about flying to the moon; I think about Skylab. The moon did not tax me, nor give me as much satisfaction as Skylab." In terms of piloting, however, Conrad had a different favorite mission. "From a flying point of view, *Gemini 11* was really it. We hand-flew our machine; . . . that was a great pilot's flight. I only got one minute of stick time landing on the moon, and there wasn't really any flying at all on Skylab."

Following his space station mission, Conrad resigned from NASA and the navy. "There was nothing left to fly," he mused. "It was very sad during the last two years that I spent at NASA, watching it continue to crumble." He began work instead with a communications company, but he never strayed far from the world of cutting-edge aerospace programs, and by 1976 was working for McDonnell Douglas. "I still have fun playing with the toys," he said at the time, "and they haven't completely shut me out of the cockpit." In 1985 he even put on a spacesuit again and practiced underwater EVA techniques for the company.

For three years starting in 1993, Conrad once again piloted a vehicle designed for space operations. This time, it was via remote control, testing the DC-X single-stage reusable spacecraft prototype. The ship took off and landed vertically, and Conrad piloted it through most of its nine-flight test program. The vehicle was the closest there has ever been to a realistic successor to the space shuttle, but NASA politics shut the program down after a minor technical mishap. In 1996, frustrated by the experience, Conrad set up his own private space launch company called Universal Space Lines, with the goal of creating ways to operate commercial reusable launch vehicles. He hoped to develop ways for rockets to launch inexpensively and manage satellites. "If somebody could begin to get space to pay commercially, then it will grow in absolute leaps and bounds," he said. "We have to make space travel affordable."

When one of the authors met Conrad on 26 June 1999, he was as busy as ever with the space business. While discussing his space achievements, he constantly checked his cell phone, following a satellite that his company had a stake in. He expressed annoyance that, while it had been possible to fly commercially forty years after the Wright brothers flew, the same opportunity in space was still not possible forty years after NASA was formed. In his opinion, the last thirty years had been wasted time, and he was eager to push the boundaries once again. Unlike John Glenn, who let NASA fly him into space in his seventies, Conrad was eager to fly again, but in his own spacecraft.

Twelve days later, Pete Conrad was dead.

Even at the age of sixty-nine, he was a self-confessed lover of "fast bikes, fast cars, and anything that moves." Conrad was riding his Harley-Davidson with a group of friends in Southern California when he made too wide a turn on a winding road and flipped his cycle. At first his chest and abdominal injuries seemed minor, but upon his arrival at a local community hospital for a checkup his condition worsened. It seemed that he had some internal bleeding, but as he went under anesthesia for some exploratory surgery he fully expected to be awake again soon and make a complete recovery. Unfortunately, his injuries were worse than originally believed, and he never woke again. He passed away five hours after the crash.

Conrad's funeral at Arlington Cemetery was one of the largest gatherings of former astronauts ever. Three Mercury astronauts, nine Gemini astronauts, and many other Apollo and Skylab astronauts paid their respects to their colleague. Neil Armstrong, struggling to fight back tears, gave an emotional farewell. "I'm not sure what he's doing right now, but I suspect he's telling some stories of the old days," Armstrong said. "Pete was the best man I ever knew. He treated me like a brother." NASA Administrator Dan Goldin also paid a fitting tribute to Conrad's irrepressible personality, saying that "at sixty-nine, he had the spirit of a thirteen-year-old." John Glenn added: "I didn't know anyone that was filled with more irrepressible enthusiasm and sense of humor and new ideas and general joy of life than Pete. He'll be missed very much."

Conrad once said that his epitaph should be "I came. I flew. I left." But his grave marker at Arlington Cemetery sums the man up just as well. Under his name are two simple words: "An Original." Looking back now on the mission he flew with Conrad, and at the entire Gemini program, Dick Gordon is justifiably proud:

The Russians were trying to get to the moon. They wanted to show the world their technology and their capabilities. So did we. The program called Gemini was created to learn the things we'd need to learn to accomplish the Apollo mission. We were planning to go to the moon in ten days—three days out, three days back, and an arbitrary four days in near-lunar vicinity. We had to learn how to maneuver our spacecraft; we had to learn how to rendezvous, how to dock. We hadn't done that yet. We had to learn how to do those things. The years 1965–1966 provided us with opportunities to learn how to do all those things. In that very, very short period of time in 1965 and 1966, we accomplished all of the individual elements that we needed to learn to go to the moon.

Yuri Gagarin, the first person in space, was in a similarly reflective mood when writing an article in 1966. "Looking back at the past five years in space," he wrote, "one can see the road as difficult, as interesting . . . entering new mysteries of space. Even five years ago it was difficult to imagine that in such a short time it would be possible to develop and launch into space multi-crewed spaceships, fulfill EVAs, and do group flights."

Bolstered by two years of stunning successes, it was time for NASA to push on ever closer to its lunar goal and debut the Apollo spacecraft. No one could imagine the profound tragedy that lay just around the corner.

4. The Risk Stuff

So far this is a thankless, risky and extremely difficult task.
It will strain our resources and talent to the limit,
and it will entail great losses too; . . .
one cannot compare space flight to aeronautics.
The latter is like a toy compared to the former.

Konstantin Tsiolkovsky

Navy Lt. Cmdr. Roger Chaffee, aged thirty-one, was a fresh-faced, dark-haired, and introspective young man, known jokingly but respectfully around the astronaut office as "The Rookie." His senior crew members, Gus Grissom and Ed White, had distinguished themselves on earlier space missions, and he was joining them on the first manned flight of the Apollo spacecraft. This was the spacecraft that, with later modifications, was scheduled to fly humans to the moon and return them safely to the Earth before the decade was out.

Chaffee came into NASA with an energy and ambition that were marked by a strong youthful desire, nurtured well before the days of manned spaceflight, to become the first person to walk on the moon. In official astronaut portraits he has the look of a gentle, sincere young man, with a wistfully shy smile and a twinkle in his eyes. It is a sad truth that the public never really came to know Roger Bruce Chaffee well, yet he would certainly have been a quiet achiever of the space age he was so thrilled to participate in. Like Ed White, he was openly and unashamedly patriotic. At the time of his astronaut selection he responded to a reporter's question by stating, "I'll be doing something for my country—something in which I can take pride." In 1963 he wrote a moving letter to the son of a man he knew while growing up in Grand Rapids, telling the eight-year-old that

"You have to love your country so much that every time you see our flag you feel warm inside."

Roger Chaffee's area of specialty within the astronaut corps was communications and navigation. In fact, he had been assigned to Apollo spacecraft communications virtually from the time he joined NASA as a member of the Group 3 astronauts in mid-October 1963. Michigan-born, his first ride in an airplane had taken him on a short but highly memorable flight over Lake Michigan at the tender age of seven. His father, Donald Lynn Chaffee, a dashing former barnstormer who had flown open-cockpit airplanes around county fairs, was at the controls. From that magical moment on, young Roger was steadfast and unstoppable—he knew his future lay in the skies. On many occasions he would enthusiastically point out aircraft flying over his Greenville home. "I'll be up there flying in one of those someday," he would tell his parents with prescient conviction. He began making model airplanes, and would proudly hang each of them from his bedroom ceiling. By the age of sixteen he had found that science and electronics held a certain fascination, but this burgeoning interest underwent a rapid evolution around the time he received his driver's license. From then on, anything to do with mechanics and engineering proved a greater and more enduring attraction for the teenager. His first automobile was paid for through his own part-time labors: a 1929 Lafayette that he bought for the grand sum of forty-nine dollars. A few months later he traded the Lafayette for a 1934 Ford Sport Coupe, which became his pride and joy. He spent many happy, instructive hours beneath the hood of the car, reassembling, tuning, and cleaning the engine until it ran like new.

Chaffee began his junior year at Purdue University in 1954, and the following year he drove out to Stair Field in Mulberry, Indiana, where he took his first flying lessons. Around that time he went on a double blind date with a reluctance that rapidly dissipated when his date turned out to be an attractive young college freshman named Martha Horn, from Oklahoma City. They soon fell deeply in love and would later marry.

The budding fighter pilot finally soloed in March 1957, and on 24 May he attained his private pilot's license. As a result, Chaffee's Purdue test administrator recommended that he undertake military flight training. In August 1957, following Chaffee's graduation from university with distinction as an aeronautical engineer, he joined the U.S. Navy. At the completion of his flight training he began a series of assignments aboard several aircraft carriers, later serving as a safety officer and quality control officer for Heavy Photographic Reconnaissance Squadron VAP 62, popularly known as the "Tigers." Stationed at the Naval Air Station in Jacksonville, Florida, he often flew RA-3B jet photoreconnaissance aircraft on covert operations high over Cuba, and would later receive the Navy Air Medal. Before joining NASA, he had been a student at the Air Force Institute of Technology at Wright-Patterson Air Force Base, Ohio, where he worked for his master's degree in reliability engineering. At the time of his selection to NASA he had logged more than twenty-five hundred hours' flying time, including twenty-two hundred hours in jet aircraft. The brand-new NASA astronaut and his wife were by now the proud parents of two children, Sheryl and Stephen.

In June 1965, Chaffee served as one of the CapComs for the historic *Gemini 4* mission of Jim McDivitt and Ed White, under Chief CapCom Gus Grissom. Like everyone else he was enthralled by White's twenty-minute walk in space. Chaffee's enthusiasm and work ethic apparently made a lasting impression on Grissom, who once said of him at a media conference, "Roger is one of the smartest boys I've ever run into. He's just a damned good engineer—there's no other way to explain it. When he starts talking to engineers about their systems, he can just tear those damn guys apart. I've never seen one like him. He's a really great boy."

On 25 March 1966, NASA announced the names of the six pilots who would serve as the prime and backup crews on the first Apollo mission, officially designated *Apollo 204*. Chaffee, who had been in training for more than two and a half years, was selected as a member of the primary crew, together with Grissom and White. This plum assignment made him for some time to come the youngest of America's astronauts ever chosen to fly.

Chaffee actually gained his seat by default. The original crew member teamed with Grissom and White was another spaceflight rookie from Chaffee's astronaut group, Donn Eisele. Just days before the crew was officially announced, Eisele threw his shoulder out for a second time, forcing him into surgery, after which he would have to wear a sling for several months. NASA could not afford to excuse a crew member from crucial mission systems training, so Eisele was reassigned to the role of pilot on Wally Schirra's backup crew, and Chaffee joined the prime crew on *Apollo 1*.

Roger Chaffee (*center*) and Ed White (*top*) pose with original backup *Apollo 1* commander Jim McDivitt in the mock-up display area at the North American plant in Downey, California. Courtesy NASA.

He may have been a young, rookie space traveler, but Chaffee was certainly not lacking any confidence in the spacecraft, or his own abilities. "Hell, I'd feel secure taking it up all by myself," he once said. "You feel secure because you know what you're doing." His unlimited passion and zeal for the task at hand, combined with a pronounced knowledge of the Apollo hardware, was infectious. Milt Radimer, former manager of the lunar module descent stage construction at Grumman, said that Chaffee's enthusiasm drove all of them to greater heights. As Radimer recalls, the astronaut once told him he wanted to meet everyone at Grumman's Plant 2, where the descent stage was under construction.

I said, "You're kidding. We have a couple thousand of them here at Plant 2. We're building the whole descent stage out there, and all of the components—a lot of components and parts and everything." He said "I want to meet them all." And he was sincere about it. And we took him through the plant. We showed him the descent stage first. I introduced him to Bob Ekenstierna, who was a supervisor, and Eky explained everything to him. This fellow was listening to everything, taking it all in. He was looking at every rivet, every part of the descent stage. And everyone in the department lined up, and he wanted to shake their hands. And he did. He told each one of them, "You know, your job is just so important to me, and to an awful lot of other people. There's no gas station on the way, no place where I can pull over, like you can in a car." He told one man, "If this line fails"—and the man is bending the line on a machine—"you know, I have no way of getting this fixed. When you're making this, you've got to make it the best you can." Roger actually changed everybody's mode of thinking. It wasn't just a piece of metal anymore. It was Roger Chaffee.

On 25 August 1966 an unmanned Saturn IB thundered off the launch pad, becoming the thirteenth successive—and successful—flight of the mighty rocket. All of the booster's major mission objectives were fulfilled, and the next scheduled Saturn flight would be a manned shakedown test of the new Apollo spacecraft from North American Aviation's production line. Designated production number 012, it was intended to carry three astronauts into Earth orbit.

As one of NASA's most experienced and respected astronauts, mission commander Grissom had been given the option of an open-ended mission. If all went well, the crew would be permitted to remain in orbit for up to two weeks. Their flight was not only scheduled to check out the Saturn and Apollo systems, but would also establish that launch operations, ground tracking, and control facilities functioned according to plan. Grissom, a renowned perfectionist in all he did, said at one press conference that he was determined to see out the full fourteen days in orbit in order to maximize test results.

The spacecraft they would occupy was the Block I version, designed for use on unmanned and manned checkout flights. It did not contain the docking tunnel through which astronauts would make their way to the lunar module on later flights, and it also lacked the advanced hatches and some of the interior changes planned for the more sophisticated Block 2 version. That December a scheduled launch date of 21 February was announced, and this was followed by a press conference featuring the crew. During the conference, Chaffee was asked to comment on a number of problems that were known to exist in their spacecraft. He brushed this off with a shrug and said he felt good about it. "I think we've got an excellent spacecraft," he stated with conviction. "I've lived and slept in it. We know it. We know that spacecraft as well as we know our own homes, you might say. Sure, we've had some developmental problems. You expect them in the first one."

CBS correspondent Nelson Benton asked Ed White whether he harbored any particular concerns about taking part in the spacecraft's maiden, shakedown flight. As always, White was laconic and introspective. "No, I don't think so," was his thoughtful response. "I think you have to understand the feeling that a pilot has—that a test pilot has—that I look forward a great deal to the first flight. There's a great deal of pride involved, in making a first flight." Benton then put the same question to Chaffee, who responded, "Oh, I don't like to say anything scary about it. There's a lot of unknowns of course, and a lot of problems that could develop or might develop, and they'll have to be solved, and that's what we're here for. This is our business, to find out if this thing will work for us."

Earlier, mission commander Grissom had been his usual pragmatic self when asked if he felt he was defying the odds by flying a third mission. His words were sadly prophetic. "No," he responded. "You sort of have to put that out of your mind. There's always a possibility that you can have a catastrophic failure, of course. This can happen on any flight. It can happen on the last one as well as the first one. You just plan as best you can to take care of all these eventualities, and you get a well-trained crew, and you go fly."

The Saturn IB (originally called the Uprated Saturn) was a stubbier forerunner to the massive Saturn V that would one day carry people to the moon. This shorter version only required sufficient thrust to place an Apollo spacecraft into Earth orbit, and therefore required far less fuel. The IB stood 224 feet tall on the launch pad, including the escape rocket tower mounted atop the Apollo spacecraft. It was powered by a Chrysler first stage and a Douglas S-4B upper stage. The mighty rocket was surrounded and protected at the pad by a 310-foot steel service tower, which provided a series of access platforms for engineers and other workers. On the uppermost level was a swing-arm catwalk, which gave access to the spacecraft hatch within a canvas-enclosed area known as the "White Room." Here, test conductors kept instrument watch on the Saturn rocket, while others monitored systems and events from within a massive concrete blockhouse, situated nearly a thousand feet from the launch pad.

On the morning of 27 January 1967, technicians at the Kennedy Space Center and Manned Spacecraft Center began to check systems for the "plugs-out" test. By this time spacecraft 012 had undergone twenty weeks of tests and checkout at the plant in Downey, California, and a further twenty-one weeks of tests and modifications at the Cape. The "plugs-out" test involved disconnecting all the electrical, environmental, and ground checkout cables. This procedure would verify that the spacecraft and launch vehicle could function on internal power only, after the service tower's umbilical lines had been disconnected at an appropriate phase of the countdown. It was not considered a particularly dangerous test of the spacecraft and its systems as the Saturn was not loaded with fuel.

Joe Shea, manager of the Apollo Spacecraft Programs Office at Houston's Manned Spacecraft Center was at the Cape that Friday morning. Because the prime and backup crews had experienced unresolved spacecraft communications problems, Wally Schirra had suggested to Shea that he consider joining the crew inside the craft for this test to better understand the communications difficulties they were experiencing. There was sufficient room for another person to squeeze in. Shea agreed that it was a good idea and tried to get an additional communications line installed in time. Technicians determined, however, that the extra communications line could only be installed if the hatch were left open. This particular test called for a closed hatch so an emergency egress simulation could be carried out afterward. As a result, Shea did not join the crew inside the spacecraft.

At 7:41 a.m., NASA technicians began powering up spacecraft 012, sending electric current surging through nearly thirty miles of wiring coiled in thick bundles around the floor as well as through enclosed recesses above and below where the astronauts would sit. Later that morning, following an early lunch, Grissom, White, and Chaffee had electrode patches placed on their skin, donned their spacesuits, and were driven out to Launch Complex 34.

Joe Shea was not the only person to be invited into the spacecraft that day. As Deke Slayton accompanied the crew on the elevator ride up the gantry to the top of the service tower, Grissom was discussing the chronic communications problems with him. He suggested that Slayton, as chief of Flight Crew Operations, might want to be sealed up with them during the test to observe firsthand some of the difficulties they faced. For a few moments Slayton actually considered doing this, but decided he would be of more overall benefit in the control room, relying on the experience of Grissom and his crew to iron out the problems.

The three astronauts began making their way into the troublesome command module at 1:00 p.m. Grissom entered first, swinging feet first through the hatch and sliding into the left couch. Chaffee, who occupied the right couch where the communication equipment was located, followed him into the spacecraft, and then White swung himself down into the center position. All three were seated and secured at 1:19 p.m. The well-rehearsed crew began to quickly run through their checklist as the hatches were sealed, ahead of the spacecraft being pressurized up to 16.7 psi with pure oxygen to simulate normal flight conditions. The closing of the hatches took a little time. First, there was the inner, or pressure, hatch. Unlike the outward-opening hatches on the Mercury and Gemini spacecraft designed and built by the McDonnell Aircraft Corporation, this was a brutally heavy hatch that opened inward, above the center couch occupied by Ed White. Once secured, it was held in place and sealed by a series of clamps. The spacecraft's middle, or heat shield, hatch was next to be secured, and then the outermost or Boost Protection Cover (BPC) hatch, part of a protective shield made from fiberglass and cork that shielded the command module during launch, was closed.

When the crew needed to exit the spacecraft, extensive ratcheting with a torque wrench, plugged into a slot in the inner hatch, was required to retract the six dogleg locking bars. In addition, some cabin pressure had to be vented using a purge valve before this hatch could be hauled inward into the cockpit. White would then pull a quick-release mechanism to unlock the outer hatch. On a good day, and under routine conditions, this procedure could take up to half a minute. The other astronauts would meanwhile be releasing their crew harnesses, then unplugging air and communication lines. With that done, and wearing cumbersome spacesuits in a mere 210 cubic feet of space crowded with instruments, they would slide backward one by one through the opened hatches and clamber across the catwalk to the gantry. It required good teamwork, familiarity with procedures, and structured haste. The simulated mission that day would be used to practice these procedures.

The astronauts were finally settled inside spacecraft 012 for a long session, expected to last several hours. As they sat in their couches, Grissom commented on a sour smell, like buttermilk, in the spacecraft's environmental control system suit-oxygen loop. The crew decided a sample should be taken of the air and then the drill could continue. At 1:20 p.m., a halt was called in the countdown to allow this test to be conducted. The odor was later determined to be unrelated to the events that would follow. Once again, problems with a faulty voice link to the Manned Operations building more than five miles from the pad now began to hamper and frustrate the crew, causing a number of holds in the test countdown. Grissom was quickly reaching his boiling point and made his feelings known with some terse remarks and a little profanity. The afternoon wore on into evening.

Another hold was called at 6:20 p.m. Although it cannot be substantiated, strong circumstantial evidence suggests that Grissom tried to tackle the communication problems by locating and changing a cobra cable inside the spacecraft. This cable was a sheathed, multiwire communication cable that connected each astronaut to the instrument panel, which the crew member could only unplug by turning a ring backward and then pulling on it with substantial force. In actions later verified by NASA's investigation, Grissom unbuckled his seat harness and disconnected his cobra cable. Reliable expert opinion says he then eased himself down into the lower equipment bay, beneath the feet of White and Chaffee, to replace the cable with another. At this time, a slight increase in White's pulse and respiratory rate was recorded, while an electrocardiogram indicated that he was performing some kind of muscular activity that lasted several seconds. Exterior TV cameras also monitored White removing his glove.

The reason for this and his increased muscular activity has never been explained, although he may have been helping Grissom remove and replace the cable. Later hearings would state that the cobra cable was definitely disconnected, and this could not have occurred accidentally. White put his glove back on again shortly afterward. Although there was no actual voice communication to confirm this, Grissom's live microphone transmitted what the official report later stated was "brushing and tapping noises which are indicative of movement." Motion sensor indicators located within the spacecraft also substantiated this movement. "The noises," the official report continued, "were similar to those transmitted earlier in the test by the live mike when the Command Pilot is known to have been moving. These sounds ended at [6:30:58.6 EDT]."

Elsewhere in the spacecraft, it is thought that a brief electrical arc flashed between two bare segments of wire. It was enough to ignite the spacecraft's fittings, which had been permeated by pure oxygen for several hours. Some flammable Raschel netting beneath and to the left of Grissom's empty couch sparked and began to smolder. It developed a hot spot, which rapidly spread, and glowed brightly in the pure oxygen environment. At 6:31 p.m., the first flickers of fire sprang into life. Within moments, voracious tongues of flame had begun to spread upward. Grissom cried out a warning, then apparently scrambled up, and leaped onto his seat on his knees. There is evidence that in doing so he banged the back of his helmet against the upper instrument panel, causing deep gouges later found in the helmet. Meanwhile, Chaffee, seated on the other side of White, had cried out, "Fire—I smell fire!" Two seconds later, either Chaffee or White shouted: "Fire in the cockpit!" White then disconnected his oxygen inlet hose and, as the fire quickly spread, went into the well-rehearsed procedures for opening the hatch, located above his head.

Some recent research, supported by the independent findings of NASA engineers, indicates that Grissom tried to initiate the prescribed procedure to purge the cabin of oxygen by thrusting his gloved arm through a solid wall of flame to activate the cabin dump valves, situated on a shelf above the left-hand equipment bay. Unfortunately, the valves did not fully engage, even though he pressed on them so hard they were later found bent. Even if this operation had been successfully completed, however, the vent orifice was far too small to cope with the rapidly mounting internal cabin pressure. Despite the fury of the fire around him, Grissom assisted White in lowering his headrest so White could reach the hatch above and behind his left shoulder, and began working with him to get it open. It was a valiant but ultimately futile effort. The flames were relentlessly consuming everything in their path, roaring over and under their couches, and they quickly progressed to the padded area beneath the hatchway sill where the two men were frantically working away. Molten fireballs of nylon were dripping everywhere, spreading the fire even more. The back of Grissom's suit began to lose integrity due to the ferocity of the flames.

Meanwhile, in the nearby blockhouse, a North American Aviation engineer named John Tribe was monitoring voice communications and spacecraft systems. He unexpectedly heard something very disturbing through his headphones and turned to the person seated beside him with a puzzled look on his face. "Did he say 'fire'?" he asked, his blood suddenly running cold.

As White desperately tried to ratchet open the hatch with Grissom's assistance, his safety harness caught fire and burned fiercely. Eventually, it separated, freeing White from the confines of his couch. Breached oxygen lines and tubes containing glycol-rich coolants were now feeding the holocaust. Chaffee was bravely trying to maintain communications, reporting that the fire was a bad one. Then he shouted in despair that they were "burning up." The last mortal sound transmitted from the fiery deathtrap was a sharp cry of pain, apparently uttered by Chaffee. Then the transmissions, which had lasted a total of seventeen seconds from Chaffee's first cry of alarm, abruptly ended. Some ongoing activity could still be seen within the spacecraft on television monitors, until the pictures became totally obscured by thick smoke. At this moment, internal atmospheric pressure had mounted to a point where extreme, unsustainable stress on the spacecraft's hull caused it to rupture, suddenly and violently. This, more than anything, sealed the fate of the crew. At the same time, the explosive rupture caused the charred back area of Grissom's suit to disintegrate, sending shards of white suit material, and even shreds of undergarments, out through the breach.

The spacecraft's right-hand hull had split open, from the wall adjacent to Chaffee's head to down below his feet, unleashing into the White Room a devastating fireball of flame, smoke, gas, and debris. Shocked technicians and engineers were thrown against the walls of the White Room by the concussive force of the rupture. With the cabin suddenly purged of oxygen, the incredible fury that had swirled through the spacecraft, devouring the Raschel netting, Velcro, and the small mattresses below the hatch, was suddenly gone—replaced by clouds of dense, poisonous smoke, which also invaded the crews' suit loop. The mattresses had been placed there to prevent damage to wires during the scheduled evacuation test, but now they only served to give off highly toxic fumes. Billows of thick, choking soot now added to the lethal environment, while an intense localized fire still burned in the area of the Environmental Control Unit, near where the fire is believed to have started. Again, there is strong circumstantial evidence that White, probably with Grissom's help, continued to work on for a few seconds after the cabin rupture and actually succeeded in breaching the inner hatch. Unfortunately, the temperature inside the spacecraft, high enough to melt stainless steel fittings, seems to have caused the hatch to expand and jam.

Grissom and Chaffee had begun to inhale the deadly fumes, while White, who had wisely unplugged his oxygen inlet hose to minimize this danger, continued to work on the hatch. As the last flames swept over the three men, a massive conflagration of fire and soot was pouring out through the breached hull and through access ports on Chaffee's side, momentarily engulfing the outside of the spacecraft and the Launch Escape System. The fire itself had lasted just less than twenty terrible seconds.

Ignoring their own burns and concussion injuries, rescuers finally managed to rip off the BPC hatch, release the outer hatch, and then get the inner hatch open. They had to repeatedly evacuate the area in order to reach some breathable air before plunging back into the thick, pungent smoke and heat. It was just under five and a half minutes from the outset of the fire before the inner hatch was finally opened and the Pad Leader reported that all three hatches were open. The spacecraft interior was still incredibly hot.

Grissom and White's bodies were found just below the hatchway, their spacesuits fused to each other and to the floor. Grissom's suit was the most badly damaged by the fire, although burns later found on his body were regarded as treatable and entirely survivable, so for the most part his protective suit had done its job. White's suit had been substantially protected by Grissom's, which had taken the brunt of the inferno. America's first astronaut to walk in space had succumbed while in the process of releasing the hatch. He was found on his side beneath it, lying transversely over the couches with his back against the wall of the spacecraft. The arms of both men were extended outward, still reaching in vain to release the jammed hatch.

Roger Chaffee, the engaging young adventurer who had dreamed of one day flying to the moon, had valiantly continued communicating up to the moment of the spacecraft's rupture. When rescuers finally managed to open the hatches, they found him still occupying his couch, seated upright with his seat restraints disconnected. He had taken his last breath while steadfastly doing his duty and carrying out his final, mortal responsibility.

The crew of *Apollo 1* was dead, but the terrible agony was just beginning for the families and the other astronauts. Most of their stories are well known, but many others involved in the space program were also swept up in the events of that tragic evening. Many of these people, routinely but proudly going about their jobs that day, would find themselves suddenly and profoundly affected by the catastrophe at Launch Pad 34.

Years after the *Apollo 1* tragedy, oceanographer Robert (Bob) Stevenson would be assigned to fly as a payload specialist crew member on space shuttle mission STS 41-G in 1984, but this selfless man withdrew from the flight when his wife became seriously ill. The subsequent loss of shuttle *Challenger* then robbed him of his only chance to make a spaceflight. Stevenson, who had trained Gemini crews in land and ocean observation, was one of the last to have contact with the three *Apollo 1* astronauts. He sent the authors the following reflection shortly before his death, at the age of eighty, in August 2001:

I met Ed White and Gus Grissom and I briefed Roger Chaffee at the Cape the day before the fire. Gus and Ed were so busy those days, getting ready for the preflight test, that Gus had designated Roger as their "Earth Obs" guy. I did it at the Cape, as they were so involved in the flight test that they had been in Florida for the previous months and would be there until the flight.

I had about thirty minutes with Roger in crew quarters. It was a fairly simple briefing as their flight was so full of engineering tests that there would be "little time for looking at the Earth," said Gus. I then got on a plane for Houston, drove to Galveston (I was the deputy director of the Fish and Wildlife Lab there), and drove to Brownsville, Texas, the next day where the Texas Shrimp Association was holding its annual meeting and I was the keynote speaker.

While relaxing in the hotel that evening I learned of the fire at the Cape. I was in total shock for a while, as were all of the guys at the meeting.

I never briefed another Apollo crew.

The astronauts' trusted nurse, Dee O'Hara, already knew that life with the astronauts was not all fun and games. Several had died in accidents, and it always hit hard; but to lose three at once was particularly painful:

I first heard the news of the Apollo 1 *fire over the radio while doing some laundry in my Texas apartment. My first reaction was one of total disbelief, and then, as it had been with other astronauts who died in earlier accidents, an overwhelming sadness.*

I think each death took a piece out of my heart, and was extremely difficult for me. There were a lot of emotional issues to deal with. Those of us who were privileged enough to work closely with them got to know them so well, and then when the accidents occurred, it was devastating. Some more than others, simply because I may have had a closer relationship with that particular individual.

It's the same with any close friend that you lose—a death really causes everyone so much grief. We had spent so much time together, and being accepted into their life and being so close with their families, each was quite heartbreaking. The job was, and is, very stressful and dangerous just under normal circumstances, and sometimes we lose sight of that. It takes a tragedy like Apollo 1, *or* Challenger, *or* Columbia, *to really bring this home to us. I've always said that spaceflight, like growing old, is not for sissies.*

These days Hank Waddell is a retired career naval officer who feels privileged to have once commanded one of his nation's destroyers. For three years beginning in 1956, he was assigned to work with the U.S. Navy's Polaris development test-launch program, based at Cape Canaveral. North American Aviation (NAA, later Rockwell International), recognizing his considerable experience in the field, hired him to help develop their Apollo/Saturn V space program. He left active naval service in 1965 with the rank of commander.

At the time of the *Apollo 1* fire Hank Waddell was a pad leader, supervising work on the Saturn V for NAA. He had also been given the responsibility for developing and preparing emergency procedures for the astronauts and NAA personnel, which gave him the opportunity to put into practice many things he had learned during his naval career. Implementing these procedures was eased by the fact that many of the workers were ex-navy and so quickly understood Waddell's purpose and methods. Sometime after 6:30 p.m. on the evening of 27 January 1967, Waddell was in the cavernous Vehicle Assembly Building (VAB) at the Kennedy Space Center:

I was on the second shift as the NAA pad leader in the VAB where my crew was preparing the command/service module section of Apollo 4, *the first Saturn V rocket. The telephone rang and I picked it up. Suddenly, there was a voice filled with tension on the other end, saying: "There's been a bad accident. Do not tell your crew yet, but I believe several of the astronauts have been injured."*

I listened in disbelief and shock and hardly remember what else was said. When the caller hung up I also hung up, and a million questions went through my head. I didn't know what to do or say, but I know I began praying for the safety of the three crew members. A short time later the phone rang again, and the same person said, "All three have been killed, and the brass are on their way out here. You must stay where you are, but try to keep it from your crew." This was rather difficult, as everyone knew by now that something was going on—there were sirens and alarms in the distance, and rumors of an accident were already circulating.

A few minutes later there was a third call, telling me that news of a serious accident at the Cape had apparently started reaching the press already, and bedlam was about to erupt. I was told to get my crew together and give them what few details I could. "Caution them not *to call home, nor to speak with anyone," was the final grim order. I did as instructed and assembled my crew. They were full of questions, which I was unable to answer, but I told them what I could, and there was silence. We then bowed our heads and prayed for the families.*

In the latter part of 1959 Sam Beddingfield left the U.S. Air Force after having served as a test pilot at Wright-Patterson AFB in Dayton, Ohio. Armed with a college degree in aeronautical engineering he drove up to NASA headquarters at Langley Field and applied for a job. The first person he ran into at Langley was Gus Grissom, with whom he had been flying at Wright-Patterson, and Grissom helped him secure a job with the fledgling space agency as a mechanical engineer. After training at Picatinny Arsenal he became the NASA engineer responsible for the pyrotechnics and recovery systems on Mercury spacecraft. In the countdown to the *Apollo 1* mission, he and a concerned Grissom had often spoken about problems with the troubled spacecraft. "The day that accident happened was my last day in Mechanical Engineering," Beddingfield remembered. "It was on a Friday, and Monday I was supposed to move to the Apollo Project Office. There was something that I really needed to have Gus do, so I came in at 7:30 in the morning so I could make sure I could communicate with him. We talked for a while." The "something" that Beddingfield wanted to discuss with Grissom was the postsimulation test of the emergency egress system, for which he was the supervising engineer. "As soon as the plugs-out test was over [Grissom] was going to be inside the Apollo spacecraft, and I could come and spend about ten minutes with him to discuss the type of problems that he'd be interested in. So the test started, and it went on and on and on. They were having so many problems it was delaying it, delaying it and delaying it until finally it was late afternoon."

Beddingfield eventually left the blockhouse and drove down to Cocoa Beach for some supper. Phoning in from the restaurant after his meal, he was told that the countdown test was still running well behind schedule. He drove back to the Cape at a leisurely pace, intending to return to the blockhouse before running the final test:

I came back and was just coming through the gate when the whole thing exploded. As I drove up the narrow access road to Pad 34, I noticed guards at the security gate waving frantically for me to pull over. Moments later an ambulance tore past, heading for the gantry area. I could hear sirens going off near the pad, and assumed there'd been an accident somewhere. My major concern was that it might further delay my evacuation systems check with the crew. When I finally reached the blockhouse, I could tell something catastrophic had happened. Then I was given the awful news.

That was on Friday, and my next trip home to change clothes was about 9:00 p.m. on Tuesday. I stayed busy just trying to see if I could figure out what went on. I lost track of days and nights.

Gerald D. ("Gerry") Griffin was a flight director based at the Johnson Space Center, Houston, for all of the Apollo manned missions, which included three assignments as lead flight director. At the time of the *Apollo 1* fire he was the Guidance, Navigation and Control flight controller—call sign "GNC"—in the Mission Control Center (MCC). Griffin would subsequently be appointed director of the Johnson Space Center and also served as the deputy director of the Kennedy Space Center in Florida and the NASA–Dryden Flight Research Center in California. As he told the authors,

The "plugs-out" test was the dress rehearsal for the actual launch, so the MCC was fully manned just as it would be on launch day. The test was troubled all day by communications problems, and at a "hold" in the test countdown people at the Kennedy Space Center [KSC] Launch Control Center [LCC] took the opportunity to fully troubleshoot the problem. When the hold was initiated, those of us in MCC took a break to stretch our legs, get a cup of coffee, or visit the men's room.

As the hold lengthened some MCC flight controllers had their headsets on. Others, like me, were at their consoles but had their headsets off. Those controllers with their headsets on heard a short burst of static followed by repeated attempts by the LCC spacecraft test conductor to contact the crew. We all noted a loss of telemetry data on our console displays in the MCC. In a matter of seconds it was clear that some kind of problem had occurred, and all of the MCC flight controllers returned immediately to their consoles and put on their headsets.

Somewhere in the first minute or two the word fire was mentioned by someone from KSC, but we had no indication of anything in the MCC other than the loss of data. At this point, the problem was being addressed by the LCC, and the MCC was in a listening mode. As we always did when there was a problem of any kind all of us in the MCC ordered up replays of the spacecraft telemetry data to see if we could find anything out of the ordinary before and at the moment of the sudden loss of data. Our early look at the spacecraft data did not indicate what had happened.

Within a few minutes it was confirmed by the LCC that a fire had occurred at the launch pad, but in the MCC *we still didn't know how bad it was. I can remember thinking that perhaps the fire had been external to the command module, and the crew would be okay. We were worried about the crew, though, and I can remember "hovering" a bit around the MCC flight surgeon (call sign, "SURGEON") console to see if he had any data which might indicate the crew's condition. He didn't. Shortly, though, I could see the* SURGEON *talking "grimly" on one of his communication channels, obviously to his counterpart in the* LCC*. Finally, the* SURGEON *looked up and told us that all three astronauts were dead.*

The shock was sudden and intense, but as we had been trained, we knew this was no time to stand around and wring our hands. We had to immediately implement our contingency plan, which involved capturing and preserving all of the information available to us in the three-story MCC*. This included preserving every bit of* MCC *paper material, as well as real-time and playback telemetry data and voice recordings.*

The task of retrieving, filing, and cataloging everything took us several hours, and gave us little time to focus on the tragic loss. That would come later, and in my own case, when I had time to focus on what had happened, it was much like my flying days in the Air Force when someone was killed in an aircraft accident. There was recognition of and a sadness with the loss of three really good friends, but it was coupled with a strong motivation to find out what happened, fix it, and get on with the mission at hand . . . to get to the moon and back. Perhaps the strongest emotions were anger and frustration with ourselves in recognizing that we all had missed something along the way that caused the fire.

Some time in the days and weeks after the Apollo fire I remembered from my flying days the old adage, "It is indeed difficult to fly without feathers." We had proved it again on January 27, 1967.

Like Gerry Griffin, his fellow MCC flight controller Gene Kranz was also a witness to the pioneering days of manned spaceflight, right through the Apollo program and beyond. Kranz achieved considerable latter-day fame as the intensely focused and dedicated controller featured in the Ron Howard film *Apollo 13*, and later wrote a best-selling autobiography, *Failure Is Not an Option*. But there were other lessons along the way to that successful *Apollo 13* rescue mission. As he told the authors, he was not in Mission Control that particular January evening in 1967, but he still felt "bloodied" by the *Apollo 1* fire:

I had tested all the night previously, and basically had about an eight-hour turnaround and was back on the console again at about 7:00 a.m. that day, just to give the next couple of teams a break. I brought the MCC *on line for the scheduled afternoon test, and handed over to John Hodge [chief, Flight Control Division] about noon. I told him the interface tests were complete, communications were solid, and the MCC was "go." I then handed the console logs to Hodge and headed to the nearby office area. Here, Chris Kraft and I used to listen to the MCC and pad voice transmissions in our offices before reporting for another shift in the control room. At 3:00 p.m., Kraft left for the MCC, for the final hours of the simulated launch countdown, and John Hodge handed over to him. Since everything seemed to be going well I left the office early and headed home.*

My wife Marta and I had decided to go out for dinner. I had just finished getting dressed when there was a knock on the door. My neighbor and fellow flight controller Jim Hannigan stood there, agitated and breathless. "Have you heard what happened?" he asked. I must have looked bewildered, because he strode over to the TV set, and turned it on. "They had a fire on the launch pad," he said. "They think the crew is dead." I watched a couple of news reports, then jumped into my car and drove flat out to the MCC, ten miles away.

Things hadn't been right that day, and I knew *they weren't right. And yet I continued on. We—the Mission Control team collectively—had numerous problems, with communications and life support systems. We had learned to work around communications problems—we had enough backups in there, we had learned to work with almost any configuration. That day, test procedures were out of whack. Around sunset at the Cape they called a halt in the countdown to allow for a little troubleshooting with voice communications between the spacecraft and the launch team. A new test sequence was written out and hoisted up to the crew. The countdown was then resumed. In retrospect, it would have been better if we had just called it off that day. The bottom line was, we were tired, the test procedures were just not as clean as they should have been for that kind of a test. But nobody stood up and said, "We're not ready; it's time to regroup."*

At 6:31 that evening, the MCC team was startled by cries coming from the spacecraft, and they listened helplessly to our crew's cries for twelve seconds as they died. From death came anger at ourselves, because we knew that we were responsible for America's first space disaster.

In 1967, Paul Haney was the head of NASA's Public Affairs Office in Houston, having inherited the job from Lt. Col. John "Shorty" Powers—the man who had become known as "The Voice of Mercury." He was working in NASA's Mission Operations Control Room (MOCR) in Houston that fateful day, keeping an eye on the simulation exercise then taking place at Cape Kennedy:

Promptly at 9:00 a.m. on January 27th, 1967, I got divorced. The day in Houston had started on a positive note for me. But that's not how it ended.

I was in and out of the simulation several times that afternoon. The simulation had all sorts of grief. Mainly, communication lines were fouling up. At one point, Grissom asked how the hell we were supposed to go to the moon when they couldn't even communicate with the blockhouse. I recall noting that it was running long on Friday night, overtime and all at the Cape. I left the MOCR to buy a car nearby—my need for a car was established by a property split earlier in the day in my divorce settlement.

At 6:31 p.m. Florida time, 5:31 p.m. Houston time, a flash of fire in a locked-down Apollo spacecraft full of pressurized oxygen killed three crew members rehearsing their planned liftoff only days away. Houston was tied to the Cape in a simulation that reached T minus 10 minutes. A cohort contacted me and said to call Chris Kraft immediately in the control center. Kraft said there had been a bad accident at the Cape and to come back immediately to the control center.

I still have a twelve-page double-sided statement titled "Chain of Events—204 Accident, by Paul Haney." On listening to a tape of the fire recorded for Chris Kraft, my notes say: "I recognized immediately White's voice saying, 'Fire in the cockpit!' I do not recall hearing Chaffee make such a report, although apparently he did. I do recall Chaffee saying, 'It's a bad fire. We are burrrrning up!' This was followed immediately by a scream lasting a second or two. It sounded like it could have been two men screaming: other persons theorized that it sounded like White as well as Chaffee. Gus never said a word."

Ed White had tried to open the hatch, the kind they used to have to secure bank vaults. He probably forgot that it took a second man working outside the hatch to open it quickly—in a minute or so. When the whole area inside and outside the spacecraft was not on fire.

My information counterpart at the Cape, Jack King, and I began working on a bulletin which went out within an hour of the accident under a Cape dateline. The bulletin said there had been an accident at the Apollo complex involving loss of life. We released a second bulletin an hour later, after the families had been notified, which named the lost crew members. At 10:00 p.m. that evening we held a news conference in Houston, covering all the detail we had.

The next morning I flew down to the Cape with half a dozen engineers from the Houston center. We saw the burned-out spacecraft up close. We talked to people working in the blockhouse at the time.

Joe Shea, the Apollo program manager who got a very special shafting out of the program from the Good Old Boys from MSC/Langley, was supposed to sit in the spacecraft the day of the fire—somewhere on a bottom step, maybe where they found Gus. He never got around to doing it, and would surely have been killed if he had done so.

I flew back to Houston that evening—Saturday night. The media had lots of questions. The funerals were on Wednesday. Gus Grissom and Roger Chaffee were buried at Arlington National Cemetery in Washington. Ed White, a graduate of West Point, was buried at the military academy cemetery. One of the Houston oil companies—Exxon, as I recall—provided several planes of its air fleet, which permitted hundreds of NASA employees to attend the Washington ceremonies.

Just before the Washington funerals, the late NASA administrator James Webb questioned me about the accident at a staging gathering at the Georgetown Inn in Washington. He wanted to be reassured that the crew died quickly. I wondered where he'd been. I told him what Ed White had said about a fire in the spacecraft, about White trying desperately to open a hatch that he knew wouldn't open, about the crew trying to find some place to get away from the flames, about Chaffee screaming that they were burning up. About that one last horrible cry of pain from within the spacecraft. "In other words, they died almost instantly," Webb said. I couldn't think of anything else to say to the man.

After the Arlington ceremony we gathered back at the Georgetown Inn for a drink. It was the only time I ever saw Al Shepard cry. "I hate those empty slot [flying] formations," Shepard said, brushing tears from his eyes. He couldn't bring himself to admit that it was Gus he missed.

After we returned to Houston, Webb put out the word that nobody was to talk to anyone about the fire, even within the ranks—not even to family members, press, you name it. I was the public affairs rep for about a week, but it was apparent if I was going to stick to that assignment I would have to move to the Cape and more or less go underground. So Jack King stepped into the spot, very capably.

Webb's instructions about not talking to anyone held up for a while. I recall going to a boxing match at the Astrodome in February or March after the fire. Joe DiMaggio walked up to me after the fight—Muhammad Ali destroyed somebody—and said, "You're the NASA *guy, right? What happened with that fire?" I told him, of course.*

Jack King, also from the Public Affairs Office, was on duty at the Cape that day:

It was a very long day. We had all kinds of communications problems between the spacecraft and the control center. I was in the blockhouse, the Saturn I blockhouse at Pad 34. It was a routine test, with all kinds of glitches; . . . the test dragged on, and then out of nowhere we heard, "Fire in the spacecraft!"

Everybody was stunned. I started scribbling notes, and listening intently—it was all I could do. Deke Slayton was in the blockhouse as well, and I agreed with Deke that I would hold off any statement until the three widows were notified. We had to sit there in the blockhouse for about an hour, because sitting on top of the Apollo spacecraft was a 155,000-pound-thrust escape tower. We couldn't send anybody up there immediately because we were afraid the escape tower would explode.

I still have written in pencil the news release announcing that the three astronauts were dead. We got it out as rapidly as we could, considering the circumstances. There was criticism of that at the time, which was understandable on the news media's part, saying I should have announced it immediately, because it was like a presidential assassination. I felt, and I feel secure in it even now, that we did it the proper way.

Don Gregory joined NASA in 1959 as a project engineer on the Mercury-Atlas program. With a bachelor of science degree in mechanical engineering from the University of Miami (1955), he had previously worked at the NASA's Langley Research Center as an aeronautical engineer. Shortly after he joined NASA, Robert Gilruth, then head of the Space Task Group, invited him to become his technical assistant. Gregory later said: "I thought he was one of the most talented and patient men I had the pleasure to know and work with. He had my utmost respect, just a wonderful human being. I often said if you looked up 'class' in Webster's, you would see a picture of Bob and Jean Gilruth." In 1964 Gregory went to work for Deke Slayton at the Manned Spacecraft Center in Houston, spending ten years as his Executive Office, Flight Crew Operations Directorate. His last position in NASA was logistics manager in their Space Shuttle Program Office. Don Gregory retired from the space agency in 1983. "Deke was an outstanding manager," he told the authors.

Our organization consisted of Astronaut Office, Aircraft Operations, and the Flight Crew Operations Office. One of the activities we had implemented was a Friday afternoon (just after the end of the work day) staff meeting. We felt it was important to review the past week's happenings and to plan for the upcoming week, coordinating the activities of our close-knit directorate. When Deke was out of town, I would conduct the meeting.

On January 27, Deke was at the Cape with Gus and his crew for spacecraft tests. Mike Collins sat in for Al Shepard, who was off somewhere. Joe Algranti (head of Aircraft Operations) and Pete Woodling (for Flight Crew Operations) attended the meeting, held in Deke's office since he had a large conference table. Shortly after we started, our red phone rang. It was a dedicated line that had been installed after Ted Freeman's aircraft accident. Occasionally they rang it for a test, but the number was only known to a few people. Deke was on the line. He told me there had been a fire in the cockpit of Spacecraft 012, and that we had lost the crew. He said he would call back as soon as he had more information.

I didn't need to say anything to the staff as they saw my expression, but I gave them a repeat of what Deke said. I asked Mike Collins to get some astronauts and wives over to Betty's, Pat's, and Martha's home straight away. He went into my office to call the Astronaut Office to get that action going, and I believe he contacted Al Bean, who helped getting the troops to the homes. I then called George Low, Bob Gilruth's deputy. Bob was also not at the office. Since Deke said he would call back, George came down to our office complex. Joe went out to the Aircraft Ops Office to make sure anyone who needed to fly that evening would have crews to support the launch of the aircraft. Pete stayed with George and me.

Deke called back a little later and provided more details, much of which has been subsequently documented. George had called the Public Affairs Office to let them know where he was in case they needed to relay any press requests to him. I understand the Mission Control Center was in what is now called "locked-down" mode. George and I eventually went down to his office and, as you may imagine, the phone was ringing constantly.

Somewhere through all this, Mike reported back to me that there were other astronauts and wives at each home. They perform an important function; in addition to support for the widow, they step in to provide a barrier to outside inquiries and well-wishers. Sometime later, maybe 10:00 p.m. or so, George and I went over to Mission Control, since they had an open line to the Cape's activities. We stayed there for several hours.

On a personal note, these were three men that were friends. We once had a lot next to the Grissom's home (never built there), and during the holidays my wife and I were at a party at Pat and Ed White's home.

It was a pretty traumatic time, but the legacy of the fire—if you could ever say anything good came from it—was that NASA ended up flying a far superior spacecraft to the moon.

Lola Morrow worked as a secretary in the Astronaut Office at the Cape in the 1960s and 1970s. This office was located in the Manned Spacecraft Operations Building (MSOB), a small two-story building housing the NASA Mercury Project staff and administration staff, along with the McDonnell contract people. Her duties were not the same as the secretaries at the Manned Spacecraft Center in Houston. In fact, she was the only secretary whose job it was to take any unnecessary stress off a crew, as well as manage the quarters and offices. She looked after her famous charges so well that she became affectionately known as the astronauts' "den mother."

On the morning of the test she was at work in her office when the three astronauts came in from crew quarters for their suit-up, wearing bathrobes and slippers that she had personally picked out for them. But that morning, as Morrow recounts, things just seemed inexplicably different:

That morning, when I saw the crew, I had a premonition, and I feel they also had a premonition. I said to them, "Are you all right? Is everything okay?" They were just not themselves.

We had a squawk box—an intercom from the test site. I listened to the crew and to everything that was happening. My premonition got worse. I just concentrated on that squawk box. My anxiety grew when I heard Gus saying, "How the hell can we get to the moon if we can't talk between two buildings?"

I tried to stop the test by calling Stuart Roosa in the blockhouse. Stu, one of the new astronauts, was communicating with the crew. He was sitting near Apollo spacecraft manager Joe Shea, Rocco Petrone, the Director of Launch Operations, and Deke Slayton, chief of Flight Crew Operations. I was unsuccessful, but Stu reassured me, by saying, "Lola, it's a 'go.' The problems are ironed out. You can go home."

I left and called my daughter Linda on the way home. "Oh Mom," she said, sounding upset. "You need to call Chuck right away!" (Charles) Chuck Friedlander was chief of the astronaut office at the Cape. When I got through he sounded awful. "Lola, there's been an accident," he said. "How soon can you get back here?" I went cold, and knew immediately my premonition had come true.

Back at the office it was both incredibly sad and chaotic. The whole world seemed to be calling us; every line on my telephone was ringing. I kept tearing up and trying to hold my emotions in check. I wanted to be brave for the crew and Deke.

I hardly remember what I was doing. Then poor Deke walked in. He was white and trembling like a leaf. He had a cigar, but he couldn't hold it to light it, he was shaking so badly. I tried to light it for him, but I was shaking as well, and couldn't do it.

That night I placed calls to the three astronauts' wives for Deke. I then called the mortician and did whatever else I could do. I left around midnight when the telephones died down. Driving home, my emotional dam broke and I cried deeply throughout the night.

Later, I helped pack up the crew's belongings, which were put on NASA*'s Gulfstream airplane headed for Houston. Days later, along with other coworkers, I watched as the three caskets were boarded onto a flight, ultimately bound for burial at Arlington and, in the case of Ed White, West Point.*

It was too heartbreaking for me to continue at that office. Deke then closed the office for four months. Later I reopened the office and greeted the Apollo 7 *crew ready for training. In the meantime, I volunteered to work with the Accident Review Board. But all of my emotions surfaced again when I transcribed the pages recording the astronauts' final words.*

On the day of the fire, astronaut Dick Gordon was attending a White House reception to celebrate the signing of the *Treaty on Principles Governing the Activities of State in the Exploration and Use of Outer Space*—more readily known as "The Space Treaty." He was there with fellow astronauts Neil Armstrong, Scott Carpenter, Gordon Cooper, and Jim Lovell. When Dick Gordon finally got back to his hotel room in the Georgetown Inn, the red message light on his phone was blinking. The front desk told him to call a number at the Manned Spacecraft Center in Houston. They stressed it was urgent. "We thought we were ready to go in January of 1967," Gordon told the authors,

and then Gus Grissom, Ed White, and Roger Chaffee lost their lives in a spacecraft under routine testing conditions. That took our breath away, set us back a bit. It was a very, very difficult lesson for us to learn. We had been all the way through Mercury and Gemini, and all of a sudden this kind of thing happened. But we had become, as a team, very complacent about the environment in which we were operating. One hundred percent oxygen. Flammable material within the spacecraft. Overpressurized: in fact, above sea-level pressure during this particular test. All that was required to have the disaster we had was a source of ignition.

Also, the hatch design was a very, very poor one. Ed White was a physically strong individual, but it would have taken three or four Ed Whites to get that hatch open.

We lost the lives of three good friends and compatriots that day. But, knowing those three individuals and what they wanted us to do, we had to continue that program. It took us more than a year and a half, to October 1968, to solve all the problems we had encountered with Apollo 1. *Between* Apollo 1 *and 7 there were some unmanned proving flights and finally,* Apollo 7 *launched. With the redesign of the Block 2 spacecraft, we had eliminated 100% oxygen. We eliminated flammable materials. We redesigned the hatch. We redesigned the wiring system within the spacecraft.*

Realistically speaking, I think the loss of those three lives allowed us to go to the moon and accomplish the mission that President Kennedy asked us to do. Without them, it is very difficult for me to say, but there may have been a few more disasters like the one we had.

That was a very trying time, not only for the families, but also for the rest of us in the program at the time.

That evening, NASA Administrator James Webb led a subdued press conference. At one point he said, with obvious pain in his voice: "Although everyone realized that some day space pilots would die, who would have thought the first tragedy would be on the ground?" Alan Shepard later echoed Webb's comments. "We . . . had gone through Mercury," he observed, "had gone through Gemini; man, we thought, we're leading. We're beating the Russians. We thought nothing could go wrong."

The Apollo program was immediately put on hold. Over the ensuing months investigators completely stripped and examined the gutted spacecraft, while others exhaustively traced its entire construction history. NASA, under considerable public and congressional criticism for conducting its own probe into the fire, would not involve itself in premature speculation and issued only brief, tersely worded reports as their inquiries into the causes of the tragedy continued. In Washington DC, NASA's chief spokesman, Julian Scheer, was adamant that there would be no discussion of the accident or its causes before the findings were released. "We're not going to answer—day by day—all the reports, rumors, and so on that come up," he stressed to the media.

In April the investigating board presented its findings. It concluded that the precise point of the fire's origin could not be positively determined, although physical evidence had been found of electric arcing from wires that had damaged, frayed insulation. Although a specific initiator could not be identified, evidence seemed to point to an area beneath and to the left of Grissom's couch, where components of the faulty environmental control system had repeatedly been taken out and replaced, both during construction and in testing. Insulation had been scraped from a wire, and this, it was felt, caused an arc that had ignited flammable material. The resultant fire in a pure oxygen environment had spread with alarming speed and intensity.

According to NASA's Apollo chronology, *Where No Man Has Gone Before*, an appalling number of factors were involved in the tragedy: "The simulation had not been considered hazardous because neither the launch vehicle nor the spacecraft contained any fuel, nor were the Saturn's pyrotechnics installed; consequently no emergency equipment or personnel were at the launch pad. Wiring carrying electrical power was not properly protected against accidental impact. Far too much flammable material—some 70 pounds of nylon netting, polyurethane foam, and Velcro fastening—had been haphazardly spread around the Command Module, creating unobstructed paths for flames."

For future missions, the pure oxygen environment was replaced with a 60/40 oxygen-nitrogen mixture, and the hatch was redesigned. It became a single-hinged door that could be unlatched in just three seconds and swung outward with only half a pound of force. Where possible, fireproof materials such as Beta Cloth were substituted for items now considered dangerously flammable, while exposed wires and plumbing were bundled, harnessed, and covered to prevent inadvertent contact or abrasion.

Dale Myers was managing the Apollo command module program for North American Aviation at the time of the fire. He said the next eighteen months were made up of "the hardest work I have ever put into anything. I can remember going to those change boards every week . . . and that was when we were changing over to the outward-opening door that North American had originally proposed . . . and the two-gas system that North American had proposed originally, and all those changes were coming into the system."

Five weeks after the fire, *Apollo 1* backup crew members Wally Schirra, Walt Cunningham, and Donn Eisele were relaxing in the living room of the Cape crew quarters after dinner when Deke Slayton walked in and joined them. At this point, no replacement crew had yet been assigned to the first manned Apollo Block 1 flight. The four men talked casually for a couple of minutes, and then Deke suddenly paused. A serious look came over his face, and he stared squarely at Schirra. "I want to let you guys know," he said, "that you have the next flight."

Walt Cunningham was well aware of the procedural and managerial changes that were systematically taking place as investigations into the fire continued, and he had few concerns about being part of the next crew to fly. "From the astros' point of view," he later wrote in his memoirs, "the changes encouraged the belief that a spacecraft would be built that could, at last, perform its intended mission. The relentless pressure to meet a schedule was gone for the first time since the program's inception. And out of the whole mess, North American was to bring forth one of the greatest machines ever built."

As author Andrew Chaikin sagely observed in his *A Man on the Moon: The Voyages of the Apollo Astronauts*, the Apollo command module was without doubt the most complex flying machine ever designed and constructed, and "an intricate package crammed with state-of-the-art equipment." It was quite inconceivable that nothing would go wrong with the prototype vehicle, he observes, but no one at NASA or North American would have ever knowingly compromised the safety of the astronauts who would fly it.

Over the next few weeks, Chaikin adds, the astronauts came to realize "that there was a hidden blessing in the disaster: the wreckage of *Apollo 1* was there for the accident board to examine, not in a silent tomb circling the Earth or drifting in the translunar void. Although three men had died, three or perhaps six more lives had probably been saved."

The hardware was not the only thing that needed to be fixed, however. As Gene Kranz admitted, people had been just as responsible for the accident. It was not a case of laying blame; rather, it was a case of learning what could be prevented, and moving on. Kranz resolved that, for his team at least, a new line would be drawn: "We wrote two more words into our mission vocabulary as flight controllers—'tough' and 'competent.' Tough, meaning that we would never get shirked from our responsibilities because we are forever accountable for what we do—and what we fail to do. Competent—we will never again take anything for granted; we will never stop learning. From now on, the teams at Mission Control will be perfect. We designed a badge, the emblem of the mission controllers. The price for admission to our ranks became discipline, morale, competence, commitment, teamwork. This team must never fail again."

It was April 1967, three months after the fatal Apollo fire, and for the second time in three years Valentina Komarova was waiting for her husband Vladimir to return home from space. His prior spaceflight, the first Voskhod mission in October 1964, had ended with a happy family reunion and the conferring of Hero of the Soviet Union status on him and his two crew members. He had obviously excelled in performing his mission, as he was the first cosmonaut selected to make two spaceflights. This time, however, the waiting would be briefer, and there would be no happy ending. Valya's husband was at the very zenith of his cosmonaut career, but he would not live to celebrate this latest accomplishment with his family.

In September 1965 plans had been set in motion for an ambitious flight schedule using the new generation Soviet 7K-OK spacecraft known as Soyuz ("Union"). Eight military cosmonauts including Komarov and Yuri Gagarin were selected at that time to begin training for Soyuz missions, while Chief Designer Sergei Korolev had initiated the formation of a parallel civilian training group, whose on-board responsibilities and duties would extend well beyond piloting operations.

No Soviet cosmonaut had been launched since the flight of *Voskhod 2* in March 1965. A third Voskhod mission due in March 1966 had been canceled in part due to Korolev's death and the need to fly Soyuz. NASA, in the meantime, had been conducting its highly successful Gemini program. Soviet space designers and planners knew they were losing their critical edge, and it was becoming increasingly apparent that they were slipping behind in the race to the moon. In fact, there was a gloomy prognosis for Soviet political leaders: the entire Gemini program of ten flights would be concluded before any Russian cosmonauts once again soared into the skies.

The untimely death of Korolev in 1966 had proved to be a monumental blow to Soviet space plans. Despite this, there was steadily mounting pressure on the administrators and engineers to continue the Soviet push to the moon, and Soyuz carried the aspirations of many as final preparations were being made for the craft's maiden flight into space. If the Soviet Union was going to do something spectacular, 1967 offered two very significant anniversaries to coincide with the event: in April ten years would have elapsed since Sputnik triumphantly led the way into space, and in November Russia would be celebrating the fiftieth anniversary of the Bolshevik Revolution.

Soviet space plans and objectives at this time were not dissimilar to those of the United States, with orbital rendezvous and docking high on the agenda followed by precursory manned flights around the moon, leading to an eventual landing. However, there were deeper-set problems. In *Starman: The Truth about the Legend of Yuri Gagarin*, coauthors Jamie Doran and Piers Bizony spoke of rampant disorganization and a lack of unity in the Soviet space program. "In contrast to the overall NASA effort, with Apollo as its privileged centerpiece," they wrote, "the Soviet lunar programs were divided, confused, and contradictory."

Soyuz was only one of many different designs being considered to achieve a number of ambitious goals: lunar flybys, lunar landings, as well as military and civilian orbital operations. Work on Soyuz had in fact stopped for much of 1964 and 1965 while design work was redirected toward a lunar goal. Increasing the complexity of the challenge the engineers faced, the Soviet manned spacecraft were not only designed to be flown fully automated but also with manual controls for a pilot. American engineers found it difficult enough to create either automatic spacecraft or manned spacecraft; Soviet engineers were trying to combine both elements in one machine. Nikolai Kamanin, head of the cosmonaut program, was one who thought the Soyuz design was "overautomated."

In 1966 the Soyuz spacecraft was undergoing construction after more than five years of design studies and development. The previously-flown Vostok spacecraft design was its model, but it would incorporate features related to Soviet plans for rendezvous and docking missions, as well as later techniques to be employed in the planned Soviet manned lunar program. The Soyuz craft had three main components. The forward section of the assembly acted as an orbital module, which was essentially a crew habitation and storage facility that would also provide EVA access. Then there was a central crew or command compartment also used as a descent module. Unlike the spherical Vostok, the descent module's underside was flattened to give it aerodynamic lift for reentry, similar to the ablative heat shield on American spacecraft of that period. Finally, there was an unpressurized aft service module. The full assembly, twenty-six feet in length, would derive its electrical power from two solar panels, folded against the spacecraft during launch, but extended once Soyuz was in orbit.

Unlike the earlier Vostok, a considerable saving in weight and volume had been gained by excluding bulky cosmonaut ejection seats. Instead, three thin couches were provided for the crew. Following reentry, as the plummeting descent module reached around 6 miles' altitude, an air pressure sensor would signal the parachute cover to be ejected. This action would also deploy two small drogue parachutes that would in turn pull out a larger braking parachute. The braking parachute would slow the descending craft to around 280 feet per second, and the main parachute would deploy around twenty seconds later as the spacecraft reached an altitude of about three miles. While the main parachute billowed out, the vehicle's base heat shield would be jettisoned, exposing the landing system. Moments before touchdown, and just six feet above the ground, the soft-landing engines would fire, slowing the craft to around seven feet per second. It was expected to give the crew a reasonably comfortable landing.

In 1966, Vladimir Komarov was appointed chief of the engineering department at the Gagarin Cosmonaut Training Center, and by September he had resumed work on the Soyuz spacecraft. He had repeatedly shown himself to be a top candidate for the first Soyuz flight, demonstrating incredible technical competence in the difficult tests. With the post-Voskhod fanfare now behind him, he was keen to make a second flight. Yuri Gagarin endorsed Komarov as one of the most experienced and technically adept of his colleagues, saying he had to "learn certain skills three times over. As backup for Popovich in *Vostok 4* he perfected control of Vostok. Then in preparation for the Voskhod flight there were a few things he had to relearn. The third time was the flight of *Soyuz 1*, which in its construction was essentially different from preceding aircraft and required a new set of handling skills. . . . Komarov performed brilliantly." For his part, Komarov was fully aware that the Soyuz vehicle had many fundamental problems. He was carefully pragmatic in all that he did and would never let his enthusiasm blind him to the sober realities of spaceflight. In 1966, while discussing future missions, he stated that "The most important thing is how to carry out everything in safety."

Komarov's concern was probably raised after the launch of the first, unmanned Soyuz craft on 28 November 1966. To allay any Western suspicions, these unmanned Soyuz flights carried a more generic identification as Kosmos series craft. Not only did a badly underperforming booster place the unmanned Soyuz into a lower-than-desired orbit, but the flight was soon beset with numerous attitude control problems. Plans had called for a second unmanned Soyuz craft to be launched the following day, with automatic rendezvous and docking the principal goal, but the attitude problems caused the second launch to be called off. Controllers then attempted to bring the Soyuz back to Earth, but the remotely signaled engine burns kept shutting down prematurely. Two days after launch the craft finally reentered, but its descent trajectory was far too flat, and it began to overshoot Soviet territory, heading toward China. An automatic self-destruct device took over and blew the craft to pieces over the Pacific Ocean, east of the Mariana Islands.

Two weeks later, on the afternoon of 14 December, the second Soyuz craft was poised on the launch pad at Tyuratam. At 2:30 p.m. the launch was initiated, but only the rocket's second-stage strap-on boosters ignited, and computers quickly terminated the launch. Once everything had appeared to stabilize, pad workers were sent out to reposition the escape frames of the pad structure to prevent the booster from toppling in gusting winds that were now sweeping across the launch area. Other service personnel were also climbing up the service tower to inspect the rocket.

Less than half an hour after the launch abort, the rocket's launch escape system suddenly and unexpectedly began to fire. Within moments, the booster was engulfed in flames, and panic-stricken workers began fleeing for their lives. Two minutes later the rocket's first and second stages exploded in a deafening conflagration of fire and debris, sending thick clouds of black smoke billowing into the air. Fortunately, most of the workers had managed to reach safety, but one unlucky person was killed. Several others were injured, and the entire launch complex was badly damaged in the devastating explosion.

A third launch attempt took place two months later on 7 February 1967, just a few days after the fatal pad fire at Cape Canaveral that had claimed the lives of three Apollo astronauts. Once again there were major orientation and attitude problems after orbital insertion, but it was felt that if a cosmonaut were on board he or she could have resolved these difficulties. This time the reentry process began well, but the unmanned Soyuz assumed the wrong ballistic trajectory and ended up well west of the planned landing area, parachuting down to a landing on the frozen surface of the Aral Sea, seven miles from Cape Shevchenko. Unfortunately, just as rescuers reached the charred spacecraft, it broke through the ice shelf and sank in thirty feet of freezing water. Divers eventually recovered the Soyuz, and investigations showed that the descent module had depressurized when a heat shield plug had burned through. If the spacecraft had been manned, the crew would not have survived.

Soviet leaders cared little for the technical and design problems now confronting Vasily Mishin, the newly appointed head of the Korolev design bureau *Opytnoe Konstructorskoe Byuro*, or OKB-1. Khrushchev's successor, Leonid Brezhnev, and the secretary of the Central Committee for Defense and Space, Dmitri Ustinov, were anxiously anticipating the next Soviet space spectacular, having seen NASA successfully complete the ten-flight Gemini program and then turn its attention to Project Apollo. Extreme political pressure was being exerted, and a reflective Mishin told *Ogonyok* magazine interviewer G. Salakhutdinov in 1990 that "there never was a time when we worked in peace, without being hurried or pressured from above."

As head of air force cosmonaut training, Kamanin had expressed in his diaries his own serious concerns about the pressure to launch Soyuz. "We must be fully convinced that the flight will be a success," he wrote. "It will be more complicated than previous flights, and the preparation will have to be appropriately longer. . . . We do not intend to rush our program. Excessive haste leads to fatal accidents, as in the case of the three American astronauts last January."

An ambitious flight plan was built around the maiden manned flight of a spacecraft that with modifications could one day carry cosmonauts to the moon. *Soyuz 1*, with Komarov as the sole pilot, was to dock in space with *Soyuz 2*, carrying cosmonauts Valery Bykovsky, Alexei Yeliseyev, and Yevgeny Khrunov. Once they had achieved a rendezvous and docking on the second craft's first orbit, Yeliseyev and Khrunov were to perform an EVA transfer across to the open hatch of *Soyuz 1* and return in that spacecraft. Bykovsky would land alone. It was a disturbingly overambitious mission for a brand-new breed of spacecraft.

Space researchers Doran and Bizony state that many lingering concerns remained as the launch deadline loomed. In their biography of Gagarin, the co-authors emphasize the fact that OKB-1 technicians had already identified 203 faults in the spacecraft hardware that required urgent attention, and backup pilot Gagarin had taken part in this evaluation. "By March 9, 1967," Doran and Bizony wrote, "he and his closest cosmonaut colleagues had produced a formal, ten-page document, with the help of the engineers, in which all the problems were outlined in detail. The trouble was, no one knew what to do with it."

In addition to Chief Designer Mishin, who later said that the craft would not have been launched unless it was safe to do so, up to fifty senior engineers were familiar with the damning report. Many had even helped

Vladimir Komarov: the first cosmonaut to fly in space twice. Courtesy Colin Burgess Collection.

prepare it. But no one, it seems, was prepared to jeopardize his or her career by making a formal approach to the Kremlin hierarchy requesting that the launch be delayed until the existing problems were rectified and others identified. Through the discreet use of unofficial channels the report did in fact reach as high as First Secretary Georgi Tsinev, a close friend of Brezhnev. Unfortunately, Tsinev was not prepared to compromise either his personal friendship or his ascendancy within the KGB, and the report apparently never left his office.

Final training for the crews of both spacecraft ended with a test on Soyuz systems, held on 30 March. All four cosmonauts passed satisfactorily and were told to prepare for a move to the launch site at Tyuratam where they would conduct final training for the flight. Whatever doubts and anxieties he may have harbored, Komarov made his preparations for the journey.

Popular press reports surrounding the flight of *Soyuz 1* suggest that Valya Komorova was not even aware that her husband was leaving her to fly into space. However, in 1997 a former colonel at the training center, Sergei Egyupov, revealed that this was not correct. He disclosed that a tradition had evolved in which the crew members would visit Moscow before leaving for Tyuratam, and he had personally accompanied the four cosmonauts and their families on a train journey into the city. Following this, and after making fond farewells to their families, the four cosmonauts had boarded a flight and were flown to the launch site.

On 15 April, Nikolai Kamanin speculated on the forthcoming mission in his personal diary. He not only expressed deep concerns about the flight, but also lamented the fact that, for the first time, Korolev would not be at the helm. "I am personally not convinced that the whole flight program will be carried out successfully," he wrote, "but having said that, there are not enough reasons weighty enough to object to the flight. On all previous flights we were confident enough about the successful outcome, but now there is no such confidence." Prior to any Soviet space mission a high-level government committee would travel from Moscow to Tyuratam to certify that the spacecraft was in good condition. In his coauthored book *Two Sides of the Moon*, Alexei Leonov reveals that the committee on this occasion "was led by Leonid Smirnov, then head of the military-industrial complex, and included, of course, Mishin, as chief designer, and Boris

Chertok, chief engineer, together with many other engineers and designers. Yuri Gagarin and I were also on the committee." Despite the report listing a total of 203 hardware problems, approval was given for the complex flight of *Soyuz 1* and *2* to proceed.

Launch dates were then set. *Soyuz 1*, carrying sole cosmonaut Komarov, was scheduled for liftoff at 3:35 a.m. Moscow time on 23 April. Then, all being well, *Soyuz 2* would be launched at 3:10 a.m. the following day to play cosmic catch-up with *Soyuz 1*.

Early on the morning of 23 April 1967, Vladimir Komarov boarded his Soyuz spacecraft in preparation for what would be the first manned launch at night. Right on schedule, in a spectacular display of white-hot fire, thunderous sound and power, *Soyuz 1* was blasted into the dark skies, achieving orbit nine minutes later. Less than half an hour after the successful launch, cosmonaut Pavel Popovich called Valentina Komarova to let her know that her husband was safely orbiting the Earth. Sadly, it was a situation that would quickly deteriorate.

"*Ya Rubin*"—"I am Ruby!" The voice of Vladimir Komarov, reprising the code name he had selected for his Voskhod mission, could be heard as *Soyuz 1* hurtled over central Asia. "*Vsyo khorosho*"—"Everything is fine!" Everything was far from fine. Ground tracking stations were receiving data indicating that the left solar panel had failed to deploy, which would effectively cut the main power in half and also prevent the deployment of the backup telemetry antenna. As scheduled, Komarov had moved into the orbital module section of *Soyuz 1* during the second orbit, and what he saw through the porthole caused him immediate concern. "The left solar panel hasn't opened," he reported. "The charging current is only 13 to 14 amps, and short-wave communications are not working. An attempt to turn the ship to the Sun did not succeed. I tried to turn the ship manually using the [attitude control thrusters]."

On the fifth orbit Komarov once again attempted to manually orient his spacecraft, but without success. The lack of available electrical power had adversely affected his guidance systems. It was reported that he even resorted to banging and kicking on the wall of the spacecraft to try and dislodge the recalcitrant solar panel. Without it, the craft's electrical power would fall to critically low levels, placing the entire mission in jeopardy.

Yuri Gagarin was immediately dispatched to the Yevpatoriya control center to communicate directly with Komarov and attempt to resolve some of the mounting difficulties.

As *Soyuz 1* entered its seventh orbit, Komarov was ordered to take a lengthy break while ground controllers and engineers worked on salvaging the crisis. Between then and the thirteenth orbit he would be out of Soviet tracking range anyway, so it was an opportune time for him to grab some sleep. It is unlikely that the highly trained and diligent Komarov actually complied with this instruction; he knew his ship was in dire trouble.

Contact was reestablished with Komarov on the thirteenth orbit, and he confirmed that he was still unable to manually orient the spacecraft. This not only had fearful implications for a successful return to Earth, in which the correct reentry orientation was crucial, but serious consideration now had to be given to scrubbing the impending launch of *Soyuz 2*.

It has always been thought that when the three-person crew of *Soyuz 2* was stood down it was because of the problems with the first craft. In an interview with researcher Rex Hall in September 2004, however, the commander of *Soyuz 2*, Valery Bykovsky, said this was not the case—in fact, they were to be launched as a mission rescue team. Chief designer Vasily Mishin and Konstantin Feoktistov, the OKB designer who had flown on the first Voskhod mission, had studied the situation and decided that Khrunov and Yeliseyev should be able to make their way to Komarov's craft and release the balky solar array during their planned EVA. The go-ahead was given to launch *Soyuz 2*, and the crew retired for the evening. That night, as the rescue crew slept, a wild thunderstorm swept over the launch area. This in itself would have been enough to postpone or cancel the flight, but Bykovsky revealed that the storm also affected the waiting booster's electrical circuitry, and the rescue mission had to be reluctantly abandoned. As it turned out, the violent storm that night was fortuitous—it actually saved the lives of the crew of *Soyuz 2*.

With the rescue EVA option gone, all the energies of the ground controllers focused on getting Komarov down, now scheduled for sometime between the seventeenth and nineteenth orbits. During the sixteenth orbit, controllers at the Yevpatoriya center transmitted data to the cosmonaut on how to use the craft's ion sensors to orient *Soyuz 1* for landing on the following orbit. Unfortunately, the suggested procedures did not work. On the seventeenth orbit Komarov reported to Gagarin at the control center that *Soyuz 1* was still deviating in pitch, and the craft's automatic systems had subsequently prevented an automatic retroburn. Things were now becoming desperate; the controllers knew they could only rely on the craft's storage batteries for another three orbits. There was a backup battery that would give them an additional three orbits, but this would cause a reentry outside Soviet territory, which was regarded as a highly undesirable option for security reasons.

An innovative but hazardous retrofire plan was devised. Komarov was to manually orient the ship using an untried procedure that had not featured in his training. With very little time available to complete this task, he had to combine his visual skills with the gyroscope system to accurately align the spacecraft. An unplanned side effect of the procedure, however, would cause Komarov to reenter the atmosphere on a ballistic, rather than guided, reentry path on the nineteenth orbit. The trajectory was still a safe one, and would bring him down in Soviet territory. The taciturn cosmonaut accepted this plan and carried out his instructions with professional calm.

Komarov's skills or motivation were never in doubt, and he finally succeeded in placing the troubled Soyuz craft into the desired orientation for reentry. He then fired the retrorocket thrusters for a precisely judged 146 seconds. Though the objectives of his mission had not been carried out, it seemed that Russia's most experienced cosmonaut had at least saved his spacecraft, as well as his own life, by successfully negotiating a complex sequence of tasks. Five minutes after retrofire, *Soyuz 1* entered a communications blackout caused by ionization as the spacecraft plummeted through the intense heat of reentry. When communications were eventually restored, Komarov apparently reported in, although the actual transcripts have never been released. According to reports from ground controllers, including Gagarin, the cosmonaut's voice was "calm, unhurried, without any nervousness."

As the descent module hurtled downward with Komarov occupying the center couch, he would have been hoping that all of the dramas associated with his mission were at an end. Soon the parachutes would deploy, and he would come to a landing near Orenburg in Central Russia. Sadly, things would go terribly wrong at this critical time. The small drogue canopy popped out as planned, but it failed to pull the larger canopy from its storage bay. A backup parachute was then deployed, but it became entangled in the first drogue. There would be no deceleration.

Karabutak is a small farming village located within the Orenburg region, nestled in the foothills of the southern Ural Mountains. As the sun rose on the morning of 24 April 1967, many farmworkers were already working the fields surrounding their village. It promised to be a perfect spring day, with clear skies and the likelihood of warmer temperatures as the sun brightened and rose slowly higher in the east. Then, at 6:24 a.m., some of the workers and farmers began pointing to the sky, shouting an excited alert to others as a large, blackened object hurtled toward the ground, trailing some sort of material. It was traveling at high speed, and within moments was lost from sight as it reached the horizon. Then there came the sound of several loud explosions. The farmers began dropping their tools and running toward the impact site. They did not yet know that they had just witnessed the last moments of one of their nation's greatest modern-day heroes.

Without any means of deceleration, the *Soyuz 1* descent module had slammed into the ground at around ninety miles per hour and split open on impact. Unused solid propellant fuel in the deceleration retrorockets had then exploded. The flattened, shattered capsule burst into flames of ferocious intensity, and a thick pall of black smoke drifted skyward, bringing not only curious villagers, but directing waiting search aircraft and rescue helicopters to the scene. Local people reached the site of devastation first and began throwing bucketfuls of dirt onto the midst of the inferno in an attempt to quell the flames. Recovery troops soon joined in the frantic effort to put out the fire, while others began transmitting urgent messages about the crash. They even suggested in some messages that the cosmonaut might require urgent medical treatment, which seemed to indicate that he had somehow survived.

Meanwhile, Nikolai Kamanin and several other members of the State Commission were aboard an Ilyushin 18 airplane flying to the airport at Orsk, about thirty miles from the projected landing site. Before landing, they had heard to their relief over the airliner's intercom system a report that *Soyuz 1* had touched down and recovery teams were already making their way to the landing site. There were smiles all around as they waited to land at the airport where they had planned to later greet the newly returned cosmonaut. But as the engines shut down and the door of their aircraft swung open, it was immediately apparent from the faces of those who met the airliner that something terrible had happened. A somber official told Kamanin that the spacecraft had crashed and burst into flames, and there was no sign of Komarov.

Kamanin was naturally devastated by the news. Despite the reported enormity of the damage, he still held a faint hope that Komarov had somehow survived the terrible impact described to him. First reports from the scene had optimistically indicated that the cosmonaut's body was not found at the crash site, suggesting he had somehow survived. Suddenly, there was more cause for hope when a telephone report indicated that the cosmonaut had been admitted to a hospital just two miles from where *Soyuz 1* had come down. This would turn out to be nothing more than a fanciful, hopeful rumor.

As head of cosmonaut training, Kamanin had received orders to remain at the airport, but he angrily dismissed them as absurd and used his considerable influence as a three-star general to commandeer a nearby helicopter, which took him to the crash site. It only took one look at the scattered, blackened devastation for Kamanin to realize that no one could possibly have survived. In fact, it would be some hours before the first physical evidence of Komarov's presence in the wreckage could be found. Identifiable human remains were finally located in the pilot's position, the center seat.

Alexei Leonov had joined a team of officials that traveled from Tyuratam to the crash site, where they found "little more than a lump of crumpled metal." In 2004 he would write in his autobiography that the cause of the malfunction was traced to evidence that "the parachute container had opened at a height of 11,000 meters and had become deformed, squeezing the canopy of the main parachute and preventing it from opening. The conclusion of subsequent tests was that the parachute container was not rigid enough. For future missions the container was strengthened and meticulously polished on the inside, so that there would be no question of a similar problem occurring."

There may have been even more to the problem, however. During their manufacture the two Soyuz craft had had an ablative thermal coating baked on the outside. Undetected problems had occurred during the intricate baking process, which polymerized the synthetic resin. Apparently, the protective covers over the parachute containers were not in place during this procedure, which would mean that the parachute containers received an internal contamination of rough resin that was not removed. This would contribute to the failed deployment of the parachutes on *Soyuz 1*.

The basic facts behind the tragic end of Komarov have been known for many years, but the full story of those final hours and his last transmissions have been dogged by turbulent and demeaning rumors, surfacing virtually from the time his death was reported. Most of these stories supposedly emanated from American National Security Agency staff monitoring the flight from a listening station near the Turkish city of Istanbul, and in particular a then-twenty-year-old U.S. Air Force radio analyst.

Causing a sensation years later, the retired analyst contended that he had eavesdropped on what he described as a "heartbreaking conversation" while on duty at the station, giving graphic descriptions of the cosmonaut pleading and begging for controllers to fix the problems, sobbing as he talked with his wife, making his will from space, cursing those who had placed him aboard the doomed craft, and even berating the Soviet leadership. Coming some eight years after the death of Komarov, these "revelations" touched off a battery of headlines in legitimate daily newspapers around the world. They also sent space historians on a misleading detour as they tried to piece together the facts and set down the real story of the tragedy. In fact, in a personal letter to Colin Burgess in 1978, astronaut Walt Cunningham stated quite emphatically that "The gruesome description of cosmonaut Komarov's death is an old and oft repeated fiction. Hopefully, you can help put a stop to it by labeling it what it is."

It was only decades later that historians had enough proof to categorically state the truth: Komarov had done an exemplary and difficult job in aligning the spacecraft for reentry and was a professional test pilot right to the end. Valya Komarova never spoke to her husband in space and was only informed of the accident afterward, at the Cosmonaut Training Center, by senior space officials.

In his book *Russians in Space*, space journalist Evgeny Riabchikov spoke of the utter grief Yuri Gagarin felt at the loss of his colleague and friend.

Gagarin was later prompted to issue a media statement, in which he said, in part: "Men have perished. But new ships have left their moorings, new aircraft have taxied to the takeoff strip, and new teams have gone off into the forest and the desert. . . . Nothing will stop us. The road to the stars is steep and dangerous. But we're not afraid. Every one of us cosmonauts is ready to carry on the work of Vladimir Komarov—our good friend, and a remarkable man."

Although he had no way of knowing it when he died, Vladimir Komarov actually saved the lives of the three-person crew of *Soyuz 2*. Valery Bykovsky recently revealed to Rex Hall that with the cancellation of their linkup rescue mission and the death of Komarov, the parachute system on their spacecraft had been carefully examined. It was determined that if *Soyuz 2* had launched, a very high probability existed that its parachute system would also have failed to deploy correctly. Then, instead of a single fatality aboard one spacecraft, the Soviet Union might have lost two Soyuz vehicles and four cosmonauts.

The remains of Vladimir Komarov that could be immediately found at the crash scene were carefully gathered together and cremated, and his ashes were placed with honor in the Kremlin Wall. It was later rumored that other human remains of the cosmonaut were subsequently found and interred at the impact site. A modest but magnificent obelisk was erected at the very spot where *Soyuz 1* came down and cosmonaut Vladimir Komarov lost his life. On 25 April 1968 an emotional memorial was held there for the lost cosmonaut. Over ten thousand people made their way out to the steppe, some driving hundreds of miles to attend.

There is little doubt that extreme political pressure caused the loss of an exemplary cosmonaut and his spacecraft. Like the three Apollo astronauts, Komarov knew that he was going to fly a defective spacecraft, but he accepted the risks and never shirked his responsibility in accepting the role of pilot. Rather than being remembered as the occupant of a doomed spacecraft, he deserves instead to be regarded as a magnificent and courageous pioneer of manned spaceflight. "Komarov gave his life on *Soyuz 1* for the future conquest of space," Gagarin wrote as a eulogy to his colleague.

Vladimir Komarov was posthumously awarded a second Hero's Star. In keeping with Soviet space policy, details of the problems that had occurred in the early part of his flight were suppressed, and have only leaked out in recent times. The loss of a highly respected space explorer would prove a vastly sobering blow to the Soviet space program, but ultimately it did little to curtail the unrelenting Soviet drive to fly Russian cosmonauts to the moon ahead of America.

The deaths of three astronauts and a cosmonaut reinforced yet again the highly unforgiving nature of spaceflight. But even as both space powers tried to find causes and grieved for those they had lost, others who dreamed of flying beyond our planet were stepping forward to take their places.

5. The Astronaut Enigma

So we'll go no more a' roving,
So late into the night,
Though the heart be still as loving,
And the moon be still as bright.

Byron

The three crew members selected to fly *Apollo 7* spent most of 1967 and early 1968 in Downey, California, working with the improved version of the Apollo command module. They pushed the engineers hard, looking for every possible weakness, every possible improvement. The engineers, however, did not need to be pushed; they were determined to learn from the fire and the spacecraft defects that the investigation had uncovered and create the best spacecraft ever built. The redesign and rebuilding process progressed carefully, but it also moved fast. Though safety was now a priority, NASA still hoped to beat the Russians to the moon, and do it before the end of the decade. For this to happen, each Apollo flight would have to go almost perfectly, right up to the first landing attempt. Valuable flying time had been lost because of the fire, but it had been used wisely. The spacecraft, and the Saturn IB booster that would launch it, had been improved and refined to an unprecedented degree. Engineers and managers took personal responsibility for the delivery of components, which they guaranteed to be as safe as possible. Not only were their reputations riding on a successful flight—so was their sense of moral pride. This flight could not afford to fail.

By the fall of 1968, following a series of unmanned Apollo test flights, the spacecraft was finally ready, and the crew was ready to fly it. This was the first time that three Americans would fly in space together. Contributing to the sense of difference from previous flights, the *Apollo 7* crewmembers were three of the most diverse characters ever assigned to fly together. Gemini crews, in many instances, had consisted of a spaceflight veteran taking a rookie astronaut on his first flight, and were almost always a couple of pilots with similar backgrounds who worked together as a tightly bonded team. With three people, the collaboration was more complicated, and there was no guarantee that the crew would be compatible. Yet, as *Apollo 7* astronaut Walt Cunningham explains, it worked:

You had thirty guys, all type A, that came in from their own little world; they were used to being a chief, and all of sudden we had to play Indian. There weren't very many important jobs in the office; Deke Slayton had one, Alan Shepard had the other, and anything else was just kind of part of the group. So you had all of these guys jostling for space, and yet you'd be put together on a crew, and you had to work together. It is something I believe that highly motivated men can do, and do all the time. We could do the job regardless of who we were flying with. Nobody did any psychological evaluation before they put us together. The media thought that somehow there was some big fancy selection process that matched crews psychologically. We didn't have time for that nonsense. Deke wouldn't have had any patience for that—if Deke said that he was going to put you there, you'd damn well better get along or your ass was out of there!

Wally Schirra, once again a crew commander and now flying his third mission, was a different man after the *Apollo 1* fire; the tragedy had changed his outlook dramatically. He'd already decided he would be leaving NASA after this flight, and he used the freedom this gave him to ensure that everyone knew his opinions on what needed to be fixed and what needed to be left alone. The Apollo spacecraft was far more complicated than its Mercury and Gemini predecessors, and Schirra spent a lot of time reining in design managers who oversaw a single system they wished to make changes to without fully appreciating the effect those changes would have on many others. Schirra saw it as his job to look at the whole picture and judge whether each change was worthwhile in the context of the entire spacecraft. This often meant saying no to people. He frequently won the day, but to do so he had to speak very bluntly, and this upset some engineers used to the more jovial Wally of Mercury and Gemini. Schirra's determination to take charge and make the mission flawless would also put him in direct conflict with some of the flight controllers, particularly Chris Kraft. But Schirra to this day insists his approach was the right one. It was, as he says, what a test pilot was trained to do—to be a communicator between the engineers and others in the design process. In criticizing Schirra's style, Kraft failed to realize that Schirra never fully considered himself part of NASA—he was officially a naval aviator on loan, and, as he told the authors, believed in the authority given to the commander of a naval vessel:

I wasn't about to stay in NASA by then; I knew what it was. I had always believed that I worked with NASA, not for NASA. There's a big *difference! By 1968, I saw a bureaucracy developing—the fun days were over. I could see that I was out of line already. If Cooper was already out of line, how the heck could I get back in again? Deke said that we of the original seven are done; there's a whole new crew now. That I even got that Apollo flight was unusual. The second group was brought in to go to the moon. We were supposed to be out of there by then. It just turned out they needed me, so I stayed for the* Apollo 7 *flight. That was unique.*

When we lost three men on the launch pad, I knew we were facing up to a real problem. I said, "We are going to do this one right." I'm afraid others didn't always like it. They didn't realize what a command was, particularly Chris Kraft. Chris decided he was going to be God and tried it on me, and it didn't work! He didn't make a big issue out of it, but he did say I was kind of grumpy. I wasn't *grumpy; I was merely asserting my authority. It helped, and it worked. The flight controllers felt like they had the right to the last word—but I was the one taking the risks. I have yet to hear of a flight controller killing himself by falling off his chair. They were younger men who had not really put themselves physically at risk. They could wear black armbands if I died, but that wouldn't help me any. By then I had responsibility for two crewmen. When you have the responsibility, it's your problem; you accept command. That's the way it goes—if you give me the ship, it's mine. Then I'll tell you what I'm going to do or not do, within the rules of the ship. That comes down from the Royal Navy, and the authority invested in the commanding officer—I was taught that at the U.S. Naval Academy as well. An admiral, even aboard ship, does not order the captain. The admiral suggests. Then the captain might accept, because the captain is responsible for the ship and the crew. The flight controllers were much younger than I at that time, had very little experience with command, and didn't understand that.*

Though Schirra may have disagreed with some of the flight controllers, he told the authors that his determination resulted in very good, close working relationships with many engineers, especially if they were people he had worked with from the start of the space program:

I arranged for Guenter Wendt to be our pad leader. He essentially had been working for McDonnell Douglas for Mercury and Gemini. After the Apollo 1 *disaster I asked North American Rockwell to hire him as our Apollo pad leader. They said, "Do you want a Barbie doll too?," or something like that. I said, "I don't think you fellows understand where I am coming from this time. You screwed up. I want a good man on the pad." I think the major thing that happened all the way through to Apollo-Soyuz is that we could communicate and become buddies with the engineers; we became very close to them. I can still name every one of the engineers that made the* Apollo 7 *flight work, and I still hear from some of them on the anniversary of my Apollo flight, saying "Thank you for working with us." That kind of bonding is what made the space program go.*

Group 3 astronaut Walt Cunningham had been chosen by NASA for his scientific background as well as his piloting skill, and like Aldrin before him he would find acceptance by the test pilot fraternity hard to achieve. Cunningham had never really dreamed of becoming an astronaut. "If I looked at the moon as a kid," he reflected, "it was just like everyone else, because there was no such thing as astronauts, no flying in space. We had to settle for reading Buck Rogers. The whole field of astronautics has come about since I became a grown man." Yet he would make contributions to spaceflight before he even joined NASA.

Born in Creston, Iowa, on 16 March 1932, Ronnie Walter Cunningham grew up in the relaxed beachfront atmosphere of Venice, California. He was always interested in science, so much so that he was an unlikely candidate for ever fitting the heroic astronaut image. "In high school I was considered kind of a square guy," he admitted. "My brother was a little embarrassed that they called me a nerd because I was reading a pocket book all the time in the five minutes between classes. But, like I now tell them, the nerds inherit the Earth!"

Although Cunningham was busy working on his academic studies, he also nursed a childhood ambition to be a lieutenant commander in the Naval Air Corps. In his freshman year at the local community college, Cunningham made a calculated decision. He wanted to become a fighter pilot, but flight school required two years of college. Yet if he waited, he risked being drafted into the army before completing the two years. He therefore dropped out of college in his first year to enlist in the navy, where he eventually took and passed a two-year college equivalency test and was admitted to naval flight training. By the age of twenty-one Cunningham had his wings and chose to take his commission in the Marine Corps. By doing so, he was guaranteed an assignment to single-engine fighters, instead of the transport aircraft to which many of his navy contemporaries were assigned.

He spent three years on active duty at various air bases around the world and was in the last squadron to fly a mission toward the end of the Korean conflict. After leaving the Corps, Cunningham decided to go back to college in Los Angeles. "I realized that in the Marine Corps, without a college education I was not going to go very far," Cunningham told NASA historian Ron Stone. "I didn't expect them to send me to test pilot school; I didn't have the qualifications they were looking for at the time." One of the jobs Cunningham held to pay his way through college was at the RAND Corporation, a nonprofit think tank researching innovative solutions for U.S. military challenges. He worked on possible defenses against submarine-launched ballistic missiles and studied the Earth's magnetosphere. His thesis project at UCLA was an experiment carried on the first orbiting geophysical observatory. His involvement in spaceflight was soon to increase, as Cunningham recalls:

I had been a Marine Corps fighter pilot, transferred to a reserve squadron, and was back working on a PhD in physics at UCLA. *I was holding down sometimes three or four jobs at the same time, trying to supplement the G.I. Bill. I'd get up very early in the morning, going to some of those jobs. On the morning of May 5th, 1961, I was driving in my little Porsche speedster down into Santa Monica,*

California. It was the day of Alan Shepard's first little suborbital flight, and I had the radio on in the car. When it got to the last couple of minutes of that countdown, I had to pull my car to the side of the road. I couldn't drive any more; I was so into what was happening. Before that little Redstone rocket had cleared the tower, I caught myself screaming out at the top of my lungs, "You lucky sonofabitch!" I looked around kind of sheepishly because I thought, did anybody hear me? At that moment, I changed from just being envious of the Mercury astronauts to deciding, hey, that's what I want to do. Sure enough, two years later, there I was sharing an office with that same Alan B. Shepard.

Walt Cunningham today is still sharp, focused, and sleekly built, and possesses a devastating sense of humor that he is unafraid to use. Of all of the living Apollo astronauts, he is the one who most closely resembles his photos from the 1960s. He hasn't aged much, nor has he mellowed—he pulls no punches when it comes to expressing his opinions and seems to enjoy being known as the astronaut who "tells it as he sees it," even if this stirs up controversy. When it comes to the hero mythology, he was always careful never to take himself too seriously. He was the first to write a book that lifted the lid on the astronaut mystique and told the truth behind the image, and *The All-American Boys* is still considered by many to be the best of the astronaut autobiographies in this regard. Today, Cunningham's candor serves as a valuable counterbalance to the bureaucratic decision-making that frustrates future space exploration. In interviews and newspaper articles, he gets to the point of an issue and gives his opinion, no matter who this may upset. As an author and radio host in the twenty-first century, this works very well. As a rookie astronaut hoping to be picked for his first flight, his personality did him no favors at all, as he recalls:

We weren't astute to the selection process; matter of fact, I never figured out a thing. I am the archetypal one that always thought, if I work hard, it is bound to be noticed, and it will all pay off in the end. I was, and still am an obstinate s.o.b., and I never figured out that I had to join the system. If I had figured out the system, I'm still so obstinate that I would not have joined it! I mean, that's the way I am today, some forty years later. Even in high school, I was selected the most obstinate in the class, and I didn't realize the significance of that at the time. We astronauts talked amongst ourselves about what it would take to be selected—Bill Anders and I were probably the closest alike on that—but in the end it didn't make any difference to me. I was going to be the best I could, wherever it was. And only in the last ten years did I understand how significant Deke's impressions were when you came in—he even knew when you came in how far he was going to let you go. It was a really capable group of people I was competing with. Some of them I am perfectly glad I am not competing with any more!

Unfortunately for Cunningham, Deke considered him one of the "scientists"—although he was a military pilot, he had also been selected partly for his scientific training. To others, this might have been seen as an advantage, but Slayton always preferred those who had been to test pilot school, Cunningham remembers: "I didn't know that at the time. I knew that being a scientist was kind of a strike against you, but I didn't realize how significant a role it played. I mean, I just set out to be the best damn pilot around. I can't really complain, because I found it very satisfying. It occupied me mentally, physically, and I benefited by it, but more importantly the space program benefited by me being there. Alan Bean says Skylab wouldn't have been half of what it was without the work I put in—that's good enough for me!"

No matter how far he went at NASA, Cunningham was just happy to be there. "It was the golden age of manned spaceflight," he recalls; "very much like the 1920s in aviation. We didn't fly biplanes and have our silk scarves trailing out of the cockpit, but it really felt like it a whole lot. Most of the time I was there, until it got towards the tail end of my tenure, we had thirty astronauts or less. We loved flying airplanes, we loved riding the rockets. We looked at NASA as the last great flying club."

Schirra and Cunningham are well known to spaceflight history. Neither of them has ever held back when asked for their opinions, and both continue to be two of the most visible ex-astronauts on the public speaking circuit. They still disagree profoundly about how the *Apollo 7* flight should have been conducted, and are more than happy to go on record criticizing each other on the subject. It is, however, a good-natured disagreement. They comment on each other in the same way that many ex-military men make jibes at each other. In fact, they seem to enjoy it. And when they do verbally joust with each other, they tend to end up complimenting each other. For example, Schirra characterizes Cunningham by saying: "Walt has lots of little fanciful ideas, once in a while. He's like a puppy dog, keep scratching him and he'll roll over and be nice, but worry about when you stop scratching him. I stopped scratching him, and boy, he got *nasty*! But he's all right, really. That crew was made up of three men who could work together beautifully, and no one could fault Walt's knowledge of the spacecraft systems." To which Cunningham counters with:

I've always had an irreverent approach to authority. Wally was willing to debate who was in charge even in a situation where nobody cared! Other than that, incidentally, he is one hell of a stick-and-rudder guy. Some of the best memories I have are of crew training, and flying with Wally and Donn. These were people that we didn't have to have warm affection for, but we had strong respect for each other. We would bet our lives on the other guy—and we did. So I came away from that with a sense of camaraderie, and respect for people that I knew would give their lives for me.

The ribbing between these two astronauts tends to overshadow the most enigmatic member of the crew, the one about whom least has been written: Donn Eisele. Schirra sums it up well: "He is the forgotten astronaut." In his book *Carrying the Fire*, Mike Collins offers insightful and often amusing insights into the characters of his fellow astronauts. When he reaches Donn Eisele, he begins with one word—"Who?" When the authors had a group conversation with Walt Cunningham, Gordon Cooper, and Charlie Duke, two of whom served on Apollo crews with Eisele, they all quizzed each other and scratched their heads to come up with reminiscences of the man. It almost appears as if Donn Eisele moved through some of the biggest, most tumultuous events in Apollo's history receiving scarcely a second glance from anyone he worked with. Even the head of NASA at the time, Jim Webb, mispronounced his name when he introduced the crew to President Johnson. It was a mistake that was repeated in 1996 by then-chief astronaut Bob Cabana when dedicating a memorial grove at the Johnson Space Center. The *Apollo 7* crew became known as "Wally, Walt and Whatsisname," and at his launch breakfast Eisele even used a coffee cup with "What's-his-name" printed on it.

Donn Eisele, however, should be remembered. It is likely that he is less well known than Schirra and Cunningham only because the two are such powerful, sometimes forceful personalities—which was far more typical of the astronauts with piloting backgrounds. As Walt Cunningham told the authors, "Donn was a capable, competent guy, and a capable aviator who I flew with a lot. He didn't make waves. I did, so people maybe noticed me more! Donn always just kind of went along: I don't ever remember him taking a strong position on anything. He didn't seem to me to be a typical fighter pilot. He was an unusual kind of guy." Gordon Cooper, who would be Eisele's commander on the *Apollo 10* backup crew, had similar memories of him: "Any team, including an Apollo crew, is always very close and becomes a kind of a family," he told the authors. "Donn was a very capable guy, and a very nice guy. He was somewhat of a prankster—he had a good sense of humor, and was all for practical jokes. Yet he was very soft-spoken, and not overly aggressive—not really aggressive at all, although I encouraged him to be, sometimes."

One of the things that Eisele is best known for is a prank of a kind, a throwaway pun. With Guenter Wendt on duty as pad leader for the *Apollo 7* launch, Donn Eisele watched as the gantry was rolled back from the command module and the White Room receded from his view. The moment was perfect, and irresistible. In a German accent, Eisele deadpanned—"I vonder vere Guenter vent?" It is a sad irony that many people later assumed this pun came from Wally Schirra—while the pun was remembered, Eisele was not. "It sounds like something that Donn would have said," Walt Cunningham remembers. "Wally, of course, was an inveterate punster, but Donn could join right in, and could also do great comedy routines." He certainly livened up their flight with some moments of mirth, even talking to the ground in a falsetto voice.

It is also most ironic that so laid-back and fun-loving a pilot as Donn Eisele would be caught in the center of two major events that threatened to shatter NASA's carefully crafted public relations image. One event damaged the perception of perfect harmony and synchronicity between the astronauts and Mission Control. The relationship had been strained a number of times in the past, yet these disturbances had remained relatively internal, NASA affairs. This time, however, it literally became front-page news. The second iconoclastic event was the shocking concept of an astronaut actually divorcing—something that upset the image of NASA's boys as uniformly happily married, All-American heroes.

In truth, the well-scrubbed astronaut image had been slipping ever since it was imposed in 1959. The America of 1968 was a much different place, and the far more liberal public of the late 1960s cared far less about such things. In an era of long hair and psychedelia, however, NASA managers were still happily stuck in the age of pressed white shirts, narrow black ties, and crew cuts. They were also intensely concerned about any waning in public interest in the space program, and didn't want anything to mar the image of their most effective public relations representatives—the astronauts. Always "an unlikely candidate to challenge the system," in Cunningham's words, Eisele became the test case to see if an astronaut could divorce and still stay on flight status. His colleagues watched what was happening very closely indeed: many were stuck in unhappy marriages they were holding together only for the sake of their careers. The prospects before 1968 had not looked encouraging. Duane Graveline, selected as a scientist-astronaut in 1965, had barely been at NASA a month when his wife filed for divorce. Forced to resign when the news broke, Graveline paid the price for becoming embroiled in a potentially damaging scandal. His astronaut career had lasted a mere fifty-one days. It was an unwritten, unspoken rule, but keeping one's marriage together, however empty and loveless it might be, seemed to be part of the price of getting into space.

In terms of his early career, Donn Fulton Eisele showed the same impressive career path as the top pilots who went on to command lunar landing missions. He was born in Columbus, Ohio, on 23 June 1930, the son of Herman Eisele, a printer who worked for the *Columbus Dispatch* newspaper. (His mother June did not want people to think his name was short for Donald, so she added a second *n* to his first name.) Reporters later seeking interesting snippets of crew information would discover that Eisele was exactly twenty years older than Wally Schirra's son. To his own parents, Donn was very close. He also adored his grandfather, who lived with the family. Not atypically for an only child, Eisele's mother doted on him, keeping him busy with piano lessons and other skill-building activities. She would also intervene whenever she felt her son was being unfairly treated. When Donn was in high school, for example, June Eisele once appeared at the school to contest what she felt was an unfair grade.

While at school in Columbus, the future astronaut excelled in the high school track and cross-country teams, and also achieved the prestigious rank of Eagle Scout, which requires years of challenging work in the community. Somewhere in those years, Eisele also found time to play bass in a musical group and work at the Columbus Zoo, where he had to contend with the occasional monkey bite. Eisele earned an appointment to the Naval Academy at Annapolis by taking the challenging examination rather than receiving a political appointment—a personal achievement in which he always took great pride. His mother still kept a watchful eye over him, visiting the academy when she learned her son had chosen to skip formations. While at Annapolis Eisele shared a room with Ed Givens, who would also become a NASA astronaut, and future astronauts Jim Lovell and Tom Stafford were in his graduating class. Eisele graduated from the academy with a bachelor of science degree in 1952.

That same year, he met his first wife, Harriet. "We met at a classmate's family picnic and boating day in Columbus in the spring of 1952," she told the authors. "Donn was the cousin of a classmate. He was cute, fun, and had a good sense of humor. Education was an important factor to me when choosing a boyfriend, the Naval Academy seemed glamorous, and I was attracted to his close-knit family." Harriet was both intrigued and impressed by Donn's family; unlike her own relatives, they seemed to do everything together. She didn't get to see too much of Donn initially—"we had a long-distance relationship," she remembers—and yet, three days before the end of 1952 the couple were married, in Harriet's hometown of Gnadenhutten, Ohio. She was delighted to have joined such an apparently close family. She soon found, however, that her initial impressions had been misleading. "I realized later that what I saw as closeness was actually smothering," she told the authors. "His mother even asked me to get him to stop flying, telling me 'If anything happened to him, my life would be over.'"

Donn Eisele did not give up his flying ambitions. He had always been interested in piloting, and decided to go into the U.S. Air Force instead of the U.S. Navy. He spent the four years between 1954 and 1958 as an interceptor pilot in South Dakota and also saw overseas service in Libya. He then returned to Ohio, becoming a student at the Air Force Institute of Technology. During these years, the young couple also had their first two children—Melinda, born in 1954, and "young Donn," in 1956. At the institute, Eisele earned an MS in astronautics in 1960. He stayed on as a rocket propulsion and aerospace weapons engineer for a year, before being assigned to the Aerospace Research Pilot School at Edwards AFB.

Harriet was growing used to the endless moves and separations that came with being a military pilot's wife and says their relationship was "pretty much" solid in those years. She remembers Edwards as being one of the better postings: "We were at Edwards for ten months. Aside from the worries about Donn flying, and venomous sidewinder snakes, it was a neat assignment. I learned to play bridge, and we met and interacted with a lot of interesting people. We made five family trips to Disneyland, and Faye Stafford and I made monthly shopping trips to Los Angeles. Our daughter Melinda would ride on a motorbike with another future astronaut, Ted Freeman, and bake cakes with his wife, Faith."

In those first years of marriage, Harriet got to know the man she had married a lot better. "Donn was capable, competent, soft-spoken, and probably avoided making waves. He also told me that he had great difficulty in making decisions, using as an example taking an hour to choose a hammer. It seemed that he needed a 'push' to be motivated. For years, the COS would tell me to get him to work on time: a duty I bought into and worked hard on for many years." The couple seemed to have a good relationship, but looking back, Harriet now realizes that they were actually avoiding many of the problems that would eventually drive them apart. "We almost never had a fight, which in retrospect was a mistake. I believe this stemmed from a history of mutual avoidance of discussing negative feelings, as well as not discussing expectations. My later attempts to change this elicited the response, 'I'm not given to introspection.'"

In 1961, the couple had a third child, Matthew. "Matt was a sweet child, and everyone loved him," Harriet remembers.

He was born with Down Syndrome, and without his right hand below the wrist. He loved books and music, and I can still remember Matt and Tom Stafford singing "Snoopy and the Red Baron" together. Having one hand never slowed Matt down. We used to joke that he would be the first baseball pitcher with Down Syndrome in the major leagues because he could throw fast, hard, and accurately with his left hand. When he wanted attention, a toy would whiz by my ear. He liked to throw his toys onto the roof of our two-story house, and a standard chore was a weekly retrieval of the toys.

Harriet found that having three children, including one who needed extra, specialized attention, took up almost all of her time and energy. Like many military wives, she could not expect much assistance from her husband, who was consumed by his flying duties. Their lives were about to get even busier, as Donn's career took an unexpected turn.

By 1963, Eisele was an experimental test pilot at the Special Weapons Center in New Mexico. While they were on leave in Columbus, Harriet remembers that Tom Stafford called. He was now an astronaut and urged Eisele to apply for the forthcoming intake of new spacefarers. Eisele followed Stafford's suggestion and was successful, chosen by NASA in October 1963 for the third group. "He was happy and excited," Harriet recalls. "I was happy for Donn, and it was exciting, but it was also scary." Eisele was one of eight test pilots selected in the fourteen-person astronaut group. Over his career, he would log more than thirty-six hundred hours flying jets. He brought to NASA an impressive test-piloting background, with all of the qualifications that decision-makers such as Deke Slayton shared and admired.

Tom Stafford knew Eisele from both the Naval Academy and Edwards, and told the authors that he was a considerable influence in helping Eisele through the selection process. Having a respected figure like Stafford as your friend in the astronaut office could do much for your career, as others found to their advantage. Deke Slayton also kept the test pilots in this group in mind as command module pilots (CMP) for future missions—they'd be responsible for the vital task of piloting the spacecraft solo. With a first mission as CMP under their belts, their chances of commanding a later lunar landing mission would be pretty good.

Harriet knew that Tom Stafford's influence was a major factor in her husband's selection as an astronaut:

Donn seemed to follow Tom's pathway. While in Tripoli, Donn applied and was accepted for grad school at Wright Patterson. When Tom later came to Tripoli for gunnery practice, he told us he had applied for test-pilot school. Donn expressed disappointment that he had not thought of that too. After grad school, Tom called and told Donn about an upcoming test-pilot selection. Donn applied and was accepted. We were in Columbus, visiting Donn's parents, when Tom called to tell of an upcoming astronaut selection. Donn again applied and was accepted. Prior to that, I had never heard him express a desire to be an astronaut.

In 1964, the year Eisele reported for duty at NASA, Slayton assigned him to the Apollo branch of the astronaut office, working under Gordon Cooper, where he joined astronauts such as Pete Conrad and Dick Gordon. His major work involved testing the spacesuit designs that would later be used by those who walked on the moon, as well as general work on the command module and lunar module. By May of 1965, Eisele was working with Dick Gordon on zero-g parabolic flights at Wright-Patterson Air Force Base in Ohio, trying out methods of transfer between docked spacecraft. Eisele and Gordon found that the Apollo hatch was difficult to move in weightlessness and suggested that more handles be provided. The same month, he worked on Grumman's prototype lunar lander, taking part in simulated landings on the moon. His work proved that an earlier problem of stability bounce had been resolved.

Despite his laid-back approach, Eisele evidently believed in hard work more than in the fame of being an astronaut. Decades later, he would state, "There are a lot of people who would go over Niagara Falls in a barrel if given the chance and enough acclaim. But that kind of attitude has no place in the space program; . . . it's too hazardous." Mike Collins, from the same astronaut group, got to know Eisele well. "He was a nice guy," he told the authors, "a very friendly guy, an easygoing guy, which was different from most of the people in that group—they were mostly overachievers, you might say. Eisele I don't think was. He was more balanced, more relaxed, very friendly, very nice, very outgoing." Scott Carpenter, another relaxed and easygoing character, agrees. "He was a fine fellow," Carpenter recalls.

Though it was an exciting year for the new astronaut, it was also a tragic one. "Donn's father had experienced a series of heart attacks," Harriet explains. "The first one was in March 1957, and the Air Force brought us back from Tripoli on emergency leave. His father subsequently died of a heart attack in 1964, while I was pregnant with our fourth child, Jon. Four days later, Donn's mother also died of a heart attack."

Life had not changed much for Harriet when Donn became an astronaut. She was still extremely busy looking after the children, especially after Jon was born. "The pediatrician told me about what was then called the Houston Council for Retarded Children, which had a branch school in a church in Pasadena, Texas. I drove Matt to school every day for a two-and-a-half-hour class, a real hardship when Jon was still a newborn." Yet Harriet understood that these difficulties were part of her role as an astronaut wife, and did her best to support Donn. "I believe that the astronaut wives accepted and adapted well to the men's schedule," she observed. "It was the children who incurred the biggest losses."

Walt Cunningham agrees. "Our wives and children suffered a lot more than we ever thought about," he told historian Ron Stone. "Many weeks, we'd leave on Sunday night or Monday morning and come back Friday night or Saturday morning. One year, I was gone all or part of 265 days. And when you came home, you had to catch up on the discipline with the kids, the lawn needed mowing; you had a lot of responsibilities. A lot of people today would probably call it stress. We ate it up: we were all proud of the fact that nothing could get us down. It was stupid!"

Harriet does remember some amusing times while Donn was at NASA, especially involving the mispronunciation of their last name, correctly pronounced "eyes-lee." They never knew what they would get when a stranger attempted their name. "That confusion happened all the time. Even in his Naval Academy days when he was aboard ship for training, the denim shirts he was issued were stamped 'DFEISELE,' without punctuation, so the letters all ran together. He became known as 'Mr. D'Feesley!' Faye Stafford and the other neighbors frequently called him by that name, and I was referred to as Mrs. D'Feesley."

On the other hand, there were times when Harriet became deeply distressed with the unending demands placed on her husband's time. She felt she was constantly losing him to both social and work commitments. "I did understand and accept the demands of his job," she explains. "But I did feel resentful over his inability to decline any social invitations, which encompassed most weekends. If I complained, he would not go—then I felt guilty. So I asked that he think about it first, accept it if it was important to him, and told him if I got mad, I would get over it." Jeannie Bassett, widow of Eisele's fellow Group 3 astronaut Charlie Bassett, remembers just how much Eisele enjoyed letting his hair down at these social events:

Donn was a really funny guy. He was a special friend to me because he was the Air Force officer assigned to assist me with all of the miserable stuff military spouses have to deal with. He had also gone through test pilot school at Edwards with Charlie. Donn had a great sense of humor, and always revealed some surprising talent, such as dropping to the floor and doing a hundred pushups when everyone thought he couldn't do five! When we moved into our house in Nassau Bay, Houston, we were invited to a party at the Cernans. We were exhausted because the move had been horrible, and were not very good company, so we went to bed, worn to a frazzle. But at 9:30 the party came to us, and being good kids we got up and let the boozy partiers come on in. Some workman had left a twelve-foot ladder in the living room. Donn had a ukulele with him, and climbed up to the top step. There, he sat and serenaded us with the Debbie Reynolds song "Abba Dabba Honeymoon." Fun memories.

In June 1965, Eisele was paired with another Group 3 astronaut, Roger Chaffee, to test the Apollo lunar surface spacesuit's life-support systems. Most importantly, the Grumman engineers wanted to determine whether the life-sustaining backpack got in the way when astronauts climbed in and out of a lunar module hatch. Eisele had no way of knowing that just over a year and a half later he would be serving as one of the pallbearers at Chaffee's funeral.

As many other astronauts found to their chagrin, Eisele discovered that working hard on a specific project was no guarantee that you would be noticed and respected. Being friends with influential decision-makers such as Slayton, Shepard, or Stafford made far more of a difference. As a former Edwards pilot and friend of Tom Stafford, Eisele seemed to be in the right circles. Yet it didn't work out that way. In 1965, Slayton tentatively assigned Eisele and Ed White to Gus Grissom's crew for the first Apollo flight. Slayton reasoned that he didn't need a very experienced crew, as there would be no lunar module to test. He had decided to assign either Eisele or Chaffee to this mission to test their capabilities, since he felt they were less impressive than many of the others. "Frankly, I thought [they] were weaker," Slayton would say of the two men in his autobiography. Somewhere along the line, perhaps through no fault of his own, Eisele had not sufficiently impressed the right people.

Then, what initially seemed to be a piece of bad luck may have saved Eisele's life. In September 1964 he had been in Ohio, participating in a

KC-135 zero-gravity training flight, when he accidentally dislocated his left shoulder. It was not serious enough to require a cast, but it did require an overnight hospital stay, then a return to Houston and placement on limited duties for three weeks. The injury did not prevent him from being placed on Grissom's *Apollo 1* crew, but it did prevent him from staying on it. In January 1966 he was admitted to a hospital when he dislocated his already-injured left shoulder once again while participating in some strenuous physical training. This time, surgery was required. Removed from flying status for a few months, he began falling behind in training, and thus lost his place on the crew to Slayton's other "test case," Roger Chaffee. "Gus reassured Donn that he would not be replaced," Harriet recalls. "However, that was not the case; Roger replaced Donn on Gus's crew. Donn was certainly disappointed."

Eisele had lost the opportunity to prove himself on the first Apollo crew. Yet Chaffee would also lose that chance, dying in the horrific *Apollo 1* fire. "Donn was deeply saddened by the loss of that crew, as was everyone else," Harriet remembers. She says that Grissom had been very unhappy about losing Donn from his crew and had tried to get him back in before eventually giving up. If Grissom had been granted his wish, Donn would have been by his side on that tragic day in January 1967. "All of this took place before any Apollo crew had been made official," Cunningham recalls. "Donn told me, while walking down a street in Mexico City in 1967, I believe, that Roger had taken his place on the *Apollo 1* crew." It is impossible to say whether an identical sequence of events would have taken place if Eisele had remained on the *Apollo 1* crew. However, Cunningham sensed in Eisele's words that he knew a terrible fate had eluded him through mere chance. If he did feel that way, he never shared any such thoughts with his wife.

Harriet was, in fact, busy dealing with the beginnings of a family tragedy. "Matt was diagnosed with leukemia at age four," she remembers. "His first remission lasted seven months, and he was able to return to school. Even during this time there were frequent doctor visits. My life was pretty difficult, as it is in any family dealing with a catastrophic illness." She remembers having to deal with this heartbreak almost alone, as Donn was always away from home, busy with his work. "Harriet had a great heart, and generous spirit," Jeannie Bassett says of those difficult years. "Very kind—we always adored her."

The *Apollo 7* crew relaxes prior to water egress training in the Gulf of Mexico. *From left*: Donn Eisele, Walt Cunningham, and Commander Wally Schirra. Courtesy NASA.

Eisele, meanwhile, had found a place on another crew—one that was planned to fly the second Apollo flight. He now joined Walt Cunningham under the command of Wally Schirra. When Deke Slayton announced the crew in September 1966, he did not have a future lunar mission in mind for any of them—especially as he knew Schirra would probably retire soon afterward. He did, however, think that they were a capable crew for testing the Apollo spacecraft in Earth orbit. After the flight, he imagined moving Eisele and Cunningham over to the Apollo Applications project, which would eventually be renamed Skylab. Already, Donn Eisele's chance of flying to the moon seemed to be over. Schirra, at least, liked him and his work ethic. "Donn and Walt were assigned to me by Deke. Donn was very compatible, very easy to work with. He just sort of melted in, wasn't really a person who would assert himself all the time. But he was great at his primary job, that flight control computer system, the inertial guidance. He had that down cold, he really was great: the people at MIT really respected him."

Cunningham remembers that he and Donn were delighted to be assigned to a mission. As he told Ron Stone, "Donn and I, being rookies, you know, we ate up everything. We just wanted everything to be absolutely correct. To us, it was the most important flight in the world."

Despite their hard work, the proposed mission never took place; in November 1966 *Apollo 2* was canceled as an unnecessary repeat of the *Apollo 1* mission. All of a sudden, Schirra and his crew found themselves without any mission at all. The crew could have been given the next planned flight—one that would test the lunar module. Slayton, however, planned to let McDivitt's crew fly that mission. He did not want a rookie astronaut flying solo in the command module on such a crucial flight. Eisele, still awaiting his first mission, was at a disadvantage. A few days later, Schirra says, they ended up as the backups to the *Apollo 1* crew:

I had a prime flight scheduled, Apollo 2. *I convinced the NASA people it was a dumb flight, to do the same thing all over again, much like the second Mercury flight. We'd finally stopped doing that in Gemini, and I asked why we were doing it in Apollo, if we were in a hurry to get to the moon and back. I didn't want to do that flight, I tried to cancel it—I* fought *to cancel it. So they made us backups. I was in a meeting after having had that mission scrubbed, with Shepard, Slayton, and Grissom. These three guys said, "Wally, you get to be backup for Gus." I said, "Oh, shit! I'm not going to be a backup again." I'd been a backup for Carpenter; I'd been a backup for Gus. Having been a Mercury backup, then a Gemini backup—this was three backups, and that was too much. Had Gus been a backup for me? Once, in Gemini. I said, "This is ridiculous—has anyone else here been backing up anybody? No." I didn't think that was fair.*

If Schirra thought that the meeting wasn't going his way, it was about to get worse. He was next told that he would probably never fly in space again:

They said, "Well, you are not going to ever get an Apollo flight anyway." I said, "What?" They said, "Well, you and Gus and your two crews are as close as anybody is to a Block One, and we'd have to move somebody from Block Two to be a backup." I said, "What happened to Gus's backup crew, McDivitt and some other guys?" They said, "We are going to move them on to another flight." Oh. So I said, "Well, I guess I am screwed. I'll see how Eisele and Cunningham handle this one, because this is a lousy deal, going from flying to backup." They said, "Well, anything you can get is good." I had spent ten years around there, and I didn't go along with that, but I decided to stick with it for a while, with great reticence. And, of course, Gus and I were very close: next-door neighbors, all that kind of stuff. Okay, Gus, I thought, I'll work with you, we'll see what we can do. I'm probably going to leave after the mission anyway, so what the hell. I was furious. We were being pushed around, and I didn't like that very much.

Unlike Schirra, Cunningham was still happy to have any flight assignment. He worked at his *Apollo 1* duties with a growing sense of confidence. As he reflected later, his confidence was misplaced: the prime crew died.

We—the backup crew, prime crew, all of the astronauts working on Apollo—we were as guilty as anybody over that particular tragedy. We knew that the spacecraft in those days was not very good. We tried to get things fixed, but didn't succeed for a number of reasons, usually because it had negative time or weight impacts. But we were just egotistical enough, I guess, to think that we would make up for it. We believed our own publicity, that we could fly the crates they shipped them in. That bit of optimism cost us the lives of three good friends. Following the fire, I was part of the accident investigation board for the Apollo 1 *fire, and we went to work to fix the spacecraft.*

One job that Eisele and Cunningham had to do was to listen to the tapes of the *Apollo 1* astronauts' final moments. As they had worked closely with the crew, they were among the very few who could identify the individual voices. The thought of that grim task, listening to their colleagues dying over and over, still makes the hair stand up on the back of Cunningham's neck.

Schirra, having been written off by Slayton and Shepard, was now suddenly needed. "They came to me with a platter saying, 'Won't you please take the first mission?,'" he told the authors. "I said, 'Aren't you cute! I thought you said nobody in my crew would fly on Apollo?'" Schirra took the job. He now had to fly a mission that, if it failed, would mean, he believed, the end of the Apollo program.

In the same year as the *Apollo 1* fire, Donn and Harriet Eisele suffered another tragic death. "Matt died at the age of six," Harriet remembers. "His second remission had been short lived, and we never achieved a third. He spent much of his last months in and out of the hospital." During this difficult time, Harriet relied on a friend of a NASA flight surgeon who had come to live with her when Jon was born. "She and I took turns staying with Matt in the hospital and being at home with the other kids. Without her help, I could not have managed. This really troubled me: neither Donn nor I were there for them consistently. Donn spent the night in the hospital with Matt once. He told me the next morning that he was so unsettled by Matt's discomfort that he couldn't remain in the room. I didn't ask him to do it again."

Donn Eisele had gone from planning to fly the first Apollo flight to readying himself for the second flight to backing up the first. Now, it had come full circle—he was going to fly the first Apollo mission after all. He would be the first astronaut to check out the Apollo navigation and guidance equipment for the spacecraft—a complex task, vital for travel between the Earth and the moon. It was an incredibly busy time for him. "A couple of times the doctors suggested I call Donn about Matt's worsening condition," Harriet says, thinking back to that difficult time. "I told them I did not want to upset him or disturb his training. In truth, I knew I wouldn't be able to find him, and didn't want to deal with that situation either. As it turned out, Donn was in town the day Matt died—he spent much of the day on the couch in Matt's room. It was very difficult for him to face, and I know that Matt's death significantly affected him. However, he was not around, and he did not talk about these feelings." Harriet had to face the tragedy of losing a child essentially without her husband. She did at least have the reassuring and helpful support of neighbors and other astronaut wives: "I felt totally supported by the wives. When Matt died, Jan Armstrong came and trimmed the grass around the trees and shrubs. They sent cards, said prayers, and kept my freezer filled with meals in disposable containers without names (so I couldn't write thank you notes)."

Donn, busy working on fixing the problems the *Apollo 1* fire had exposed, was also feeling the same frustrations Schirra had encountered. Decades later, in an interview for *People* magazine, he discussed some of those concerns: "I asked the Apollo managers before the fire, 'What's the hurry?' We had until 1970 to land on the moon. They had it worked out that they were going to do it in 1968, two years early. The trouble was, they were forcing the pace. I wouldn't call NASA negligent . . . too strong a word. They had a blind spot." Just as Schirra did, and perhaps because of his influence, Eisele felt that the concerns of the pilots were being overlooked. When disputes arose, Mike Collins observed that Eisele often played the role of a "good-natured referee" between Schirra and others. For a rookie astronaut, siding with Schirra in the battles with management was risky. But Eisele was standing up for something he strongly believed in: "We thought that we astronauts were the pilots and that we were going to have to get into this thing and fly it. NASA officials, on the other hand, tended to see the astronauts as passengers, a sort of cargo. They regarded us as troublemakers. We raised issues they didn't want to hear about. What happens, I believe, is that they fall into a pattern of group think; . . . they decide to perceive something in a certain way. From then on, all information that doesn't jibe with the mindset of the group is rejected."

As he got to know his crewmates better during the demanding training, Walt Cunningham never quite figured out what made Donn Eisele tick. He noticed that Eisele seemed to adopt the style of his commander very readily and enjoyed the astronaut lifestyle of working hard and playing hard. Beyond that he remained somewhat of an enigma, even to Cunningham. Eisele certainly liked to play hard, and this became even more evident during the long weeks of training at the Cape. The Cocoa Beach party scene had always embraced the astronauts, and it warmly embraced Eisele. Traditionally, when launch dates were approaching the crews spent a large part of their time down there, and the social scene surrounding the astronauts grew accordingly. Cunningham remembers one big party that ended with Eisele trying to swim the three miles back to his hotel along a canal. "He wasn't unusual in that respect," Cunningham remembers. "A lot of guys had a 'home away from home,' if you will: a life away from home."

For those who were less careful, or didn't care to be careful, there was even more on offer for the astronauts, as Cunningham explains:

There was a little hanky-panky that went on every once in a while. In the sixties, being an astronaut was literally a license to steal in that area. And being cool, an astronaut, having your own T-38—picture this. Arriving in a strange airport, with a strange blind date meeting you, and you pull up in a T-38, come leaping down in a flight suit—here I am. Now that was cool! In fact, I can remember, the very first lecture I ever had, when I reported to the office as an astronaut—I think it was Wally Schirra and Deke Slayton took all of us new guys aside, fourteen of us. All of us were going to go down to the Cape, as none of us had ever even seen a rocket launch. They told us to really mind our Ps and Qs, because for the last three months the National Enquirer *has had a reporter assigned down there just to watch for astronauts coming and going at the Holiday Inn. So, whatever you do, you had better be awfully careful. That was the first lecture I ever had at NASA.*

The lecture did not stop many of the astronauts from enjoying the illicit pleasures the Cape offered. Although having a girlfriend in Florida and a wife in Houston was officially against the rules, it does seem that unofficial rules governed this behavior. Most importantly, the affairs should never become serious. If the girlfriend knew the score, she would enjoy the astronaut's company at the Cape but never expect him to leave his wife. The wives were not supposed to know, and if they found out, they were supposed to take it well, knowing that the affairs would never affect their Houston home lives. It was a finely balanced situation, and one that could not last. In the case of Donn Eisele, both his wife and the woman he began an affair with were not going to play within those rules. Harriet, for one, would confront the husband she felt was slipping away from her.

Donn's "second life" began long before, but apparently it was nothing lasting. I knew at the time of Matt's death that Donn was involved with someone else, although he denied it to me. He would leave me clues, deny when confronted, leave more clues and tell me I was crazy. By the time of the mission, he had almost convinced me I was crazy and imagining things. Finally, the uncertainties led me to tell Donn that, if I was really crazy, then I should go and see a psychiatrist. But he told me, "You can't go see a psychiatrist—I'll lose my job!"

During the summer after Matt's death the kids and I went to the Cape. Even then, Donn did not come home or spend much time with us. After that experience, I felt certain that something was going on, and asked to be quietly let out of the Apollo 7 *spotlight. I told him that I wouldn't rock the boat or disclose anything, but I didn't want to have the flight publicity in October followed by a divorce in November. Donn tried to convince me that I was wrong, appealed to Faye Stafford and others for help, told them that I was upsetting him and that this could impact the flight or get him killed. They almost convinced me that I was imagining things.*

But Harriet was not imagining things. Donn was dating other women. At first, they were casual affairs, but as the flight of *Apollo 7* neared, he met a woman who was never going to fit the traditional mold of "astronaut girlfriend." Susie Hearn was a professional career woman and a mother, who told the authors that she had never been part of the astronaut social scene at the Cape:

I was a single mother and the controller and treasurer of the Cape Kennedy Savings and Loan Association. The only astronaut I had ever met was Wally Schirra and that was at the Cocoa Beach municipal tennis courts. I was taking lessons from a pro, and when Wally came in one evening the pro asked us to practice hitting together. We did for maybe an hour—that was my extent of knowing other astronauts. I was never part of that scene. But one of my neighbors was in the social group at the Cape, and she became very friendly with some of the astronauts.

I met Donn by accident. One evening in June of 1968, I was having my hair done, and my neighbor Betty was too. As I was leaving, she called me over and asked me to do her a favor. She had a date that evening with a man called Donn Eisele, and she was expecting him at her door in fifteen minutes. She asked if I could please be there when he arrived, to let him in; she would be there as quickly as possible. I told her that my mother and sister were visiting me, so I could not stay long, but I could do it if it would not be for more than thirty minutes. I met him, fixed him a drink, and entertained him with jokes until Betty came home, at which point I went to my apartment for dinner with my family. About a month later I had a call from Donn, asking me out for a drink—I said yes.

Donn and Susie would go out and socialize, usually with his coworkers and their women friends. At first, Susie believed that Donn was single—after all, he had been dating her neighbor. She says that she did not know about Donn's disintegrating marriage. But as the relationship grew more serious, Susie believed the secret could be kept no longer:

Donn told me he was married around the end of August. I went to Portugal by myself for two weeks; I really wasn't interested in a married man. But things got out of hand. I knew I was getting in over my head by the middle of September, and I wanted to break things off with Donn after I came back from Portugal. Donn was very upset about that. Deke Slayton called me and asked me if I would hang in there until after Apollo 7 *flew. So I did, but I fully expected not to see Donn after the* Apollo 7 *flight.*

Walt Cunningham believes that the awkward situation did not affect Eisele's work until the last few weeks before launch. At that time, any concerns outside of the mission were supposed to be put aside, in order to focus solely on the upcoming flight. However, Cunningham watched his crew member disappear to Cocoa Beach at every opportunity during this time. Schirra also worried about Eisele and gave him a stern lecture about his personal life. Compounding the problem, Eisele was also maintaining his hobby of water-skiing, even in the last three weeks before launch, and Schirra felt that risking injury so close to launch was irresponsible. Schirra and Tom Stafford were concerned that a lack of focus might affect Eisele's future career, and eventually decided that a heart-to-heart talk with their colleague was needed. With Slayton trying to keep Susie in check with phone calls, Stafford tried to persuade Donn to stay in his marriage. According to Schirra, the difficult situation Eisele was in had begun to affect his work:

"It was kind of strange. We saw him slipping just before launch, starting to slide during the mission training. He wasn't really totally committed. He was falling in love with his Susie, that's why. He was already entranced with her. And so as a result, he wasn't really putting out the way he was supposed to. I made note of it, not to a great degree, but I made note of it."

What Donn Eisele was actually thinking at that time can now never be known. One part of the puzzle Susie is adamant about—if Donn was neglecting his training, it was not because he was spending time with her. She takes particular offense at a suggestion made in one account of the *Apollo 7* flight that Donn even snuck out to see her the night before the launch. "The crew is locked down the night before a flight," she maintains. "If Donn got out, it wasn't to see me."

The day finally came for NASA to show that all the lessons of the *Apollo 1* fire had been learned and to complete the mission that Gus Grissom, Ed

White, and Roger Chaffee had hoped to perform. On 11 October 1968, twenty-one months after that terrible fire, there was another burst of flame on Pad 34. This time, however, it was precisely directed, precisely controlled, as the Saturn IB's eight first-stage engines thundered into life. The crew could not feel the exact moment of liftoff; the booster shook the spacecraft so fiercely in its buildup to full power that the upward rise was hard to detect. This rocket was almost four times as powerful as the one used in the last Gemini launch, and Schirra immediately noticed the difference in sheer brute strength. Eisele later related his impressions of the moment of ignition: "You can tell things are happening; . . . you don't know what the sequence is, but you can tell, feel and hear, at very low levels some fluids running around, and you can hear the engines thumping and feel the whole stack shake a little, and you know something is going on down there. . . . It's very evident that the whole vehicle sways. . . . We could see the whole vehicle sway; at least, I could. I could look out and watch the White Room structure go back and forth." Schirra, who could compare this with previous launches, expressed his feelings on launch even more vividly. "It sure was vibrating up a storm," he would relate postflight. "I thought the world was coming to an end." As the powerful forces thrust the spacecraft and booster on a steadily accelerating path into space, the first manned Apollo flight was underway.

Despite the successful liftoff, Schirra was not happy. He had been fighting a safety issue up until the hour of launch, and it did not go his way. The couches that the crew would sit in during the mission were of the old *Apollo 1* design, and a newer, safer version had not been readied in time for this flight. The old couches were not designed to adequately protect the crew should the spacecraft abort during launch and come down on land. Schirra had insisted that the rules on wind conditions for launch be revised, so it would not take place if there was a chance the spacecraft could be blown back onto land during an abort. As the countdown had neared the moment of launch, it became obvious to Schirra that the winds were outside of the safe margins agreed upon—but no one was planning to stop the countdown.

That was not the time to launch that day, and I didn't want to. Those were the wrong conditions: they broke the mission rule that we had established. We were not to launch if the wind could blow us back over the beach, which would then force a land landing if we had to abort. That would essentially have been a death penalty. The winds on launch day were such that they would have blown us back over the beach. There was no problem with which day we launched. It was really a case of someone wanting to go. I fought that, until I became rather difficult, and I finally yielded, with great concern. I conceded when we got to about T minus an hour and counting, when I realized that this could be a hard one to redo. But they were the ones who should have called the shutdown, not me. I tried to play it light. We launched because everything else seemed to be in good shape. It was one of those things, you say, "Okay, we'll take a go." But I was furious.

During ascent, the boost protective cover blasted away as scheduled, revealing a spectacular view as they climbed ever higher. Although he was monitoring their trajectory on the spacecraft computer, Eisele had less to do during the boost phase than Schirra and Cunningham. Fascinated, he took the time to take in the impressive sight. Schirra watched his two crewmates with amusement as the rocket went through staging—he'd done it twice and thoroughly enjoyed observing their surprised reactions as they were slammed forward. Eleven minutes after launch, *Apollo 7* was in orbit.

During the mission, the spacecraft's data storage equipment taped everything the astronauts said, even if they were out of communication range with the ground. The previously classified transcripts make for fascinating reading. During the first orbit, out of touch with the ground for much of the time, the transcripts reveal an experienced astronaut, used to the sensations of launch and of orbit, talking two rookies through their first awe-inspiring glimpses of the space environment. Half an hour into the flight, Eisele told Schirra, "You know, I haven't even hardly looked out yet," to which Schirra replied, "Okay, why don't you take a minute and look out. It's much fun." Soon, Eisele and Cunningham had both spotted fires on the shadowed side of the planet far below them—beautiful "orange spots," according to Eisele—and something else that "looks like stars if I were looking at a simulator star ball . . . except I can't see stars because I'm looking at the ground, and they're not moving." Schirra, who had seen it all before, explained that Eisele was probably seeing lightning from above.

Both rookies laughed at how much they were enjoying being in orbit, with Eisele adding, "Oh, it's a good feeling; . . . it's too bad that everybody can't see this." Schirra, clearly enjoying the role of mentor, quipped, "Few nice guys like me want to hang around and go again!"

Eisele would soon be busy—he was scheduled to be first out of his seat, to head down to the guidance and navigation station, ready for the rendezvous. Cunningham had suggested that Eisele perhaps unstrap before they were in contact with the ground—but Schirra suggested that they wait until approval was given by the flight directors. "I'm sure we are go—but let's play honest," he told Cunningham. With ten minutes to go until they were in touch with the ground again, Schirra suggested the two rookies continue to relish their new environment and the views it afforded. "I'm telling you something, guys," he said with a laugh, "let's enjoy these few minutes." Eisele took in the ever-changing views, which at first were hard for him to decipher. "That's weird," he would say, "I see two points of light on the ground. Looks like the stars are going through the Earth again." In those first hours of flight, Schirra pointed out thunderclouds and sunrises to them, adding comments such as, "There you go: that's the thrill of this business." He even took the time to repeat and chuckle once more over Eisele's pun about Guenter Wendt.

One of the crew's first tests was to separate the command and service module (CSM) from the Saturn booster's second stage and then turn around and fly back to it, simulating the extraction of the lunar module that would take place on later flights. Schirra, pulsing the thrusters to stay in place with the spent rocket stage, came within four feet of the booster. He was delighted; back in space for the first time in almost three years, he was doing some real flying again.

When he first jettisoned the module's exterior dust covers for the optics system, Eisele was captivated by the beauty of what he saw. "Boy, was that ever pretty!" he explained. "I never saw anything like it; . . . the two covers, sitting out there twirling around, still reflecting sunlight." It had also been a moment of some tension, Eisele later explained. "The first surprise I got from the optics . . . I was told that twenty degrees of shaft rotation would kick the covers off, and it took about 180 degrees before they let go. I had my heart in my mouth there for a few seconds until they finally went, because I could envision the thing not coming off. . . . However, they did pop off, and I could see the little pieces with sunlight reflected off them disappearing as they drifted away from the spacecraft." Using a combination of telescope, sextant, charts, and support from ground controllers, Eisele took on the demanding task of testing the guidance and navigation system to align the spacecraft. It was a job that would need to be done solo during lunar missions, so Eisele worked as an independent unit. It was quite a responsibility. With every test, he learned something new; sometimes, it was a lesson in the limitations of the equipment. In the second hour of the mission, he looked too closely toward the sun. "That sun is brighter than hell," he discovered. "I just caught a corner of it in my eyeball." He also learned that looking outside for a long time required him to adapt to the darkness. "I am having a hell of a time adjusting," he stated. It would in fact take him four or five minutes for his eyes to adjust to space. At first, he could see only the brightest stars, but as his eyes adapted further he could discern the dimmer ones as well.

It was planned that the *Apollo 7* crew would use the spacecraft thrusters to move into a slightly different orbit than the booster, then undertake a number of rendezvous exercises with it. This meant firing the CSM's big engine for the first time, to enter a new elliptical orbit. The engine burn only lasted ten seconds, but it pushed the three men back into their couches with unexpected force. "We didn't know quite what to expect," Eisele recalled after the flight, "but we got more than we expected. . . . That thing really comes up in a hurry, and it's a real boot in the ass when it lights off. It's not a smooth thrust rise . . . man, it comes all at once, smacks you right in the tail and just plasters you to the seat." Schirra described it as "a catapult shot . . . you know you have got to have your head back, or you're going to get snapped." He was delighted with the power this new-generation spacecraft gave him. He later described it as being able to fly "a ballet in space."

Firing the big engine again, and using the smaller thrusters for minor corrections and braking maneuvers, *Apollo 7* rendezvoused with the booster stage, which was now tumbling and steadily losing orbital height. The booster was large enough so that the faint traces of atmosphere 140 miles above the Earth had reduced its altitude by over four miles in just one day. Eisele was delighted that they had pulled off the rendezvous. "Break out the champagne," he shouted. "Goddamn it, we made it!" He also noted that the close maneuvering with the booster was "rather like one car overtaking something, but a car with very weak brakes and not much acceleration."

Schirra was pleased at how much easier it was to rendezvous using an Apollo spacecraft than it had been flying Gemini. This far more sophisticated machine allowed a much wider range of options when it came to navigation and rendezvous. Even so, some of the magic was gone for him. Piloting a spacecraft far bigger than the others he'd flown, he noticed something else that only an aviator would naturally sense:

Apollo was a big, unwieldy vehicle, like flying a big transport plane, which fighter pilots don't really revere. It was a pretty big vehicle, like driving a big bus. I had a problem with the flight controllers over that. I said, "I am tired of changing attitude up here, we are being affected by the atmosphere." They said, "What do you mean?" It was such a large vehicle, it would try to fare its way like an airplane, causing the spacecraft to try and get the trimmest attitude. They didn't ever anticipate that, and there we were, very sensitive to anything that caused the vehicle to move. You don't read that on instruments, you have to feel it.

With the rendezvous maneuvers completed, the crew undertook other tasks to give the new spacecraft a thorough shakedown—navigation, attitude control, and guidance. They would also continue to thoroughly test the propulsion systems for the remainder of the flight. The majority of the spacecraft tests were of the environmental, electrical, and thermal control systems, and these also checked out extremely well. The redesigned Apollo spacecraft was better than anyone had dared to hope. It was also far roomier than the Mercury and Gemini spacecraft, and in space the lack of gravity made it seem even larger. Its entire volume could be used. It was even possible for an astronaut to float into the lower equipment bay and be out of sight of his colleagues.

Eisele found that it was not as easy to navigate the spacecraft using the stars as had been hoped, however. One persistent problem occurred whenever the spacecraft dumped water; the glittering frost particles made star identification difficult. "At certain times, notably near sunset and sunrise," he explained, "if you happened to be dumping fluids through the overboard dump at the same time you are looking through the optics, you will see snowflakes out there which look very much like stars. You see a whole field of them, and it just obscures the entire star field. You can't tell the stars from the flakes of frozen particles." When looking out the window, Eisele could even see the entire spacecraft's shadow silhouetted on the drifting field of ice.

Because Earth's horizon was indistinct in such a low orbit, he also found it was hard to use rising and setting stars to obtain any kind of navigational fix. "The Earth horizon is a fuzzy kind of a thing," he explained after the flight. "It does not appear as if there are any precise lines or delineation that you could use reliably. . . . I just could not pick anything out in the airglow. I could not pick out what looked like a horizon at all." Nevertheless, Eisele worked diligently to test the equipment, learning what was possible and what could be improved as he went. He was also still getting used to weightlessness. "I picked this box up," he told his colleagues on the first day of the mission, "and I'd swear there is nothing in it, because it doesn't weigh anything." In an admission that made all three of them laugh, Eisele told them he even opened the box and looked inside, still not believing his senses.

There was one aspect of the mission that was not well planned: to have at least one astronaut awake at all times. The spacecraft was roomy, but not *that* roomy; it was impossible for astronauts to work effectively without disturbing another crew member sleeping just inches away. As a result, none of the astronauts slept well for much of the flight. "It was the first time that we ever had this spacecraft up," Cunningham remembers, "and we were in contact with the ground 5 percent of the time. So it was important to have somebody on watch. It was logical for us to do that. We couldn't sleep for the first couple of days anyway, so it was really a tortured sleep schedule. I don't think there was a complete appreciation at that stage of the space program of the business of not jacking the schedule around too much. They were pushing our daily cycles all over the place."

The plan for this mission had Schirra and Cunningham sleeping at the same time while Eisele remained awake. It was hard for Eisele to keep awake in a quiet spacecraft with his crewmates asleep, however, and he took an amphetamine medication on one occasion to try and stay alert. "Everything was quiet," he later described, "particularly on the dark passes, on the dark side at night." When it was his turn to sleep, the others working made enough noise to keep Eisele awake. Cunningham remembers waking up to see Eisele floating, pencil in hand, looking like he was working—when in fact he had fallen fast asleep in mid-task. "It was harder to stay awake when you were by yourself," Cunningham remembers. "He complained when the two of us were up, because of the noise. We whispered—but trying to whisper for eight hours? Impossible." When he did sleep, Eisele dreamed, Harriet remembers. "He did tell me that during the flight, he dreamed about Matt."

On the first day of the flight, Schirra came down with a bad head cold, and soon Eisele caught it too. "I went down in the LEB [lower equipment bay] and blew my nose for a couple of hours," Eisele would later joke. The pure oxygen atmosphere did not help, nor did the weightless environment, meaning stuffed noses and sinuses stayed stuffed. Though some accounts of the flight state that the entire crew caught a cold, Cunningham says he never had it. This is borne out by the mission transcripts, which have him stating that "so far I have been able to resist . . . getting these things, but Donn's coming down; . . . if there is some way I can hold it off, I would just as soon take . . . pills." He even joked about it during the flight, saying to the ground, "Tell the doctors not to worry about the cold; . . . it takes a week to get rid of if you treat it, and only seven days if you don't."

"The only reason they said everyone caught the cold," a smiling Cunningham told the authors, "is because if Wally had a cold, *everyone* had a cold! Wally really was miserable—but I never caught a cold. I think at the time Wally felt that having a cold might justify some pretty damn obnoxious behavior, in my opinion, and if everybody had a cold that it wouldn't be a reflection just on him. Donn was not feeling too good either, but he didn't have what I'd consider a bad case of the sniffles or anything. Other people don't know; a lot of it is hearsay, from what was said right after we came back."

The lack of sleep and bad colds probably contributed to what happened next.

Schirra believed in sticking as much as possible to the flight plan, the one he had agreed to during the mission planning. During the flight, Mission Control was asking that some tests be moved in the flight schedule, and this was beginning to irritate him. His exasperation first reared its head when the ground asked him to reschedule a television transmission from the second day of flight to the first. The transmission would be a great boost for Apollo's public relations, but Schirra was more concerned with the engineering safety aspects of the mission, he relates:

I eventually said, "Okay, if you're going to be violating rules, guess what I'm going to be doing! I am going to judge these rules from now on. If you are going to break that rule and not give me a chance—then I am going to break some of the rules that you have given me problems with." I didn't want to do things that hadn't been tested in their proper sequence. We were to test the circuit of the television set prior to using it. That was scheduled for one day; then the next day we were to turn it on. But the day they requested us to play games with the television, I was trying to do a rendezvous with the booster. I didn't want to mix that up with something else that was not important. They wanted the television on a particular day, and it wasn't scheduled for that day. I said, "We'll put it on tomorrow." That made sense to me—but not to them. I was pretty annoyed with Chris Kraft and the flight directors, who were ordering me. I said, "You don't order a commander!" That's why we called the position a commander, after all, not pilot and copilot. The pushing around should only be done by the engines and thrusters, not the flight controllers. But having said all that, I felt a lot for the flight controllers, and worked with them, not against them.

Certainly, the vast majority of Schirra's conversations with the ground were polite and professional and he explained any problems clearly. "I believe that television should be left as the last low-priority test objective in relation to any other event that may occur simultaneously," he explained from orbit. "We were paying much too much attention to the TV camera and not to the spacecraft." When this point is considered objectively—that in a front-loaded mission the rendezvous, alignment, and engine tests should be done before television shows—it is hard to argue with him.

Schirra did it his way, but he did it well. Despite feeling lousy because of his cold, he rose to the occasion every time the television camera was switched on, and with his trademark humor gave the viewers quite a show. Unfortunately, Eisele was the one who had to give Mission Control's original request about the television to Schirra, and so, as Schirra explains, "Donn sort of got caught in the middle of that one. I heard him say to the ground, 'Wally's not going to like it.'" Eisele had been placed in a difficult position between two different people who believed they had control over the flight objectives. It was a no-win situation for him.

Meanwhile, Eisele was providing valuable feedback about the navigation systems. On the third day he told the ground that "In general, the optics drive is very smooth and much better than we have seen in any of the simulators. The visibility, however, in the telescope is no better: in fact, a little worse than what we experienced; . . . it was most disappointing to find that you really do have to bury your eye in that thing for a good five to ten minutes . . . to get dark-adapted before you can recognize anything. . . . Without a well-defined pattern, it's pretty hard to find anything by itself."

Gene Kranz, on the night shift for this mission, felt he was getting the best deal of any of the flight controllers. He did not experience the disagreements with Schirra heard on other shifts. Instead, he enjoyed listening to the near-whisper of Eisele, the only person awake in space, as he identified stars and described what he could see from his unique vantage point. Kranz was surprised to hear what Schirra was doing during other shifts. Until that flight, he says, Schirra was one of his favorites to work with: "Wally, up until *Apollo 7*, was as good a pilot as anybody has ever had. He was the first one to really close the loop with the ground. He was also the first one that spent enough time on his flight plan . . . to understand what his plan was, how he would fly it, and what aberrations to the plan could possibly occur. He got on with it—Wally did a great job!" Kranz's initial good opinion of Eisele would waver a little during the mission. As the flight continued, he was saddened to note that the previously mild-mannered Eisele began to change his attitude, complaining about a test concerning a horizon navigation fix, for example. According to Cunningham, Eisele was simply following the style set by his commander, got "caught up in the swing of it," and started to "throw a little weight around." A reading of the mission transcripts reveals a clear difference in the way that Eisele talks about the tests as the mission continues. On day five, he explains why a test did not work, and follows it up with "the gist of it all is that I don't think it was too worthwhile or realistic a way to perform that program. It wasn't designed to be used that way."

By day eight, his tone had changed, and Schirra warned the ground during one transmission that they were about to "get some sweet remarks" from his colleague. "This horizon jazz doesn't mean anything," Eisele grumbled. "Hell, it's all going to be one. . . . We didn't get the results that you were after. We didn't get a damn thing, in fact; . . . you bet your ass . . . as far as we're concerned, somebody down there screwed up royally when he laid that one on us." Schirra believes that this was the one major occasion when Eisele upset the ground. "I think Donn got caught on that one, where they screwed up our onboard computer," he recalls. "Donn said, 'Who in the hell sent that up? What idiot sent that command up?'" It is important to look at why Eisele was annoyed. In a situation where the crew's lives might depend on their alignment, Mission Control had decided to try a new, untested procedure, which had the effect of calling up program alarms and forcing a system restart, as the computer tried to perform using square roots with negative numbers. As NASA's official mission report concluded, "erroneous procedures were given to the crew." Eisele had spent years developing procedures to test the spacecraft in orbit, and now last-minute ideas were ruining his efforts. Anyone who cared about his or her work would be angry in such a situation.

Harriet, listening on the "squawk box" NASA had provided her, which allowed families to listen to the in-flight communications in their homes, was disturbed by what she heard. "I had mixed emotions during the mission: concerned and scared for Donn and the crew, sadness over the loss of Matthew just six months earlier, and sadness over suspicions of Donn's 'life' at the Cape. Listening to the squawk box during the flight, I knew something was drastically wrong. I can't cite specifics; it was just not the Donn I knew."

During the mission debriefing, Eisele calmly explained what happened: "This was a rather hastily conceived test, I think, and was not properly checked out on the ground before it was read up to us. When I performed the exercise, we got a program alarm, a restart light . . . and the computer hung up, froze such that we could not make any entries into it with the keyboard; . . . that indicated that there was something badly wrong with trying to do landmarks when they are out of Earth's horizon." Eisele knew the system thoroughly, however. He worked out a procedure to reset the computer without any help from the ground and was able to bring the system back without any damage. Yet it was the tone of his voice, and not his actions, that seemed to be noted on the ground.

His testiness was not only confined to conversations between the crew and the ground. According to Cunningham, one moment when Eisele was working on verifying a program on the guidance computer Schirra became anxious about moving on to the next mission objective. This developed into an argument in which Schirra accused Eisele of "threatening mutiny." The row soon blew over. Cooped up in a spacecraft together, the crew had little choice but to make up. "I was flabbergasted when he finally resisted one of Wally's ideas," Cunningham recalls. "It's a shame that this was the first time he'd stood up to Wally, because if he had before, there would have been two of us, and we'd have been in control of the situation a lot better. But Donn always just kind of went along with things. That was just Donn's character." Cunningham believed that if he and Eisele had really focused their efforts on Schirra, they could have changed the way he was reacting to the ground's demands during the flight. But Cunningham later admitted that was never going to happen:

Have you ever been in the military? You don't say no. It really depended on the commander's personality. With some guys, you could do it: for example, you could handle things differently with Gordo than you could with Wally. It was very tough to influence Wally Schirra. I think it was because Wally, in spite of everything, had to demonstrate that he was in charge. Pete Conrad was always in charge, but he never had to demonstrate that he was in charge. He was just in charge. With Wally, it was important for him not just to call all the shots, but for the ground to know he was doing it. Wally still stands behind his particular attitude, but it wasn't an attitude that we shared at all. I don't think that that there was disagreement between the Apollo 7 *crew and the ground; I think there was a lot of disagreement between Wally Schirra and the ground. I didn't agree with it, I just tried to stay out of the controversy; after all, I was a rookie astronaut. Wally, bless his heart, to this day thinks he was right. I have heard him with others defending his position. Wally and I have just agreed to disagree on that particular point.*

Certainly, after the flight, any disagreement between Eisele and Schirra seemed forgotten. In fact, Schirra praised Eisele for this work, saying

Donn was our navigation expert, and he made the guidance apparatus sing. Although he'd started sliding during the mission training, this did not affect

the mission itself. During the flight he came back on again, no problem. The mission went fine, he didn't do anything wrong. In fact, he saved the computer a couple of times, when the ground screwed it up. One time the computer went down, he fixed it between ground stations. I love that! The ground said, how did you fix that? I said—you find out! It took them two orbits to work it out! So I was pretty pleased with that.

In the postflight technical debriefing, Eisele described what he said was a good working relationship that he and Schirra had shared during the flight. "We both checked each other," he explained, "and we did it on a simulator. One or the other would start to do something, start the wrong way or forget to do it when you should, and we've always just kind of overlapped each other that way." Cunningham agrees, noting that Eisele "knew the guidance and control system and had a good grasp of the mission and its objectives." He also says that Eisele was sufficiently cross-trained to fill any of the three crew positions, including that of commander.

Over the decades, the *Apollo 7* flight has been unfairly tarnished as one on which the crew "mutinied," as if the crew had been constantly hurling insults at the ground. The truth, demonstrated in the transcripts of the space-to-ground transmissions, is very different. They show a crew working carefully and diligently to carry out a thorough test flight of a new machine. There are very few disagreements, and those few were usually accompanied by a clear explanation of why there was an issue. It was not the military precision and exact conformance to orders that had characterized earlier missions. It was the human side of spaceflight: three men working to complete a flight schedule in which they had invested a huge amount of effort. Though each of the crew members sometimes spoke very frankly about something they disagreed with, it was always about a specific point—usually the addition of a new experiment to the flight schedule. This manipulation of the plan is something that Kranz freely admits the mission controllers were doing to the crew: "We really put the spacecraft through its paces; . . . we had a single flight test to do it, and . . . we kept piling a lot of stuff on Wally, because with only one test to get the job done, every time we saw an opportunity . . . we'd go for it, we'd press it, we'd try to get some more testing in there."

When reading the transcripts, joking and general levity between the ground and the spacecraft are far more common than testiness. "The problem is," Cunningham explains, "for eleven days, with people calling down, back and forth, you don't always know who is saying what. Because you never identify yourself. I've gone back and looked at the transcripts, and I never found anything I'd ever said that could have been interpreted as nasty, with the exception of one comment. Given the nature of what was going on, someone said 'Well, there's another example.' Well, it wasn't."

One thing is certain: after a couple of days, Schirra had grown bored. The mission was going so well that the primary test objectives had all been accomplished at the beginning. "That was not uncommon," Cunningham remembers: "They front-load everything." *Apollo 7* would be the first "open-ended" mission, NASA managers had agreed before the flight. This new philosophy was designed to obtain maximum return from a mission. It meant that a flight could still be successful if it had to return early, but it also meant that the end of the mission would drag on interminably. For Schirra, a fighter pilot who saw an hour-and-a-half jet flight as a long mission, flying for eleven days soon began to feel like a chore.

By day four, their onboard exercise equipment was beginning to fray, cloth footwear was starting to unravel, and microphones and ear tubes were cracking due to heavy use. "Some of our equipment is starting to show signs of age," Schirra told the ground. His cold was not going away either, and this soon led to the biggest disagreement between the commander and the flight controllers. At the end of the mission, the crew was supposed to reenter wearing not only their spacesuits but also their spacesuit helmets, in case of a loss of cabin pressure. But Schirra was concerned that, if he was not able to blow his nose to equalize pressure, he could damage his eardrums. "We feel the risk of rupturing our eardrums is higher than the risk of injury from not having the suits on," Schirra explained to the ground. It had been an ongoing discussion throughout the flight, in which the ground had even admitted, "I think that's all subject to some discussion. You guys have got a better feel for that than anyone else." This conciliatory attitude did not last.

Toward the end of the flight, when Chief Astronaut Deke Slayton came on the radio to request that Schirra and his crew put their helmets on, Schirra must have known it was serious. But Schirra had made his decision, and the helmets stayed off. At the ground's insistence, the crew taped food bags to their headrests, then put towels both in front of and behind their necks to ensure their heads would be adequately supported. "We tried to figure out every way we could turn our neck or head with those helmets off," Eisele later explained, "and the only thing we could see was perhaps banging our chin on the front of the neck ring."

It is a command decision that Schirra still strongly defends:

We had severe head colds, and we agreed to come back with our helmets off, so that it avoided the mucus causing tremendous ear pressure. I mean, the command module is the most beautiful structure ever known. All of those structures that were designed by Max Faget were outstanding. We were not going to lose pressure in there: we had just lived in there for eleven days. We could get to the helmets if something happened slowly; we could put the helmets on. The ground was really giving me hell—but they couldn't come up there and make us. Houston, you have a problem!

For the last time, the engine that had been tested so thoroughly over eleven days made another precise burn, and the crew began their reentry sequence, separating from the service module soon afterward with a sharp jolt. Eisele was surprised at how early the heat of reentry became visible out of the window. A "pretty . . . very strange" effect, was how he described it. "A light pink, like a pink sherbet. It was very pastel." As the spacecraft plowed further into the atmosphere, Eisele watched the ride become increasingly rough. Chunks of the shield appeared to be breaking away and disappearing into a glowing, twisting, tube-like vortex in *Apollo 7*'s wake: "The spacecraft was flying itself like mad back and forth; . . . we'd been pulsing pretty heavy in pitch and yaw, . . . and the system was fighting all of these movements that were going on. All of a sudden, this big hunk of whatever it was went, pow. I was looking out of the hatch window, and this big hunk of yellow. . . . I sat over there saying 'Look at that fireball!'"

For a moment, the disintegration was so jarring that Eisele thought a pitch thruster had exploded. Schirra, also surprised, wondered if the outer pane of his window might have blown out. To their relief, it turned out to be just a larger part of the heat shield melting away. "Apparently, what happened," Eisele recalled, "is that one or more chunks of the ablator came off rather explosively. I don't know how it happened; they just blew off. At the same time, we were into the atmosphere far enough that we had aerodynamic standing. . . . I think that's what happened, and we got faked out."

The heat shield held perfectly, and almost eleven days after they had launched, *Apollo 7* splashed down in the Atlantic, where they were recovered under rainy conditions by the USS *Essex*. The spacecraft pressure had stayed well within safe limits during reentry; the helmets had never been needed. Eisele had been nervous about the splashdown—the mission controllers, while trying to persuade the crew to put their helmets back on, had stressed the dangers of a hard landing on water. "From the time they started giving us that," he explained, "I kind of quit worrying about the chutes opening; . . . we were all sitting there . . . scared to death, hanging on to our shoulder straps, heads tucked in, and just waiting for the worst possible happening." After the flight, Schirra and Eisele were angry about being put on their guard by the ground. "That is absolutely ridiculous to put a crew in that frame of mind," Schirra fumed in the postflight debriefing. He was ultimately proved right—when the *Apollo 11* crew came back from the moon they didn't even wear their spacesuits, never mind their helmets, for reentry.

Splashdown had also been a moment when Cunningham felt anything but an astronaut hero. "There is no question," he joked, "that when the space shuttle comes back and lands on a runway, and they walk out—that's a much better way to return as a hero than to be dropped in the ocean like a bag of cats!" Eisele wasn't feeling too good, either. "Queasy, hell!" he explained after the flight. "I got dog sick after about five minutes." Instead of announcing he was about to throw up, he asked his crewmates if they were feeling sick too. "I was looking for company," Eisele later quipped. "I felt very green, but not with envy."

Despite the bad weather outside, it was a relief for Eisele to exit the spacecraft and feel the rain on his face. "Boy, did that feel good!" he recounted. "I got sopping wet just from getting out of the spacecraft and getting in the raft. I was just about shot. If I had had to put on one of those airtight suits that they are talking about for the lunar returns, I would probably have passed out." Schirra, concerned about his crew member, was pleased to see that Eisele felt much better once he was out of the confining spacecraft. "Donn was in bad shape and needed some fresh air," he explained. "We got into the chopper; . . . right away Donn recovered, almost immediately, and I think he said, 'I'm now back in my element.'"

The flight had been virtually faultless from an engineering standpoint. The service module engine had been test-fired eight times, and the flight had tested rendezvous techniques planned for both Earth and lunar orbit. All of the spacecraft systems had been tested for a duration equal to a lunar mission, and they had held up very well. In all, the crew had traveled four and a half million miles and splashed down a precise quarter of a mile from the target point. Not only had all the test objectives been completed, but extra objectives not planned before the flight had also been achieved. The spacecraft faults discovered had been minor. Problems such as biomedical harness wiring being too thin and easy to break, moisture forming on coolant lines, and fuel cells with high temperatures were easy to fix. Dale Myers, the program manager for the command module at the time, remembers Schirra presenting him with a small piece of Teflon debris that Schirra had found floating in the cabin. The mission had been a great success, but Schirra was a perfectionist. Schirra's gift was his way of telling Myers—not quite perfect, but good enough.

Donn Eisele had certainly done an outstanding job testing the guidance and navigation systems. He had repeatedly powered down the system, then used the scanning telescope to find constellations to realign the spacecraft platform. Using the telescope and sextant, he'd outlined the limits of useful star visibility while the spacecraft was on the sunlit side of the Earth, and tracked landmarks on the Earth below to check the accuracy of the alignments. He'd discovered that it was very hard to use the Earth's horizon as a way to gauge star sightings, but had found the moon was a highly accurate measuring tool. The craft's rendezvous maneuvers were based on his sightings using the guidance systems. During the flight, Eisele had tested many backup options, in case systems failed on a future flight. He'd even successfully come up with independent solutions when ground controllers had sent him incorrect information and crashed the computer.

He had accomplished all this despite an unrealistic sleep schedule, which had never let him adequately rest. Immediately after the flight, Eisele underwent a physical that concluded that he was suffering from "excessive fatigue." He had given his all to contribute to the success of the mission. Despite all of the difficulties, it had been an incredible experience for him, as he later reflected in a speech at the Florida State Museum:

The first thing I noticed was that there weren't any political boundaries. You couldn't tell where Morocco ended and Algeria began. You couldn't see the border between China and India, or between North and South Vietnam. But you could see the colors and the beauty of Earth. You could see the pale blue and the pale greens of the ocean. You could see cloud forms swirling and moving about, thunderstorms which light up at night like light bulbs. You could see the land masses and the continents, the browns, the grays, and rust colors of the deserts and mountains and the deep dark green of the valleys.

From outer space one is keenly aware of the beauty of the Earth; . . . from the spacecraft you can see the Earth is small, beautiful, fragile, and unique in the cosmos, unique because it is the only place you can go back to when you have finished your mission. It's all you have; . . . the Earth itself is nothing more than a large spacecraft making its way through the universe and we humans are traveling on board.

Apollo 7 was such a successful flight that it boosted plans, already in place, to change the destination of the next Apollo flight. *Apollo 7* allowed humankind to feel confident enough to journey to the moon, as Cunningham remembers:

Apollo 1, *as a spacecraft goes, really was a piece of junk. But we went from that piece of junk to the most magnificent flying machine, and we went eleven days on* Apollo 7. *I don't believe anybody is ever going to schedule an eleven-day first flight again. It seemed ridiculous—we never expected it to last the full eleven days. We thought something was bound to go wrong and cause us to come back early: but it didn't. I think we had one minor emergency in the whole eleven days; some electric power problem on one of the buses or something, it didn't take too long to fix up. It was a wonderful machine.*

With all the successes that followed it, the first Apollo flight is now relatively forgotten, and Cunningham accepts this as part of the natural course of history:

We had a mission that had never been done before. Everybody was excited, and there was a lot of national pride and a lot of money behind it. It was a very important flight, and it was a near-perfect mission. The most important thing? It's hard to say, because every system was important, but the most important thing was to demonstrate that the engine worked like it was supposed to, and the spacecraft could do a good job for the duration of a lunar mission. If Wally had not had the attitude that he had, it might not have been "a lost mission" as much. But Apollo 8 *was such a dramatic mission occurring right after it, that it was really easy to forget about 7. Now, it is really easy to forget about almost all of them. When was the last time that you heard* Apollo 9 *mentioned, for example? Yet it was one of the most technically challenging missions we ever did. Unless you went to the moon, people don't pay a whole lot of attention.*

But you can't hang up on all that stuff. It's just a case of being able to look back, and put it in perspective for what it is. I don't know how it looks to the outside world, but to me it seems perfectly natural. It's nice when people do remember, however, because we really do think that it was a tremendous mission. When I think back on it, I don't think many of us were very introspective about it. We were fighter pilots; we had the good fortune to be up there flying. All we really wanted to do more than anything else was not screw up. We wanted to be able to come back and have our peers think that we did a good job. Every single astronaut that went into space I think must have had this same feeling—if this mission is going to fail, it is not going to fail because of me. So I didn't think a whole lot about it during the time. And I may be just less sensitive than some of them, but I don't think so. We were physicists, engineers and we just kind of did our job.

Some of the other astronauts also wondered what the fuss was about when it came to the crew's attitude. Bill Pogue was assigned to the *Apollo 7* support crew, and also served as a CapCom for the mission. When asked by NASA historian Kevin Rusnak what the crew was like to work with, he replied: "Very good. Wally, of course, you couldn't help but love the guy. . . . I've heard disparaging remarks about him, but he just sits, tells it like it is. Everybody liked working with him, unless you got crosswise with him. With Walt, it was very efficient, and Donn Eisele was a sweetheart to work with. I liked them all. That was probably one of the coolest crews we ever launched." In 1998, Gene Kranz also told reporter Roy Neal that "we all look back now with a longer perspective. Schirra really wasn't on us as bad as it seemed at the time. . . . Bottom line was, even with a grumpy commander, we got the job done as a team."

Nevertheless, after the flight, Schirra was given a dressing down by a very disgruntled Deke Slayton. Slayton was evidently far more annoyed at what Schirra may have done to Eisele and Cunningham's careers than at his own actions. Chris Kraft had made it clear to him that he would not work with a future crew that included any of the *Apollo 7* members. He felt that Cunningham and Eisele should be included with Schirra in this ultimatum, despite the fact that this was their first flight. Slayton was not going to use a crew that Mission Control would not work with. He would offer backup slots to both men, but never formally offered either of them another prime crew position. "There was a considerable shakeout after that flight regarding the subservient crew position," Pogue recalled with a laugh. "We were working for Kraft or Gilruth or whatever. Going off and exerting that much individualism probably wasn't too popular. So we got the message." As Schirra had predicted, the bureaucrats were putting an end to independent-minded commanders once and for all—and anyone associated with them. Rusty Schweickart thought that the situation was unfair: "It was a very difficult position for Donn Eisele and Walt to be in. I think that partly because they didn't counter or contradict what Wally was doing during the mission, people assumed they had agreed with him."

Cunningham and Eisele were not invited to Slayton's talk with Schirra—but they were reprimanded in more subtle, yet ultimately just as final, ways. Both were given assignments that made them believe they had a good chance of flying in space again. Neither of them did. As Jim Lovell told the authors,

Apollo 7 *was a very successful flight—they did an excellent job—but it was a very contentious flight. They all teed off the ground people quite considerably, and I think that kind of put a stop on future flights. Wally was finished anyway, he wasn't going to fly anymore, but for Donn Eisele and Walt Cunningham, that slowed down their future spaceflight careers, which is I think unfortunate. But that was the reason, I think, more than a pending divorce. I don't think Donn's married life had as much to do with it.*

Schirra agrees that Eisele and Cunningham suffered because they were perceived as following his line during the flight: "Oh, of course, they did. Unfortunately, Walt and Donn suffered the consequences, because Chris Kraft at that point was getting much more power than he should have had. I felt trapped by those people, and I really feel that Walt and Donn did suffer by my having made that decision to do what I was going to do.

They were a good crew. Donn was very much alert. . . . I was pleased with the crew; we all did a good job."

At first, Cunningham did not realize that his active astronaut career was already over. In fact, he had been reassured by some NASA managers that the facts of the flight were understood and that he was to be held blameless. He'd hoped to be assigned to a lunar mission, but was not overly surprised to instead be given responsibility for the Skylab program. At the time, that program was, as Cunningham remembers, "a mess . . . the poor relation of the manned spaceflight program." Budget cuts and lack of focus meant the program was in sore need of shaping up. Cunningham remembers that every crazy, untried idea the engineers could think of was planned for Skylab; someone was needed who could work with test pilots, engineers, and scientists. Cunningham took on the new task with vigor, and the program gradually melded into a coherent whole. "I coordinated the operational development, system integration and habitability of all Skylab space station hardware, including five manned space modules, two launch vehicles and fifty-six major experiments," Cunningham states with pride. "I had about forty percent of the astronaut office working for me at one time." He took on the project on the understanding that he would command the first mission. However, the promise did not hold. Cunningham was eventually told that Pete Conrad and Alan Bean had been given command of the first two missions. Instead, he was offered the position of backup commander. It wasn't what he wanted—he wanted to fly in space again. Instead of taking the job, he resigned in August 1971.

"When I left NASA," he remembers,

I was somewhat self-conscious about being an astronaut, because all of a sudden I was out in the business world, and it was hard to get past being an astronaut. I felt like, you know, don't these people realize I am more than just an astronaut? It took me probably a dozen years to get past that—I had just a little bit of resentment that I couldn't be seen simply for myself. That was probably not right, on my part, because for the last dozen years I have finally begun to put the whole thing in perspective. I realize now that I had an opportunity, as very few people did, to play a very small role in a truly historical event. When I was an astronaut I was just living it, and I think all of the other guys were probably the same way. We had flown airplanes, and we loved them, we were still flying nice airplanes, and we had a chance to fly higher, farther and faster, in the best things they had going. We just didn't question it; we worked hard, we played hard, we didn't think anything about it. So we had few thoughts about the significance of it: we took it for granted. Until it became time to land on the moon. We knew that was a moment in history, but other than that we just did our jobs. Five hundred years from now, there is only going to be one thing that they remember the twentieth century for, and this is that man first left one body in the solar system and landed on another body in the solar system.

Having said that, Cunningham would like to be recognized for having done more than be an astronaut. "It's funny," he says, "I like to think that I have done a lot of things in my life, but I am only remembered for eight years, and that is 1963 to 1971." Cunningham has enjoyed a very busy career in the engineering and venture capital world ever since, and even serves as a talk show radio host in Houston. Though he has received tempting offers to move back to California, Houston had become his home, and he has lived there ever since. He attended Harvard Business School in the 1970s, became a licensed real estate broker in Texas, and has been involved in the startup and development of approximately thirty venture-financed companies. Perhaps most importantly, his continuing work as an author and media commentator has ensured the presence of an always frank and refreshing view of NASA, past, present, and future.

Wally Schirra left NASA and the navy at the beginning of July 1969, just in time to become a television commentator for the first moon landing. He wasn't interested in becoming a NASA manager, although he was offered the opportunity. Nor did he want to stay in the navy, quickly realizing that any command he was offered would be as a showpiece—a "potted palm," as he puts it. In many ways Schirra was the only Mercury astronaut to have had a "normal" astronaut career. Shepard and Slayton were still waiting to make their first orbital flights following their medical sidelining, Cooper was still hoping for an Apollo flight he would never get, Glenn and Carpenter had moved on to other ventures, and Grissom was dead. In the end, Schirra turned out to be the only one who would command a Mercury, Gemini, and Apollo flight. After NASA, he went straight into the financial investment business, taking on directorships of a number of banks and other financial service companies. He fully admitted in interviews that he had no formal business training at all—his entire adult life had been spent as a naval officer. Yet, as he explained, he had put his life at risk in aerospace industry products, and he felt this gave him great insight into the workings of industry. Schirra also felt that he could leave the smaller financial details to the "bean counters," as he called them, while concentrating on the bigger picture. Having made multimillion-dollar decisions in partnership with aerospace companies as an astronaut, he felt he had more than enough experience in this area.

It didn't quite work out as Schirra planned. His first company presidency lasted only six months, by which time he realized that his good reputation was being used to cover somewhat shady dealings. He got out just in time; the company head was later imprisoned for fraud. Though his intuition had eventually saved him, it was a sobering lesson that experience and good judgment in one field does not always translate into another. After that experience, Schirra's business dealings were more cautious, and he fared better. Within a decade, he had begun his own consultancy business, and these days he and his wife Jo enjoy a relaxed life at their homes in San Diego and Hawaii. Essentially unchanged in character, Schirra adds a jovial presence to the many aerospace gatherings he attends around the country every year, while staying as serious and resolute as ever when it comes to discussing the more serious aspects of his career. He loves to recount the fun he and his colleagues had, but he never forgets the sacrifices that were made, too.

Wally Schirra spent an enormous amount of energy ensuring that *Apollo 7* not only flew safely, but also quickly, getting the United States back into the space race for the moon. In retrospect, he is not sure if all the hurrying was worth it—especially as the haste contributed to the death of his colleague, Gus Grissom:

In essence, as I look back on it, the timeframe was that we had a real beautiful Cold War going on. The challenge was that Kennedy had made a mess in Cuba at the Bay of Pigs, and he had to do something to look good. The Apollo program concept of going to the moon and back before the decade was out was quite a goal, which we all accepted, because we all loved the man. He was the only young, committed president we've ever had. We've lost that kind of commitment since. And yet, in fact, if you think of the inherent risk that we had,

A relaxed-looking Wally Schirra in 2004. Courtesy Francis French.

and the amount of effort the country went to, it's quite apparent that we were somewhat set up, I would say. We should have done it, but we didn't need to do it in that big a hurry. It would have been a much better program to have real piloted vehicles all the way through. In fact, von Braun made quite an issue at one point about having an Earth orbit rendezvous, then going from Earth orbit with the vehicle to the moon, then back to Earth orbit rendezvous. We wouldn't have got there so quick, but we'd have something left for it.

The aftermath of the *Apollo 7* flight proved to be tumultuous for Donn Eisele, and it began the day he returned to the Cape. As Susie explains, it was a time for them both to make some difficult choices:

I will never forget that day of the flight. I did not go to the launch, but watched it at home. It was that day that I found out how truly married Donn was. I had no idea about his children; now, every day that he was in space, the television treated me to their family outings. That finished it for me. The day they arrived back at the Cape after the mission, I received a call from Donn—could he come to see me that night? Deke Slayton sneaked him out of the Cape, and he came to have dinner with me. I told him that I was not going to live that life; if he was unhappy at home, that was really not my affair, and he would have to fix that problem. I think that he found it hard that I was not going to play ball, like the rest of the girlfriends. There were a lot of women who accepted that kind of arrangement then, but I was not one of them. I had a daughter and a career, and I needed a future. This affair looked like a dead-end street to me. So we said goodbye.

I truly think that if Harriet had not divorced Donn, our relationship would have faded away. Donn went back to Houston and performed his postflight duties, but he was not happy with what was going on, and neither was I. We had both asked for it, by getting so emotionally involved when he was not available.

Donn Eisele was, on the surface, maintaining the NASA astronaut public image. In November 1968 he returned to Columbus, and as part of a homecoming ceremony he stood on the steps of his old high school while teachers and students honored him by praising his accomplishments. The town was named for Christopher Columbus, and now they were determined to shower their returning explorer hero with accolades. Yet, that same month, Eisele would finally begin to break the astronaut image once and for all. Though Harriet took part in the postflight visit to the White House, an experience she thoroughly enjoyed, she also knew that the marriage was over. "In November," she recalls, "he told me the truth regarding Susie. The truth brought relief, in that I knew what I was facing and that I was not crazy. Donn did not arrange for a divorce; I did, at the urging of my minister. In fact, when I told him I had filed, he asked if he could go talk to my lawyer. When I told him to get his own lawyer, he told me he didn't know how."

Susie, who had not seen Donn since his return to the Cape after his flight three months earlier, finally received a phone call with the news she had expected never to hear. "Some time in January he called me," she recalls, "said that he and Harriet had separated, and that he was going to be at the Cape, training as backup on *Apollo 10*. Would I please see him? Well, that was quite a time."

Walt Cunningham remembers being surprised that Donn was actually going through with the divorce. His quiet colleague was doing something guaranteed to stir up NASA and the media. "It seemed so out of character that Eisele was at the center of this divorce thing, the first guy to do it. It would be like John Glenn getting a divorce! He was an unlikely case to be the one to test the system—it wasn't like him to be out front like that." On the surface, it looked like Eisele was going to survive both the fallout from the *Apollo 7* flight and the scandal of divorce: he was assigned to the next possible crew assignment. Just as his old friend Tom Stafford had been a backup for *Apollo 7*, so Eisele was now backing up Stafford's *Apollo 10* flight, with Gordon Cooper as his commander and Ed Mitchell as lunar module pilot. Susie sees this as a sign that Stafford still had confidence in his friend:

All of the astronauts had their own circle that they confided in; that went back to their military postings and flight schools. That was Tom and Donn's relationship—the best of friends, back to when they were classmates at Annapolis together, and their first wives were the best of friends too. Tom was backup commander for Donn's Apollo 7 *flight, and then Tom picked Donn to be on the* Apollo 10 *backup crew. Why would he do this if he thought that Donn had lost interest in the work? If Donn had been derelict in his work, they would never have put him on the* Apollo 10 *crew.*

The full story, it seems, was not as simple. By now, it seemed a certainty that Gordon Cooper would never get to fly an Apollo mission and that his crew was merely a placeholder for another to step in. Eisele and Mitchell were therefore in a precarious position, facing a disturbing likelihood that they would not automatically rotate to a prime mission position. Gordon Cooper told the authors that he was unaware of this at the time, but nevertheless when he learned that Eisele was assigned to his *Apollo 10* backup crew, his heart sank: "When they assigned Eisele to my crew, and it was also my second backup assignment, that was a clue that something was wrong. Too bad he had to be involved in a divorce at that period of time. Because he was in the midst of getting a divorce, that meant that they were going to break up my crew; they wouldn't put Eisele on a prime crew if he was involved in a divorce. I was hoping I'd get someone else assigned to my crew."

No one ever directly told Cooper that a divorced astronaut would never fly, but he knew it was an unwritten rule, and not only because of possible adverse publicity. He believed management also imagined the pilot might make errors due to stress. As a result, Cooper believed the chances of keeping this crew together to fly a mission to the moon were very unlikely. The realization had an impact on how hard he trained himself. He suspected it was only a matter of time before Shepard and Slayton broke up his crew. His mixed feelings toward working with a divorcing crew member did not, however, have a bearing on his view of Eisele as a diligent and hardworking astronaut. He was, in fact, impressed by the work Eisele put into this new assignment: "*Apollo 10* was a very important mission; it did everything but land. It was the first time the LM was used in descending to the moon. Donn was not just marking time in a backup assignment; he was very capable, and could have been a very good solo command module pilot."

Ed Mitchell, the only member of the *Apollo 10* backup crew who would rotate to a prime assignment, remembers Eisele at that time as "a personable and easygoing personality." When asked if he believes Eisele had already decided to leave NASA, he responds, "I do not know for sure—but I don't think so. He was competent, and did his assigned job well." Outside of his working relationship with his crew, however, Susie remembers that Donn was meeting with quite a chilly reception from NASA. Having upset the rules about keeping marriages together and girlfriends at arm's length, he was being frozen out by his other colleagues: "My relationship with Donn was seen as perfectly okay," she told the authors, "as long as it remained as that of the other woman. The treatment Donn received from some other astronauts at the Cape was outrageous; he was now persona non grata. That really threw him closer to me; he had to have someone to talk to. So by the time his divorce was final, Donn and I were as close as ever."

Apollo 10 flew in May 1969. On the morning of launch Eisele was in the command module for several hours, thoroughly checking all the systems. But then he was required to carefully climb out, to allow the prime crew to enter and journey to the moon. He would never fly in space again.

By the time of the next mission, *Apollo 11*, the divorce became front-page news. "It was rough when the rumors of the divorce triggered constant phone calls from reporters," Harriet remembers, "with some even knocking on the door. The divorce became final during the *Apollo 11* mission. I watched the news come over the wire in the CBS office in Houston where I was working. I felt publicly humiliated, being plastered in newspapers, magazines, and TV news reports all over the country." The news was, in fact, plastered in the press all over the world. Britain's *Daily Express* ran the headline "Astronaut Donn sued in first 'space divorce'—and more homes may break under strain." Accompanying the story was a small photo of Donn—and a large shot of Harriet. The stories contained quotations from NASA officials, forced to comment on divorce for the first time, who resorted to generalities about men in middle age often suffering marital difficulties. Donn himself was quoted, saying "The astronaut life is a great strain on marriage. It is like a circus. To be avoided."

Looking back after three decades, Harriet can afford to be gracious about her former husband:

I am sure I have forgotten the many times in that marriage where I, too, was negative or angry. As such, I accept my share of the blame for the marriage problems. I feel a great sadness over what might have been, and especially for my children. After the truth emerged, and before he left for good, we talked more intimately than we ever had. He told me, "You know me better than I know myself." To this day, I do not believe he had plans to leave the space program.

Walt Cunningham believes that Eisele might have been able to survive the divorce issue with no effect on his career if it had been a divorce only. However, he was not only divorcing—Susie was giving him an ultimatum, as she explains:

By midsummer, men had landed on the moon, and I was preparing my daughter for first grade. Donn wanted me to move to Houston. I said, "Not a chance, unless we are married." I would not uproot my daughter on the chance of a maybe commitment. We were married in a small, private ceremony in Cocoa Beach. We had a small reception, with mostly my friends. Donn's friends didn't show up. Al Bishop, a friend of many of the astronauts, stood up for him. From then on, Donn treated my daughter Kristy as his own.

Cunningham believes this was the key difference: Eisele immediately remarried someone who had been a girlfriend at the Cape. "He might have survived otherwise," Cunningham believes, "because he and Stafford were friends." But this was too much. By leaving the astronaut office the next year, Eisele was "keeping faith with his friends," Cunningham believes, and allowing the others to "keep family skeletons in the closet at the Cape."

Tom Stafford felt that Eisele had done a fair job up until *Apollo 7* and before his divorce, but then, as he states in Deke Slayton's autobiography, "Donn quit doing the job. His professional skills just went to hell. He just lost interest." Harriet concurs: "I had been told that Donn was not completing his assignments and had been warned prior to them letting him go." Slayton and Stafford told him to move on—and gave him two months to do it. Susie Eisele believes that there was far more to why Donn left the astronaut office. As she explains, she and Donn were still being frozen out of NASA life, even after they married. Very few of his colleagues were willing to accept Donn's new wife, and the pressure on the couple mounted:

I moved to Houston and got a job at a savings and loan association there. We rented a townhouse in Nassau Bay. Donn went to work, and so did I. We would have his children every other weekend. It was a living inferno. I figured if we minded our own business it would be okay. But I did not know how radical this group was. Some of the wives were very kind, and I will be eternally grateful. Others were absolutely awful. Some were told not to talk to me by their husbands. The ones that were the worst were the ones whose husbands were the biggest womanizers; they knew, but they didn't want the truth in their face. And here I was. It was a very tight community, with many egos, multiple personalities, and a lot of hidden agendas involved. If you wanted to do anything that was outside this community, you had to ask permission—even if a wife wanted to appear in a community play. It was a very immature lifestyle, and nothing that I was interested in. We lived through it.

Beth Williams, who was a dear, wanted me to attend the astronaut wives' club with her, but then I received a call, and was told there were wives who would walk out if I came. Beth insisted that I go; but I was not interested in causing a scene, and I stayed home. This went on for a year. It was an interesting time; I made some friends, some of the families were great, and I still count them as friends. I don't think Donn was fired. But I think Tom Stafford felt that Donn had deserted him.

In such a situation—turning up for work to be met with stony indifference—it would have been surprising if Donn Eisele had worked his hardest or wanted to stay around. Every day, an unspoken but unsubtle message was being sent: move on. With some years to go before his retirement from the air force, Eisele stayed on active duty while becoming the technical assistant for manned space flight at NASA's Langley Research Center. It seemed like the best move, Susie explains—one that brought them happiness:

Deke called Donn into his office and said he thought Donn's career would be better served if he took the position at Langley: some research for Skylab. We had dinner with the director of Langley, and decided it was probably the right thing to do. The Apollo flights were all assigned; management had already decided that the crew of Apollo 7 *was in the dog house, so Virginia looked pretty good to us. It was a time to move on when everything was at a lull, money was tight and Donn really had other things to do. We didn't lose anything by leaving Houston; in fact, it turned out that Donn had a nice relationship with the men and women of Langley.*

If Donn Eisele had been the needed test case, it seems that he successfully removed the stigma from astronaut divorce. Not long after he left the astronaut office John Young divorced and remarried, and *Apollo 15*'s Al Worden also divorced in December 1969. Their careers remained unaffected, and both flew their missions to the moon as planned. Jim Lovell explains:

We astronauts led a very unique type of life. We were away all the time; we were in the public spotlight all the time—which was very conducive to having a lot of groupies around, I guess. It made married life kind of difficult if you weren't careful! But up until the time that Donn got a divorce, no matter what you thought of your wife, or what the relationship between man and wife was, no one got a divorce, because the men especially thought it would ruin any kind of a position for them to make a flight. It was just not done. And so we had probably the lowest divorce rate of any group; zero, as a matter of fact. Then Donn finally got divorced. I guess he saw that he would just take a chance, and see if it affected his career or not. As it turned out, it did not affect the flight aspects of astronauts that much, because as more and more people got in the program, there was a certain attrition rate amongst husbands and wives, which is probably general throughout the country. But it sure opened up the doors for a lot in our group. By the time Donn's divorce had gone though, there were a lot of other people who were thinking the very same thing, and were just looking for an opening to see what would happen. So as you look at the future after Donn Eisele, and you look at the astronauts in the first groups—many of them were divorced. In the Apollo program, the Apollo 8 *crew is the only crew that still has our original wives!*

The change in the social rules that governed the astronauts was inevitable, it seems. Certainly, Donn Eisele was not one to be bound by such institutional guidelines, as he explained in 1971:

People are simply adapting new standards of values for their own lives. They decide, for instance, that it's more important to establish and preserve one's identity as an individual than to seek one's proper place in whatever hierarchies he finds himself. . . . It amounts to a state of mind that is oblivious to social orders, political structures, and hierarchies; . . . as these organizational entities cease to be relevant, they will change or wither away, to be replaced by entities more rational and responsive to human needs.

The three men who flew into space with Wally Schirra—Eisele, Cunningham, and Stafford—all eventually divorced. Schirra is the only one still married to his original wife, and today he reflects on this statistic with a touch of sadness:

It started breaking apart during the Apollo days. Eisele divorced his wife after our Apollo flight, and then the flood came, the dam broke. Donn was the stalking horse. At first, Cooper was essentially living apart from his wife when he came in, but they came back together, and finally divorced after he left the space program. Walt Cunningham divorced after he left NASA, and remarried. I was looking at a book called Astronauts and Their Families *just the other day, which was published just before my Gemini flight. I was really shocked how few of those guys are married to those women anymore.*

After the divorce, Harriet had to pick up the pieces of her life and return to work. She found that the astronaut family support system was still there for her:

After the initial flurry of publicity, things calmed down, and I was able to begin moving on with my life. When returning to nursing for the first time in seventeen years, astronaut Joe Kerwin offered to be my guinea pig so I could learn to start IVs. *During my first year of work, which was Jon's first year in a nursery, Jon caught every illness that came around. Faye Stafford and Barbara Young babysat him and took him to the doctor. Without their help, I would have certainly lost my job. I was back to work with little free time, but although I saw the other wives less, their emotional support increased. I was invited to several get-togethers at other couples' homes. Al Shepard had his secretary call to inform me of astronaut office get-togethers. Dr. Mueller sent me an invitation to the* Apollo 12 *launch, which I attended, and I went to the Cape with the Staffords for Apollo-Soyuz. These gestures helped my beaten-down ego and were a nice transition to life beyond* NASA. *Later, following the parade of divorces, several of us met regularly in a support group. In 1991, we started gathering as the original wives for reunions.*

Donn Eisele resigned from NASA and the air force in July 1972. He spent his first two post-NASA years as director of the Peace Corps in Thailand, beginning right after he and Susie had Andrew, their first child together. Susie remembers it as an intriguing period in their lives:

We went to Thailand for two years, and had a life-altering experience. Donn had 175 great volunteers, who ranged from twenty-one-year-old teachers teaching English as a second language, to a seventy-five-year-old doctor working in family planning. In the meantime, I raised our children. Kristy went to the British school in Bangkok. Andy was only three months old when he went over, and by the time he came back his Thai was better than any of ours. Donn did some other work for the government while I was over there; I didn't ask. He traveled a lot, and I traveled with him. The Vietnam War was very much a part of our lives out there, because Thailand housed our air force bases. There were volunteers around these bases, a lot of drugs, and every once in a while one of our volunteers would get into trouble with that.

While head of the Peace Corps, Donn appeared in comedian Bob Hope's last Christmas shows. "We were not directly involved with the military," Susie explained, "but when Bob Hope arrived that Christmas of 1972, there was such an anti-war sentiment in the U.S. that he had a hard time finding people to entertain the troops. Bob Hope asked Donn if he would go out, and Donn said, of course—the troops needed to be supported whether you believed in the war or not." Eisele was indeed against the war by this point, having said in a speech the year before, "In this country we have come to see the folly of our Southeastern Asian venture and are now withdrawing—although perhaps not as fast and completely as we ought." Nevertheless, he helped give the soldiers a sorely needed piece of entertainment, Susie recalls: "Donn went to several bases with the troop show. I went along to a base called Nam Pong, a secret Marine base somewhere in North-East Thailand, where we landed on a pipe runway. There were thousands of men and women there sitting under the hot sun, desperate for anything from home. They were sad times."

By 1974, Eisele had been in the service of his country all of his adult life. As Susie says, it was time for him at last to become a private citizen:

After we came back from Thailand we settled in Williamsburg, Virginia, for six years. Donn tried to find himself and what he wanted to do for the rest of his life. With Donn retired from NASA and the Air Force we considered ourselves private citizens, and really wanted to keep a low profile. Donn did keep in touch with some of the other astronauts—we would arrange to meet at various group meetings. Walt and Donn kept up their friendship. But Donn was really a loner, and we preferred keeping a private, low-profile life. We even had a doormat, as a joke, that said "Go Away"—Donn was a great jokester.

Eisele's work in the private sector included a stint as a sales manager for a power shovel company and an attempt to start a small airline. After two years, Eisele realized that the airline was never going to be a reality, and it was time to move into a different sector. "We moved to Fort Lauderdale in 1980," Susie recalls. "Donn went to work as an account executive for Oppenheimer. He got a degree in financial planning and started a consulting company with a Japanese partner, called Space Age America." Donn Eisele had always loved Florida. "Florida has a long and colorful history that is rich with tradition, culture, drama, and adventure," he once said of the state. It was a place he came to cherish and, in the environmentally conscious atmosphere of the early 1970s, sought to protect. Environmental issues were something he had begun thinking about during his spaceflight, he would later relate:

There were a few ugly sights [from space] too—the mud and silt and the crud that spews out from the mouths of rivers, the Ganges for instance into the Indian Ocean, or the Mississippi flowing into the Gulf of Mexico. And you could see smog obliterating large cities and entire valleys, like Los Angeles, where the smog backed up against the mountains and then reached across into the desert to the East. From outer space one . . . can see how much damage is being done. . . . Man needs the Earth, but the Earth does not need man.

Take, for instance, the Everglades right here in Florida. The land developers would like to go in and carve it up into small lots . . . but in the long run we might destroy the delicate balance, the fragile beauty of the life systems that now exist there.

Proud of his work in the space program, Eisele hoped that it might provide some of the answers to these problems:

We have just now begun to reap the benefits, in real human terms, of spaceflight. Our moon flights have produced a wealth of scientific information, but the real payoff is yet to come. In studying the moon's history and geologic processes we learn, by extension, a great deal about our own Earth's formative processes. This should lead, in the future, to a better understanding of our natural resources—where to find them and how to use them best. . . . I believe that we can, through our gifts of reason and intelligence, and of emotions and feelings, find the means to employ our space research and other technologies for the betterment of mankind everywhere.

For the meantime, he would have more human concerns to busy himself with. Sadly, as often happens following a divorce involving children, Harriet and Donn's only contact for many years was negative and combative. Harriet remembers the court battles over child support and arguments over how the children were being brought up. Things improved between them, however, when Jon had a difficult freshman year, and Donn agreed to take him in for a year and help him refocus. At the end of a year at a new school, Jon was back on track with his education and his personal life.

Once the children were all older than the age required for child support, Harriet found that any lingering quarrels finally dissipated, and the families could reconnect somewhat. "When Susie and Donn's son Andy was a teenager, Donn called and said Andy wanted to meet me. We had a

A 1987 portrait of Donn Eisele. Courtesy Susan Eisele Black.

pleasant visit—Andy is a sweet boy." In later years, both of Donn Eisele's wives even attended family wedding celebrations together, and neither wife holds any hard feelings toward the other.

Donn Eisele stepped back into the limelight a little in the year before he died. "Donn wrote a rather long article about the *Challenger* disaster for a British newspaper," Susie recalls, "comparing it to the environment at NASA with the Apollo fire. It was that article that attracted *People* magazine to Donn." When talking to the *People* reporter, Eisele explained that he saw a great deal in common between the two spacecraft accidents. "I'm astonished at what I hear and see," he said of the *Challenger* investigation, "and yet, in a way, I have seen it all before. There are striking parallels between what seems to have been going on in the shuttle program and what went on in the Apollo development before the fire." He was upset to see the same frustrations he had witnessed, with managers not listening to astronaut concerns, seemingly played out again. He believed the major cause of both accidents was the same: "Go fever: trying to keep the flights on schedule." Eisele thought NASA needed to admit that space was still a dangerous frontier. They were still in effect undertaking what they had asked him to do all those decades ago. Just like *Apollo 7*, the shuttle was still "an experimental flight test program; . . . every shuttle flight is an experiment."

Since he had flown the first flight after *Apollo 1*, the reporter asked Eisele if he would be willing to fly the next shuttle flight after *Challenger*. "Oh yes," the former astronaut replied. "The one after the accident is always the best one. They will spare no effort to make it the safest flight there ever was."

December 1987 brought Donn Eisele back into contact with one of the manager astronauts who had cut him loose all those years ago—Alan Shepard. The two of them journeyed to Tokyo to open a Japanese space camp sponsored by Nippon Steel. It was part of Eisele's work with Space Age America, Susie recalls; Donn had negotiated the contract. However, while he was alone in his hotel room, a heart attack took his life at the relatively young age of fifty-seven. Wally Schirra remembers hearing the surprising news: "Donn and Al were in Tokyo at a space camp, and apparently Donn was out jogging, getting some exercise. He went back to his room and had a stroke, and died. It was a shock to me; he was much younger than I, and my first thought was, 'What's he doing out there in Tokyo?' I couldn't even remember the last time I saw him."

Susie received the tragic news by telephone. "It was a great shock when the American Embassy called me from Japan and told me that Donn was dead. Andy was just getting ready to leave for school when the call came in. He died unexpectedly at fifty-seven, leaving me with a fifteen-year-old boy to raise, a big mortgage, and a little insurance money. There wasn't much there. All I could think of was, how do I get a body out of Japan?" Susie was faced with the difficult task of trying to bring Donn's remains home to America. It would have taken up to a month for his body to be returned by the usual official American Embassy methods. To her profound disappointment, Alan Shepard not only chose not to assist, but also decided never to call Susie and tell her about her husband's last days on Earth.

Luckily, Donn's Japanese business partner and family friend was able to help. Once Susie had signed the remains over to him, he was able to have the former astronaut cremated at his family temple in Japan. Then he personally carried the urn back by airplane from Tokyo to Washington DC. In the meantime, Susie had had some assistance from the NASA Public Affairs Office to arrange for a military funeral at Arlington National Cemetery in Washington. She also had some much-needed help from other colleagues at this difficult time: "Al Bishop called when he heard of Donn's dying and arranged for a limo service for us in Washington for the Arlington burial. Our former neighbors in Thailand arranged everything at Arlington—I couldn't focus on anything." At Arlington, Donn Eisele's remains were buried in the same section as two members of the *Apollo 1* crew: Gus Grissom and Roger Chaffee, the man who had replaced him on that crew. His second marriage had lasted until the day he died.

Harriet had seen Donn only four months before, at the wedding of their son, Jon. At the time, she had noticed how thin he looked. She subsequently discovered that he had been on a self-imposed regimen of exercise and weight loss to counter a suspected heart problem. "It never occurred to me to go to the funeral," Harriet remembers,

although someone called and offered to fly me there. Rather, I babysat for my children, so they could attend. I remember talking with Susie and giving her the names of some NASA personnel who might be able to help her. Donn's death was a surprise to me; I was unaware of any heart problems. After Japan, Donn had planned to go on to Hawaii to see young Donn, so this greatly increased the sadness and distress. Walt Cunningham spoke at the service in Fort Lauderdale, and Tom Stafford at the service in Arlington. Tom spoke of our families, and what our families had done together, and the children found this comforting.

Susie agrees that the funeral and its aftermath were not only a fitting sendoff, but also a way to bring the children from the two marriages together—and to put to rest the rift between Tom Stafford and Donn that the divorce had brought: "Many friends rallied around, and we had a beautiful ceremony on 7 December in Fort Lauderdale, and the burial at

Arlington. I tried to make things right, by having Tom Stafford give the eulogy for Donn at Arlington. I asked him because of his relationship to Donn's children, and his former life. I also tried to do the right thing for his children. I took Andy and Kristy to Hawaii for Donny's wedding, and gave them personal items I thought they would want. For example, I sent them old photos of their dad as a boy."

Schirra attended Eisele's funeral service at Arlington National Cemetery. The day before the service, Schirra met with John Healey and Fred Peters, two of the engineers he had worked closely with in preparation for the *Apollo 7* flight. During the conversation, the two engineers told Schirra that he and his crew were due far more gratitude than they ever received at the time, both for the flight and for their dedicated participation in the redesign. "The three of us were indeed fiercely proud of what we accomplished on that mission," Schirra remembers. To hear such praise from those who knew best, he adds, was perhaps the best eulogy that Donn Eisele could have had.

Susie Eisele remains a fierce defender of her husband's reputation and honor, and strongly believes that he was a victim of circumstance, who never neglected his duties:

Donn talked about the "splash down" after Apollo 7 *with me. He felt Wally knew he wouldn't fly again, so really didn't care if he ticked off the management about the reentry without helmets. Wally was the commander—he ruled. I believe that Donn would have flown again if it hadn't been for his personal life. There were no divorces before Donn's, and afterwards the flood gates opened. Now, they are just about all divorced with a couple of exceptions.*

I am very conscious of Donn's contribution to space history, and really am very concerned about the wrong that has been written about him. But try as they may, they can never take away the fact that Donn flew in space.

In 1971, Donn Eisele had reflected on his own mortality and linked it to the survival of humankind as a species. He ended a speech with a message of hope for the twenty-first century, a new millennium that he would never live to see:

We all know . . . in the long run, we all are dead. Now, the notion of our nonexistence is appalling and quite intolerable to most of us, so we choose to believe that we are immortal. . . . I'd like to point out that if we succeed in destroying ourselves as a species then the issue of personal immortality has no meaning.

Man differs from other animals principally in that he has a sense of his own destiny. . . . Will man survive? I'll say he will! He'll not merely survive, he will flourish. Stay around another thirty years or so. I think we'll all be surprised.

Perhaps the last word on Donn Eisele should be left to the *Apollo 8* crew members, who benefited most from the work he did checking out the Apollo spacecraft. Frank Borman speaks glowingly of *Apollo 7*, saying *Apollo 8* could not have happened without the crew having done "a perfect job." Jim Lovell, when talking to the authors, also paid tribute to Donn Eisele and his flight: "Donn Eisele was a classmate of mine: Naval Academy class of 1952. He was very conscientious—an easygoing guy, but very competent. Of course, his real claim to fame was that he was on *Apollo 7*, and did an excellent job. Donn and the crew were so successful on *Apollo 7*, they allowed *Apollo 8* to go to the moon."

6. Starting Over

Our greatest glory consists not in never falling,
but in rising every time we fall.

Confucius

It is rare for parents to bring a new child into the world and not hope anything but the best for them. Most mothers and fathers naturally want to help their children have a bright, happy future, free of any unnecessary obstacles. Any Soviet citizen with a new child in the early 1930s, however, must have paused and wondered if this would ever be possible. Under the dictatorship of Joseph Stalin, a long series of show trials and purges was only the most visible tip of what was a terrifying time for many of the country's citizens. Stalin's despotic desire to remove any possible enemies around him, both real or imagined, would lead to the demise of millions of Soviet citizens in a period that came to be known as the Great Terror. Stalin's favorite saying at this time was rumored to be "Est chelovek, est problema; net cheloveka, net problemy," or "A person, a problem; no person, no problem." He harbored a particular hatred for Soviet Jews that originated from long years of struggle with Leon Trotsky and his supporters.

Around 1934, Stalin introduced what became known as the "Line 5 rule"—part of a nationwide registration program that issued internal passports to all citizens of Russia, who had to carry them at all times. On the fifth line of the registration form was a compulsory requirement for the applicant to reveal his or her ethnicity. No matter where they had been born and whether they practiced religion or not, if the person had at least one Jewish parent, the entry would read "Jew." This ethnic information allowed bullying officials to practice discriminatory behavior, especially when dealing with Jewish citizens, and to closely regulate any person's movement. This onerous regulation would remain in place long after Stalin's death, and, sadly, anti-Semitism still affects Russian social and political life to this day.

Stalin also instituted what he felt was a practical solution to his imagined "Jewish question," by creating a new homeland for Russian Jews in a frozen tract of near-impenetrable swampland in the far east of Siberia, five thousand miles from Moscow on the border with China. In 1934 the area was officially designated as a "Jewish Autonomous Region," which became known as Birobidzhan. Yet even here, its hard-working citizens were still not free of Stalin's unrelenting purges, as political leaders were summarily arrested and imprisoned.

On 18 December of that same year, Boris Valentinovich Volynov was born into a Jewish family during a fierce blizzard in the city of Irkutsk, the capital of East Siberia. While most of the country was struggling to keep up with Stalin's ambitious industrialization plans, the hard, spartan life of the Irkutians gave them little time for politics. Here, in a relatively secluded place once cut off from central Russia by the rocky Ural Mountains, survival and food on the table were the main priorities. Divided by the torrential Angara river and situated on the Trans-Siberian Railroad, Irkutsk had been home to many wealthy merchants in its colorful past, and they had contributed greatly to the area, building wonderfully elaborate houses and places of worship that included churches, mosques, and a Jewish synagogue. However, the vast majority of Irkutians, including the Volynov family, lived in petite but attractive wooden houses, which even today give the city a unique character.

Boris Volynov was only three years old when his father, Valentin Spiridonovich, died. The bereaved widow, Yevgeniya Izrailyevna Volynova, was a qualified pediatrician, so she and her son managed to live in reasonable comfort, in what was essentially an intellectual, middle-class family. In the prewar years, when Boris was still quite young, she took on a position as a traumatologist in the rundown coal-mining and metallurgical town of Prokopyevsk. Located in the center of the Kuznetsk Basin region (popularly known as Kuzbass), Prokopyevsk was home to the largest coal field in the world. Coal-mining accidents and associated injury trauma were commonplace events, so Yevgeniya Volynova maintained a vital medical practice in the town. A popular person in a working town where being of Jewish descent was totally irrelevant to one's community standing, she would later remarry. Boris then became the stepson of war veteran Ivan Dimitrievich Korich.

Prokopyevsk was neither an attractive nor peaceful place in which to grow up. Grimy pithead derricks and vast pyramids of rock dumps dotted the landscape, while clinging dust and constant noise permeated every part of daily life. Volynov's grandmother often told him that the city might have slept at night, but the miners never did; they toiled underground in shifts day and night. When World War II began, Yevgeniya undertook training in emergency medicine and worked as a surgeon through the war years.

Most of the boys in the village were destined to become miners, and young Boris believed this would also be his future, especially after he went on his first, exciting, underground school excursion. He and some of the boys returned later for a little "*subbotnik*" or volunteer labor, and the pit boss kindly handed them a few rubles for helping out. Boris was delighted with this newfound wealth, and decided to spend some of his wages on a couple of books he wanted to read. One was a biography of the dashing fighter pilot Anatoli Serov, which told of his experiences in testing aircraft and how, under an assumed name, he flew nocturnal combat missions over Madrid in 1937, helping to defend the Spanish Republic. Serov had died in a flying accident in 1939, but Boris was thrilled by the descriptions of combat flying. He decided he would like to emulate his hero and become a fighter pilot.

For a school project, Boris designed a small propeller-driven engine, which he tested in the gymnasium under the watchful eye of his science teacher, Vladimir Usanov, and a skeptical group of children. To everyone's amazement the rudimentary engine and propeller fired into life, but also filled the gymnasium with clouds of pungent smoke. With this new-found zeal for propulsion and engineering, Volynov began designing rockets and jet engines, and would often give talks on ballistic flight to members of a small group of amateur rocketeers he had joined. His mother, however, did not support his wish to become a fighter pilot. She urged him to go into medicine like her, but she could not know that her son would one day become a famed cosmonaut, and the first space explorer of Jewish descent.

Volynov could hardly wait to finish school so he could apply to join the air force. Finally, with ten years of education behind him at Prokopyevsk, he secretly contacted the regional Komsomol committee, pleading for a letter of recommendation to an air force school. The committee granted his request, but he felt bad about leaving his mother and could not bring himself to tell her that he had gone against her wishes. On the appointed day he quietly packed his bags and made his way to the train station, where he boarded a train along with some other young recruits. He agonized over his decision to leave home without saying goodbye.

During a long stop to replenish the locomotive's water tank at the next station, Volynov hurriedly disembarked and ran to the post office, where he bought some paper, an envelope and a stamp, and dashed off a quick letter. "Dear Mama," it read, "You must understand that I am already a grown man. Everything will be well, please believe me. I want no other profession. I shall only be a flier." On arrival, the young hopefuls were disappointed: the flying school they had applied to join did not yet exist. However, the air force officer who met them quickly reassured the recruits that they would be participants in the origin and development of the school. "I warn you at once, we are beginning from scratch," he told them. "We shall set up the school ourselves: myself, the flight instructors and you. The first year will be very hard; the second will be easier. The main difficulty will be that we shall combine work with studies. One thing I promise firmly; you will become good pilots. Meanwhile, be patient, brave and disciplined. If anyone wishes, he may go away." Volynov stood firm.

As time passed, the primary flying school evolved, as promised, before their eyes. An airfield was constructed and soon airplanes began to arrive. The cadets were given a solid classroom background in flight basics, and before long they began to take elementary flying lessons. Combined with this, however, they also had to learn the routine associated with military life, participating in drills, monotonous guard duty, and endless kitchen details. But Volynov loved the life, and wrote of this in his regular letters to his mother. "Mama, I am flying!" he said in one. "Today we took the oath of allegiance. I would like to say that it is not as simple to pledge to work and serve as the manuals demand. Army life is difficult. But we took the oath readily, without hesitation or doubts. It was as simple as breathing. We swore to our banner and to ourselves." As well as his mother, Volynov was also corresponding regularly with a childhood sweetheart from school, Tamara Fedorovna Savina, who was studying to become a metallurgical engineer. In their letters they professed their affection for each other, and had already decided to marry sometime in the future.

Volynov's instructor at the preparatory school was a young pilot not much older than him, named Veniamin Reshetov. Despite his youth, Reshetov had already accumulated considerable experience in the air. He readily imparted his enthusiasm and love of flying to his students, spurring Volynov on to put even more vigor into his studies and practical exercises. Graduating from flying school in 1952, Volynov was then enrolled at the Stalingrad Military Aviation School for advanced training, where he served under Squadron Commander Major Viktor Ivanov, a decorated war veteran. One of those who trained with him at the school was another future cosmonaut, Gherman Titov. Four years later, Volynov proudly received his wings, and then spent another four years as a fighter pilot with the 133rd Air Division of the Soviet Air Defense Forces, based at Yarolslavl, in the northern part of Central Russia.

In 1957 Volynov finally asked Tamara to join him. She came to the air base bearing a small suitcase and her diploma as an engineer, and the young couple set up their first home together in a small rented attic near the airfield. Tamara soon found a position as assistant foreman at a local engineering plant, and would later be made foreman, while Boris was promoted to full lieutenant on 14 March 1958. More exciting news came when they found out that Tamara was having a baby. On 9 May they were married in a simple ceremony, and Tamara returned to more comfortable surroundings at her mother's house in Prokopyevsk to await the arrival of their first child. When the birth became imminent, Volynov would eagerly check every day at the local post office for a telegram from Prokopyevsk, and finally he received the news he had been waiting for: "Congratulations on the birth of your son Andrei." Tamara and baby Andrei joined her husband at the Yaroslavl base late in 1958 and they resumed their life together, this time as new parents.

Citations noting excellent progress and flying skills, calmness under pressure, and leadership qualities followed Volynov throughout his training, including a glowing commendation from Major Ivanov. This would continue in his four years as a fighter pilot serving with the Air Defense Forces, during which time he became a senior lieutenant. In the fall of 1959, these recommendations from his unit commanders took him to Moscow under mysterious secret orders, where he was told that he was under serious consideration to join the first cadre of Soviet space explorers.

A thorough background check had taken place, including reports on his parents. Though Volynov has never disclosed how his father died, it is highly unlikely he had been a victim of the brutal Stalin-run purges of that era, as space historian Rex Hall points out: "The cosmonauts were subject to extensive vetting by the KGB, and some candidates were certainly discriminated against because they had relatives abroad or a questionable background. It is my understanding that if his father had been purged, Volynov's name would have been removed from the list of candidates."

On 25 February 1960, just twenty-five years old and already a member of the Communist Party of the Soviet Union, Boris Volynov became one of twenty young pilots selected to become cosmonauts. In fact, he became the second to be appointed to the group, after Pavel Popovich. "Once I got to know everyone," he recalled, "I soon realized that I was back among members of the piloting family. Constant joking around and harmless practical jokes—all the things that went along with Air Force life—were brought by the boys from the squadrons they had served with."

Volynov might even have been considered for the very first flights if it were not for one physical characteristic—his shoulders were too broad. Physical limitations were understandable, but decisions based on ethnicity should have been banished from the new era of space flight. Unfortunately, the mere fact that he was of Jewish descent would cause Volynov many difficulties during his thirty-year term as a cosmonaut, with undisguised bigotry impacting several times on a career that was otherwise exemplary.

The tragic death of Vladimir Komarov at the end of his *Soyuz 1* mission in April 1967 and the subsequent reevaluation of procedures and equipment had an immediate impact on the Soviet space program. Early the following month, Col. Georgi Beregovoi was tentatively assigned to fly the *Soyuz 3* mission as sole pilot; he planned to carry out the originally scheduled assignment of Komarov to dock with another Soyuz craft carrying three cosmonauts. In securing this flight, however, he had to fight off determined bids by Konstantin Feoktistov, who felt he was far better qualified to make this vital test flight, and Yuri Gagarin. It led to a simmering hostility between Gagarin and Beregovoi, which one day exploded into a vicious shouting match, in which both men hurled pent-up insults at each other.

Meanwhile, the composition of the three-person Soyuz rendezvous crew selected to train with him underwent several changes, and by February the following year it had been narrowed down to Boris Volynov, Yevgeny Khrunov, and Alexei Yeliseyev. However, there would be many alterations to the mission schedule before Soyuz flights could safely be resumed.

It had been eight long and often vexing years for Volynov, watching and waiting patiently as his colleagues from the first cosmonaut detachment carried out their pioneering space missions. One can only imagine his feelings when the influential Colonel Beregevoi, who had been selected in 1964, managed to leapfrog several of the first detachment members and receive an assignment to the *Voskhod 3* mission just a year later—a mission subsequently canceled when Sergei Korolev died. In fact, there had been a simmering rivalry between the first group of cosmonauts and the second Air Force group, selected in January 1963. Overall, the members of the second group were more qualified and senior in rank than the first group, creating an understandable tension at the training center.

Like the proverbial bridesmaid, Volynov had been involved in several missions and crews since his selection, but had never quite made it to the top of the list. Early on, having been added to the Vostok training group after Titov's flight, he was named second backup pilot for *Vostok 3* and *4*, and later the backup pilot for Valery Bykovsky on *Vostok 5*. He was assigned as one of two possible commanders training for *Voskhod 1* in 1964, but here he became embroiled in a battle between Korolev and the air force over crew selection. Volynov was one of the air force choices to fly—until the State Committee discovered that his mother was Jewish. Korolev was not able to argue against the weight of this institutional discrimination, and Volynov lost his chance to fly.

In a May 2001 interview with researcher Bert Vis in London, Volynov claimed that he and his crew, Georgi Katys and Boris Yegorov, were unexpectedly dumped within days of the scheduled launch. "We were the *prime* crew," he emphasized, "and the backup crew was Komarov, Feoktistov, Sorokin [first reserve], and Lazarev. Three days before the launch the crews were changed: the backup became the prime and vice versa." The only exception was Yegorov, who retained his place in the prime crew ahead of

Lazarev and Sorokin. This crew change had also been a topic of discussion during Bert Vis's interview with Georgi Katys in April 2001 in Moscow. When asked if Volynov had lost his command specifically because he was Jewish, Katys confirmed this. "Yes, by the State Commission," he stated. "Volynov's mother was Jewish; that's why he was [taken] out of the program at that time. He [and I were] moved to the second crew . . . three or four days before the launch. When Korolev learned about this he was angry . . . furious . . . he even wanted to complain to the Central Committee of the Communist Party. Khrushchev called Korolev and told him, 'Don't rock the boat—it's not worth it!' He promised to give us a chance later."

Having lost his opportunity to fly the first Voskhod mission, Volynov swallowed his disappointment and spent a further year training for the planned *Voskhod 3* mission together with a succession of partners. First there was Georgi Katys, but he was completely dropped from flight status soon after, when the KGB discovered his father had been denounced, imprisoned, and shot in one of Stalin's purges. Volynov was then teamed with Viktor Gorbatko and finally Georgi Shonin, but fate would intercede with the sudden, tragic death of Chief Designer Sergei Korolev on 14 January 1966. It was "an unfortunate piece of timing," Volynov told Vis. "Although the space vehicle was ready on the launch pad, just ten days before our flight to the cosmodrome for the space flight, they cancelled it. There was no Sergei Pavlovich Korolev; . . . his deputy and subsequent replacement was Mishin. And Vasili Pavlovich [Mishin] made the decision to speed up preparations for the flight of Soyuz."

Originally, six Voskhod missions had been planned, but following the appointment of Mishin all four of the remaining missions were canceled, and plans for the Soyuz series of spacecraft brought forward. "That was a way for him to show himself and to open roads for his perspectives," Volynov added. Soon after, Volynov was transferred to the Soyuz group, and later given a backup-support assignment for *Soyuz 3*. Choosing his words carefully, Volynov told Vis that he had become "fed up with being a backup," but he offered brief characterizations of the men who he trained with for *Soyuz 3*, curiously referring to himself in the third person. "One spacecraft commander was Shatalov and [in] the second space vehicle was Volynov and Beregovoi. Shatalov was the newer; he was also a lieutenant colonel but a bit older than me. Beregovoi was a test pilot, a colonel and a Hero of the Soviet Union. But he came to cosmonautics late and many questions were quite unknown for him; . . . nevertheless they decided to make him a commander, being emotionally stable after Komarov's death. So, again, it turned out that Shatalov and I were the backups."

Yuri Gagarin had been sympathetic toward his colleague, Volynov, and was aware of the frustration he felt in being a perennial backup. "Boris," he once told him, "you know that the longer the flight, the more complex it is. You have been trained for the most complex flight, and you'll get it. Remember what I said." Before he died, Korolev had also made a promise to Volynov. "We'll give you a very special mission," he said. "You're strong, and you'll bear up." Like many candidates before and since, Volynov returned to the sidelines and patiently hoped his day would come to fly into space.

Meanwhile, Georgi Beregovoi trained for and flew his *Soyuz 3* mission. At the time of the flight, it was trumpeted as a great success by the Soviets, but the mission was in fact a flight that failed in its major objective, with the blame attributed to a less-than-satisfactory performance by the forty-seven-year-old pilot.

Georgi Beregovoi was born in the small village of Fedorovka near Karlovka in the Russian Ukraine, on 15 April 1921. The following year, his parents, Timofey Nikolayevich and Mariya Semenovna Beregovoi, gathered up their young family and moved to the industrial city of Yenakiyevo, in the Donetsk region of eastern Ukraine. The youngest of three brothers, Georgi grew up reading the stories of Soviet aviators with great interest. His elder brother Viktor was a gliding instructor at the nearby airfield, and he would often allow Georgi to sit in the cockpit of a glider and pretend he was flying.

On leaving school at the age of seventeen Georgi took a job at a steel plant, but he had higher aspirations. With each passing year the budding aviator had grown closer to his goal of joining the aero club his brother belonged to, and finally he had his wish. The chief of the flying school and gliding school, Vasili Zaryvalov, finally gave in to the boy's persistence, and though Georgi was still too young, Zaryvalov told him to turn up in the fall and he would be enrolled in the gliding group.

Beregovoi joined the aero club the first day he was permitted to do so, and the following year, 1938, he made his first solo powered flight in a Polikarpov biplane. There was no stopping the eager youngster now; that December he enrolled at the military flying school in Voroshilovgrad, later renamed Luhansk. It was a close thing, as the instructors immediately stated that he was too small to fly, but they soon changed their minds when they watched Beregovoi take a training aircraft through its paces.

In 1941, within a week of Germany invading Russia, Beregovoi left the flying school and joined a squadron equipped with Yakovlev BB-22 bombers. Later that year he was sent to train on other reconnaissance aircraft such as the Petlyakov Pe-3 heavy fighter and the Ilyushin Il-2 Sturmovik ground attack bomber. He returned to combat duties with the 4th Air Assault Division, 5th Army, 2nd Ukrainian Front in the spring of 1942, where he flew several missions in Il-2 aircraft targeting German rail transport. His division was led by Maj. Gen. Nikolai Kamanin—a person who would one day prove influential in the Soviet space program and Beregovoi's future.

On his sixth bombing sortie, supporting tanks and infantry assaults, Beregovoi's airplane was hit by flak, but he managed to nurse it to a safe landing. In August 1943, while airborne during the height of the Battle of Kursk, his airplane was again hit by concentrated ground fire, causing it to erupt in flames. He and his gunner, whose clothes were on fire, just managed to leap clear of the blazing Sturmovik at an altitude of around one thousand feet and parachute to safety. Shot up a third time in March 1944, Beregovoi was spared by fate yet again when he was able to land his crippled airplane behind Russian lines. He was earning himself something of an enviable reputation for survival within his squadron after being posted three times as "missing, presumed dead."

Beregovoi received several awards and honors during the war years, and on 26 October 1944 received the title of Hero of the Soviet Union, with the Order of Lenin. He flew his last combat mission, his 185th, on 9 May 1945 near Brno in Czechoslovakia, ending the war as a squadron commander. He was then assigned to an advanced officers' flight tactics course, and subsequently joined a Bell P-63 King Cobra fighter unit as the regiment's navigator. His request for assignment to a test pilot school was accepted, and he completed his training at the Chkalov school in Shchelkovo, graduating in 1948. Over the next sixteen years he flew sixty-three different aircraft types and became one of the Soviet Union's leading test pilots.

"Every new aircraft is a joy to a test pilot," he would state. "At times I took risks—but for the sake of the job."

Among his many accomplishments as a test pilot, Beregovoi participated in rocket-assisted takeoffs of the MiG-19. Eventually, he rose to the rank of colonel, and became deputy chief of the test department. During this time Beregovoi married Lidiya Matveyevna, and they would eventually have two children: Viktor, born in 1951, and Lyudmilla, born in 1956.

Beregovoi's ambitions did not falter, and in 1953 he undertook a correspondence course with the Red Banner Air Force Academy, graduating three years later. In 1961, he was named Merited Test Pilot of the USSR. Two years later, at the personal invitation of his former division commander, Nikolai Kamanin, Beregovoi and three other top test pilots were hired as flight instructors at the Cosmonaut Training Center outside of Moscow. Sergei Korolev showed the four pilots around the center's facilities, and they even sat in a Vostok trainer. Korolev particularly encouraged Beregovoi to think about becoming a cosmonaut and taking the medical tests, as he and Kamanin both agreed the renowned test pilot had much to offer the Soviet space program. Beregovoi felt he was too old to qualify; he was thirteen years older than Gagarin, who later said of him, "When I had just started school, he was already fighting the Nazis." Despite this, he requested an assignment to the Center for Cosmonaut Preparation. In February 1964, at the age of forty-three, Beregovoi was selected for the cosmonaut team.

Despite his impressive background, many did not believe that Beregovoi would make the grade as a cosmonaut, and sometimes even he wondered if he was fit enough for the demands of spaceflight. However, Yuri Surinov, the Moscow center's director of physical training, placed Beregovoi on a rigorous six-month exercise program. At the end of this personal development training he was in excellent physical condition and had lost fourteen pounds in weight. In April 1965 Beregovoi's first mission assignment came along when he and Lev Demin were chosen as the backup crew for *Voskhod 3*, then slated for early 1966. Soon after, Demin was dropped from the crew, and his place was taken by Vladimir Shatalov. As well as training as a backup, Beregovoi had also been chosen to fly the *Voskhod 4* mission that would hopefully follow. In the end, he would never fly that spacecraft. Following the cancellation of the remaining Voskhod program, Beregovoi was

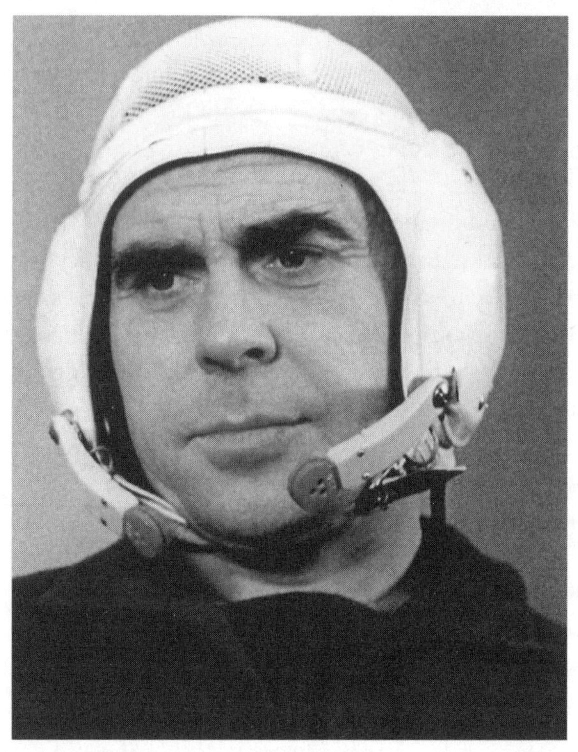

Col. Georgi Beregovoi, sole crew member of *Soyuz 3*. Courtesy Colin Burgess Collection.

reassigned to the group of cosmonauts in training for the first flights of the new generation of spacecraft, named Soyuz.

By August 1968, Beregovoi and his backups, Volynov and Shatalov, had completed their training for the *Soyuz 3* mission. The earlier mission plans had changed. Now, they no longer called for a docking with a second manned spacecraft. Instead, an unmanned craft would be launched, named *Soyuz 2*, followed the next day by the manned *Soyuz 3*; the two would then dock. The Soviets were anxious to resume their spaceflight program, particularly as NASA had finished investigating the *Apollo 1* fire, their three-person spacecraft had been fully redesigned, and *Apollo 7* was ready to be launched in October as the prelude to manned lunar missions.

On 23 October, the day after the crew of *Apollo 7* successfully splashed down at the end of their mission, Kamanin made a presentation before the State Commission at the Baikonur Cosmodrome, in which he introduced Georgi Beregovoi as the prime pilot for the first Soyuz flight since the death of Komarov, eighteen months before. There was only one small problem—the cosmonaut had failed his prelaunch examinations. According to space researcher Asif Siddiqi, this news was revealed in issues 23 and 24 of the Soviet spaceflight magazine *Novosti kosmonautiki* in 1998, in an article by I. Izvekov and I. Afanasayev titled "How from a Failure Was Forged the Next Victory." Siddiqi cites the article when writing that Beregovoi had received only two out of five possible points in his examination, which would normally have disqualified him and placed the prime backup, Shatalov, in the pilot's seat. However, Kamanin somehow arranged for Beregovoi to sit a second examination, and this time, unsurprisingly, he got four out of a possible five, which was deemed good enough.

The unmanned *Soyuz 2* lifted off from Baikonur just two days later on 25 October and achieved the desired orbit, though no immediate formal announcement of its name was made. The following day, as the unoccupied craft flew over, *Soyuz 3* lifted from the launch pad with Beregovoi on board, successfully achieving orbit just six miles from the target vehicle. "The liftoff was smooth—even gentle," Beregovoi later reported. "Then came the G-loads. But everything was normal; everything was fine." Soon after, the Soviet press issued releases stating that *Soyuz 3* would rendezvous with an unmanned craft launched the previous day, now identified as *Soyuz 2*.

As planned, the Igla rendezvous system on board *Soyuz 3* locked onto the other vessel and guided it to within two hundred yards of *Soyuz 2*, at which time Beregovoi took over manual control for the docking. When radio contact was lost with the two craft as they moved out of ground station range, they were only forty yards apart. An hour later, when contact had been reestablished, ground controllers received an unexpected shock. Not only had the docking not taken place, but telemetry also indicated that *Soyuz 3* had deviated from its intended flight path and had consumed most of its fuel.

Beregovoi, using the call sign Argon, reported that the ships had been aligned incorrectly, which prevented a docking, and that the approach and orientation engines had used up nearly 180 pounds of fuel, precluding a second linkup attempt. Beregovoi would need the remaining twenty pounds of maneuvering fuel to align *Soyuz 3* for de-orbit. The cosmonaut expressed his extreme disappointment at not being able to fulfill the major objective of his flight, but offered no further explanation for the failed docking.

The Soviet propaganda machine was cranked up once again, and the flight was lauded as a great success. However, some astute Western observers were very skeptical, and headlines such as "Russian Flight a Mystery" appeared around the world amid mounting speculation that *Soyuz 3* had failed in its mission. This was significantly fueled when the official Tass News Agency failed to disclose how close the two ships had been. Investigative observers saw no reason why the Soviets would have attempted a simple approach maneuver, which would not appear to appreciably advance Soviet space research. Tass did quote a message from Kremlin leaders to the cosmonaut, saying that Colonel Beregovoi "had so far fulfilled the program of the flight." Later reports from Moscow would also state that there was never any intention of docking the two orbiting spacecraft, which might have put the life of the cosmonaut at risk. It was all a cover-up to disguise the fact that a senior cosmonaut had performed well below expectations.

One could possibly excuse Beregovoi because the view ports were fogged and dusty and he had to attempt the docking by himself, without the aid of an onboard engineer. Yet he was later at a loss to explain why, as he closed in on *Soyuz 2*, his spacecraft had suddenly banked 180 degrees in relation to the target vehicle, despite his best efforts to compensate for what he suggested was a fault in the Igla guidance system. He would only remark that "the flight plan was difficult and saturated."

On the fourth day of the flight, the S5.35 main engine located in the service module of *Soyuz 3* fired for 145 seconds. Beregovoi could feel the deceleration process begin and closely monitored his instruments as he readied himself and the descent module for a fiery plunge through the atmosphere. Six seconds after retrofire had been successfully completed he heard the dull thud of precision pyrotechnics as the orbital and instrument modules on either side of his descent module were jettisoned, freeing it for reentry.

While Soviet controllers monitored the progress of the *Soyuz 3* descent module, feelings of relief were tinged with understandable anxiety as the first manned Soviet spacecraft since Komarov's death arced earthward. The reentry was following the expected track, seemingly without any problems.

Overall, the spacecraft had functioned well, and this was welcome news to Soviet space planners. What was nowhere near as satisfactory was the poor performance of the cosmonaut. There was a gloomy realization that the wrong person had been sent to do a job that should have been entrusted to a better-trained pilot. Although he would be accorded praise in all the newspapers and would receive the usual homecoming rituals, backslaps, and medals, the sadly underperforming Beregovoi had just flown his only space mission.

Soyuz 3 made a safe touchdown in the planned landing area in a remote area of Kazakhstan, and helicopters patroling for Beregovoi's arrival were quickly on the scene after watching the last stages of the parachute descent. Despite this, they still did not beat some eager local people to the craft, including a small boy on a donkey, who watched in awe as Beregovoi clambered out of his spacecraft. The weather on the ground was typically cold—just a few degrees above zero with gusting winds and snow flurries—and Beregovoi's first gesture was to touch some snow on the ground, as if to reassure himself that his four-day flight was truly over. "It's good to be here," he told the villagers, who had slogged across the snow-covered fields to greet him, "but it's cold."

The small crowd swarmed around Beregovoi as the rescue helicopters moved in, while some adventurous children took the opportunity to ask the cosmonaut for his autograph and posed for photographs, which he took with one of his still cameras. In answer to their questions, he merely said he felt excellent, but a little hungry. They told him a ring of stones would later be placed around the landing site, as a lasting monument to his flight.

The recovery crews then moved in, and Beregovoi gratefully slipped on the thick fur coat they gave him. "From summer into winter," he joked as he donned the coat over the lightweight jacket he had worn on his flight. He finally climbed aboard one of the helicopters, waved goodbye to the villagers, and was whisked away for three days of debriefing, as well as scientific and medical evaluations. On his arrival at Moscow Airport on Saturday, 2 November, Beregovoi was accorded a red-carpet welcome and greeted by Soviet party chief Leonid Brezhnev. During an hour-long ceremony, he was made an air force major general, presented with a second Order of Lenin, and awarded a second Hero of the Soviet Union gold star.

Outwardly, it may have seemed that Georgi Beregovoi's flight was a great success, but the planners knew it had fallen well short of its objectives. The designer of the Igla system, Armen Mnatsakanyan, was insistent that the rendezvous guidance system had functioned as expected, unlike the cosmonaut. Later analysis would prove him right, and pilot error was determined to be the cause of the failure to dock. Beregovoi had not stabilized *Soyuz 3* along the right guidance axis to the target craft, nor had he used subtle corrections to achieve the correct orientation to *Soyuz 2*. Instead, he had used strong firings of the maneuvering rockets, which placed *Soyuz 3* into a completely incorrect orientation to his target. Meanwhile, the reciprocal system on *Soyuz 2* had sensed the misalignment of the approaching active craft and had automatically tilted its docking nose to prevent what it correctly assessed to be an erroneous approach. Beregovoi, puzzled by this, had then completed a fuel-expensive fly-around of the recalcitrant vehicle and made a second approach, again without success. By this time he had used almost all the fuel allocated to the docking procedure, and was reluctantly forced to abandon the task.

In an interview conducted by the Bulgarian *Trud* newspaper in April 2002, spacecraft designer and cosmonaut Konstantin Feoktistov emphatically stated that Beregovoi "committed the grossest error. . . . He did not turn his attention to the fact that the ship to which he was meant to dock was overturned [upside-down in relation to *Soyuz 3*]. The flashing lights of the unmanned ship proved to be on top, and should have been below. Therefore, the approach of *Soyuz 3* [caused] the pilotless object to turn away. In these erroneous maneuvers, Beregovoi consumed all the fuel intended for the ship docking." During a postflight report to the State Committee on 31 October, Beregovoi was critical of the sensitivity of the manual controls he used in trying to conduct the failed dockings, and said improvement was needed in this area. Later, at a press conference, the cosmonaut was asked whether his age had caused him any problems in being selected for the *Soyuz 3* flight. He carefully replied that his height (just over six feet) had proved a bigger obstacle.

The problems with the failed linkup between the two Soyuz craft now meant far greater pressure than ever would be applied to the four crew members of *Soyuz 4* and *5*. After two successive manned mission failures, this time a docking *had* to succeed.

At 10:38 a.m. on the morning of 14 January 1969, *Soyuz 4* blasted aloft.

With the use of models, *Soyuz 4* mission commander Vladimir Shatalov demonstrates docking procedures between two Soyuz spacecraft. Courtesy Colin Burgess Collection.

This created immediate speculation in the West that another manned craft would soon join it in orbit and that this could be the start of a plan to assemble a huge orbiting workshop. As it came soon after NASA's spectacularly successful *Apollo 8* flight, there was also some speculation that *Soyuz 4* might be a prelude to a Soviet moon flight.

The Soviet news agency Tass reported that the sole cosmonaut on board was Lt. Col. Vladimir Shatalov, an air force pilot who had joined the cosmonaut detachment in 1963. The bulletin stated that he and his wife Muza had two children: a sixteen-year-old son called Igor, and a ten-year-old daughter, Yelena. The son of a railway engineer who had been a pilot during the Russian Civil War of 1918–21, Shatalov was born in Petropavlovsk, a city in southwestern Siberia, on 8 December 1927, but had spent his early years in Leningrad. His childhood hero was legendary test pilot and polar flight pioneer Valery Chkalov, and he dreamed of one day emulating him.

He was eventually recommended for admission to a special air force flying school in Lipetsk in 1945, and later underwent advanced flight instruction at the Red Banner Air Force Academy in Kacha, graduating with distinction in 1956. He then served in various air force units, rising in rank and office until his selection to the cosmonaut team. He was by far the most outstanding member of his selection group. There was little surprise when he was not only the first member of his group to fly but also flew ahead of some cosmonauts selected before him. Like Beregovoi, he had been a candidate to fly a Voskhod mission before the extended program was canceled.

Postlaunch reports stated that Shatalov was "feeling fine," as his spacecraft, *Soyuz 4*, orbited the Earth once every 88.5 minutes with an apogee of about 141 miles and a perigee of around 108 miles. All systems were said to be functioning normally, and Shatalov had begun to implement his flight program.

Soviet citizens first learned of the launching when Radio Moscow interrupted its normal programs to announce the event in laudatory terms. Later, just ninety minutes after the launch, those with television sets were able to watch images of the liftoff, the first time a Soviet manned launch had been shown on the same day. Shatalov was also shown in file footage before launch. Prior to boarding the elevator that would carry him up to his waiting spacecraft, he said, "Leaving on a space trip, I assure the Central Committee of the Communist Party of the Soviet Union and the Soviet Government that I shall give all my strength and knowledge to carrying out the tasks set before me. Goodbye, dear friends. Until I see you soon on dear Earth."

Western space experts were intrigued by images of Shatalov's cabin taken from on board the orbiting spacecraft, which looked large enough to hold more than a single cosmonaut—and it was fitted with two spare couches. They also noted that *Soyuz 4* was equipped with two television cameras inside the vehicle and another two outside. This further fueled talk of a manned linkup and possible crew transfer. By 3:00 p.m., some four and a half hours after liftoff, Shatalov had completed his third orbit of Earth. He had also assumed manual control of *Soyuz 4* in order to align the ship's solar panels to the sun. Using the code name Amur (after a Russian river) he reported back to ground control, ate lunch, carried out "various

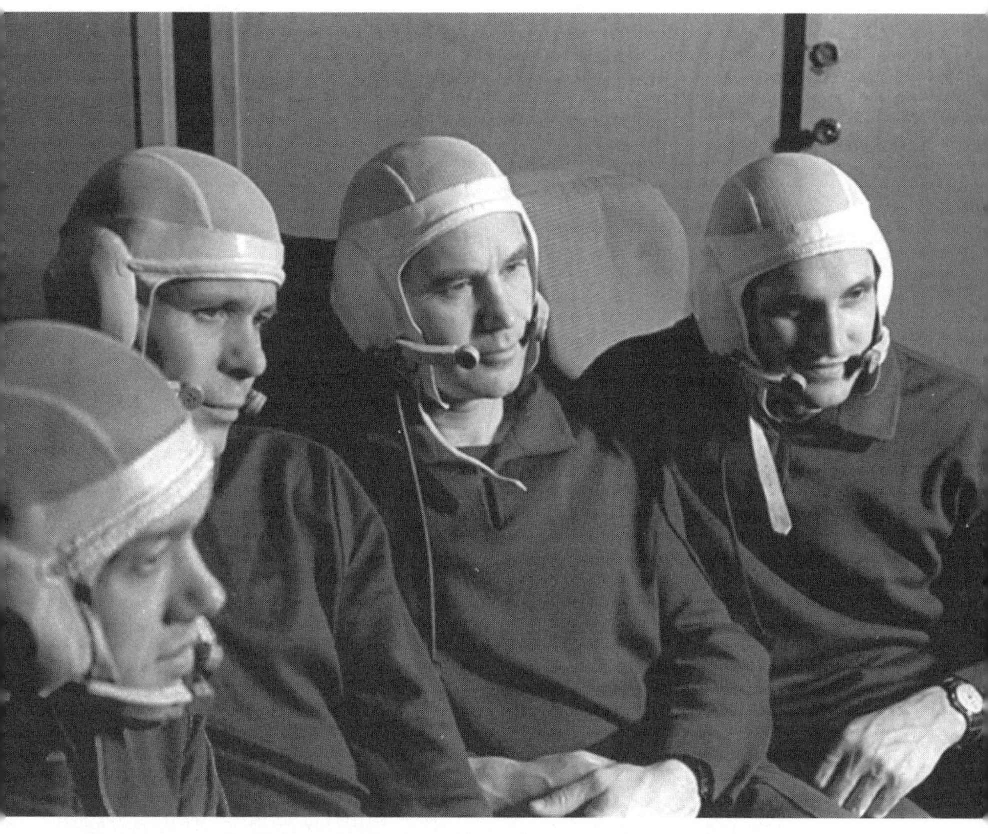

Soyuz 4 commander Vladimir Shatalov (*second from right*) sits with the crew of *Soyuz 5*. *From left*: Alexei Yeliseyev, Yevgeny Khrunov, Shatalov, and Boris Volynov. Yeliseyev and Khrunov would become the first people to return to Earth in a craft other than the one in which they were launched. Courtesy Colin Burgess Collection.

experiments," and was shown on television smiling, wearing a comfortable lightweight suit. On his fifth orbit, Shatalov fired his main engine, kicking *Soyuz 4* into an almost circular orbit. He then began preparing for his first night's sleep in space.

The following morning, three men rode the elevator up to the top deck of the launch platform, where they were inserted into their spacecraft. A delighted Boris Volynov was the commander of this mission, finally making his first spaceflight nine years after being selected as a cosmonaut. Once he reached the boarding level, Volynov took one last look around the Kazakh steppes before technicians helped insert him through the hatch into the orbital module of *Soyuz 5*. After checking out the equipment there he made his way down through another hatch into the descent module, where he and his fellow "space rookies" strapped themselves in for liftoff. Volynov was followed into the spacecraft cabin by flight engineer Alexei Yeliseyev and research engineer Lt. Col. Yevgeny Khrunov.

Alexei Yeliseyev was a civilian engineer-cosmonaut who had worked in Sergei Korolev's design bureau after graduating from the N. E. Bauman Moscow Higher Technical School in 1957, eventually specializing in space engineering in 1963. "My work is not easy," he would later explain. "That is why it is interesting." He had proved himself to be an outstanding engineer in his work designing the Soyuz spacecraft.

The future cosmonaut was born on 13 July 1934 in the city of Zhizdra, not far from Kaluga. When he was seven years old, with war imminent, his family was evacuated from Zhizdra and traveled to the Soviet Far East. At the war's end they moved to Moscow where Alexei looked after his younger brother while their parents worked. His mother Valentina was a chemist, and she instilled in him an interest in science and mathematics. Alexei completed secondary school in 1951 and was enrolled at the Bauman School. Here, he would prove to be not only an outstanding student, but would also become a "Master of Sport," as a two-time fencing champion for the Soviet Union. Just fifteen years later he was selected to join the cosmonaut team, although his selection only came after a bitter factional dispute led by Chief Designer Mishin, who wanted to form an auxiliary cosmonaut group drawn from civilian engineers, and the Soviet Air Force, represented by Lt. Gen. Kamanin. At the time, the air force enjoyed a complete monopoly on cosmonaut selection and training, and was firmly set against the presence of nonmilitary candidates at its cosmonaut training facilities. Eventually, a compromise was reached, and Yeliseyev became a member of an eight-person civilian cosmonaut group. It was a close thing; his father had spent five years in prison around the time of Alexei's birth, accused of agitating against the state. Yeliseyev was even using his wife's last name, rather than that of his father, Kureytis, to try and disassociate himself from this family episode. That he was chosen to fly such an important propaganda flight given such a background was surprising.

One of eight children raised by his widowed mother, Yevgeny Khrunov, like Volynov, had been selected in the first cosmonaut detachment in 1960. His parents Vasily and Agrafena had lived in the village of Prudy in the

Tula region, one hundred miles south of Moscow, but peasant life in the small remote village had become even harder for the large family when Vasily died in 1948. Because of the war, Yevgeny began school late, but soon caught up with the other students. He attended the Kashira Agricultural School where he studied farm machinery. Another type of machinery—airplanes—fascinated him even more, however, and he decided to become a pilot. Khrunov was accepted into a flying school in 1952 and the following year enrolled at the Bataysk military flying school. In 1956 he graduated and became a pilot in the Soviet Air Force. Four years later Khrunov was back in Moscow, this time as a cosmonaut in training for spaceflight. Like Volynov, Khrunov had also served as a backup for an earlier mission, as prime backup for Alexei Leonov on the flight of *Voskhod 2*.

Khrunov and Yeliseyev had been in this position before. With Bykovsky as their commander, they had been ready to fly the original *Soyuz 2*, to join Komarov in orbit in April 1967. They would have made the same EVA that was now planned for *Soyuz 5*. Less than a day before they were due to launch the flight was canceled, and they later listened in horror to the ongoing reports as their colleague fought and finally died trying to bring his troubled craft back to Earth. Now they would attempt to carry out the same mission, and could only hope that all the spacecraft system problems had been rectified.

With *Soyuz 5* safely in orbit, both commanders set about preparing for the rendezvous and docking planned for the following day. Mission planners had decided to take things slowly following the problems on Beregovoi's flight. On his fifth orbit, like Shatalov before him, Volynov (using the call sign Baikal, after a lake in eastern Siberia) fired his main engine in order to circularize their orbit and be more in line with the path of *Soyuz 4*.

On the morning of 16 January, ground controllers determined that all was ready for the linkup, and, following a second, smaller maneuver by Shatalov over the South Pacific, they switched on the Igla approach control system. Over the next thirty minutes, the two manned vehicles closed to within a hundred yards of each other. They were identical craft, apart from the probe and receptacle in their respective docking ports. Shatalov would now assume the active role of the two craft, while Volynov's *Soyuz 5* became the passive vehicle. Over Africa, they approached to within ninety feet of each other and then held off until they came back within range of Soviet ground controllers, which, according to Volynov, "was not a simple thing to do: I controlled the spacecraft manually, and kept it at a steady distance—thirty meters between the spacecraft. I controlled that distance myself all the time. The orbits were a bit different so the velocities were also a little bit different; . . . you could easily keep the distance but you constantly had to work to control the ship."

Once they were back in radio range, both commanders maneuvered manually and closed in even further, ensuring that the docking units of both spacecraft were properly oriented toward each other. A videotape of the actual operation was later shown on Moscow television, carrying the sights and sounds of the historic feat into millions of homes. Television cameras mounted inside both Soyuz vehicles had relayed periodic reports to anxious audiences throughout the mission, part of a propaganda campaign to help counter some of the impact of the recent *Apollo 8* mission.

The final alignment and maneuvering of the two craft was shown via the television cameras mounted on Shatalov's *Soyuz 4*. They showed *Soyuz 5* approaching, with the strong reflected sunshine giving it the appearance of glowing brightly in the blackness of its surroundings. The thin gullwing solar panels were extended outward; to Shatalov *Soyuz 5* resembled an approaching bird. The voices of the cosmonauts on both craft could be heard throughout the approach.

"I now have Baikal on my screen," Shatalov reported. "Velocity 0.82 feet per second. We'll proceed." It was a procedure the cosmonaut had practiced almost eight hundred times in the training center's simulator, and he later said he felt assured of success. "When you feel the machine is under your command, the feeling of confidence comes by itself."

"Beautiful, very beautiful," Khrunov responded. "Just magnificent. Amur is flying like a bird from a fairy tale. She's coming up like an aircraft."

Volynov was a little more pragmatic in his dialogue. "Everything is fine . . . waiting for contact."

"I'm moving in," commented Shatalov. "Everything is normal. Contact light! Linkup firm. Docking!"

Though Russian viewers were exultant at witnessing this feat soon after it had happened, they were rather stunned to hear at the moment of docking a voice from *Soyuz 5* crying out, "We're being raped! We're being raped!" When the tape was broadcast again, minutes later, this portion of the dialogue had been diplomatically edited out. After the flight, the crew would be mildly reprimanded for its inappropriate choice of words. "When you have worked very hard for something, and at last it happens," Shatalov said, shrugging off the incident, "then extreme happiness overtakes you. So it was with the meeting of our spacecraft."

During the nineteenth orbit of *Soyuz 5*, Yeliseyev and Khrunov began initial preparations for their EVA transfer to *Soyuz 4* by making their way into the orbital module of their craft, where they extracted two Yastreb spacesuits from a storage locker. Volynov was now free to help them, so he also made his way into the living compartment, assisting his crew members into their suits and with their life support systems, which he later said included "the backpacks with the oxygen supplies, ventilation, communications, and so on; switching them on, checking them out . . . many, many important details." They were also carrying a few small gifts for Shatalov, including some mail and newspapers announcing his successful launch the previous day.

Volynov then hugged both men before retreating into the descent module and closing the inner hatch. The orbital module was now effectively an air lock, so the two cosmonauts carefully checked everything before opening a side hatch as the linked spacecraft swept over South America. Unlike Apollo spacecraft, there were no direct transfer hatches incorporated into the docking apparatus; an EVA was the only means of crew transfer. Once again they conducted spacesuit pressurization and communications checks before venturing out into raw space. Both men were tethered to the assembly by safety umbilical lines, which contained their means of communicating and also wires monitoring their physiological well-being during the transfer. Volynov, assured that everything was in order, gave his permission to proceed. At first Khrunov just poked his head out and gazed around, then slowly moved out of *Soyuz 5*. He would later recall that incredible moment:

The sun was unbearably bright and scorching. Only the thick, filtering visor saved my eyes. I saw the Earth, the horizon, and the black sky and had the same feeling I had experienced before my first parachute jumps. And I'll freely admit I felt all the anxiety of an athlete at the starting line. But I overcame that anxiety. I became completely absorbed in my work and thought only of carrying out my assignment.

I emerged from the spacecraft without difficulty, and looked around. I was amazed by the marvelous, magnificent spectacle of the two spacecraft linked together high above the Earth. I could make out every tiny detail on their surfaces. They glittered brilliantly as they reflected the sunlight. Right in front of my eyes was Soyuz 4, *looking very much like an aircraft. The big, long spacecraft was like a fuselage, and the solar panels were like wings.*

Although it was strictly against orders, Khrunov briefly lifted the filtering visor on his spacesuit helmet to get a clearer look at his surroundings. The blinding sunlight soon made him lower it again. Slowly and cautiously, he then began moving toward the docking unit of *Soyuz 5*, easing himself along the spacecraft using rails that had been secured on the exterior for this purpose. "Particularly great effort had to be made for turning the body around some point on which your hands had been placed," he recalled. "That was not easy."

There was one other task to perform: he removed an external television camera from a bracket, switched off its power supply, and transferred it to a similar bracket on the outside of *Soyuz 4* to record Yeliseyev's transfer. Once this had been done he continued moving toward the hatch on the orbital module of *Soyuz 4*, which had been mechanically opened by Shatalov. Unfortunately, Khrunov lost his still camera on the journey; it floated away, taking some priceless photos of the historic transfer with it. He entered the hatch feet first, followed several minutes later by Yeliseyev. The entire transfer had taken just over an hour; Khrunov had made his passage over Soviet territory and Yeliseyev above South America. Khrunov was elated with the success of the operation—the first ever crew transfer in space—but there was still some work to be done, as he described: "We closed the hatch, cut in the pressure feed, creating normal pressure, and helped each other take off our spacesuits. Then Shatalov came swimming in toward us through the hatch. We all hugged and kissed each other, talking fast and not making much sense. Shatalov thanked us for the mail we had brought him and held up the newspapers and letters for the [television] audience to see. . . . Jokingly, we begged forgiveness of Volodya Shatalov for keeping him waiting twenty-four hours."

With the crew transfer now completed and everything in order aboard both craft, it was soon time for both commanders to prepare for the undocking maneuver. Just three orbits and four hours and thirty-four minutes after they had first linked up, the two spacecraft successfully disengaged and separated. Slowly the gap between them widened.

The next morning, *Soyuz 4* would become the first of the two craft to return to Earth. Tass described the craft's reentry in the following way:

The braking engine was fired at a pre-calculated point in orbit and, as it followed up the present impulse, it reduced orbital speed and the ship went into the descent trajectory. After the braking engine was switched off, the landing capsule with the cosmonauts detached from the orbital module. The landing capsule made a controlled descent in the atmosphere, making use of its aerodynamic features, and appeared over the preset landing area, where a parachute system and a soft-landing engine ensured a smooth touchdown of the apparatus.

Soyuz 4 and the three cosmonauts came to rest on the ground at 9:53 a.m. Moscow time, twenty-five miles northwest of Karaganda in the steppes of Kazakhstan. Shatalov had even returned to Earth with a small, unauthorized souvenir—the celestial globe that he had used for navigation. Strictly, he was supposed to leave all of the orbital equipment behind in the abandoned orbital module, but Shatalov did not want to part with something that had proved so useful during the docking.

In reporting the safe touchdown, Tass stated that *Soyuz 5* was "continuing its flight under Colonel Boris Volynov's command," and that "all was well." Volynov was overjoyed to hear that his colleagues had landed safely and began preparations for his own reentry the next day. Like Gagarin before him, he was orbiting the Earth alone—the last cosmonaut to do so before the era of manned lunar landings began. In fact, he would be the last person ever to return from orbit alone that century. He was not to know it, but he was also about to undertake the most perilous descent ever survived by anyone in the history of space exploration.

At 10:20 a.m. the next day, as *Soyuz 5* passed over the Gulf of Guinea near Africa, the landing sequence began. Retrofire took place just as planned, and the main engine shut down after firing for just under two and a half minutes. Explosive bolts designed to separate the orbital and instrument modules detonated, but seconds later a cold shiver ran through Volynov—

he could still see the whip antennas extending from the solar panels, which signified that the separation had not been fully effected. Although the orbital module had been discarded, the instrument module—containing the rocket engines and power supplies—was still attached, covering the heat shield of the descent module. There was no way that Volynov could abort the reentry phase. It was a potentially catastrophic failure, and he was in serious trouble.

Once the spacecraft reached the upper fringes of the atmosphere it began tumbling as it automatically sought the most aerodynamically stable position. It quickly stabilized, but headed into the descent nose first. This meant the instrument module and its flared base were at the rear of the assembly, while the heavier descent module with its light metal hatch was facing to the front. The heat shield with its six inches of ablative material, meant to protect the craft during descent, was now at the wrong end of the spacecraft. The spacecraft was effectively reentering back to front, and it seemed inevitable that it would burn up in the terrible heat of atmospheric friction.

As *Soyuz 5* began to plummet through the thickening air, a shimmering heat wave began to build up around the poorly protected nose of the craft. There was only one inch of insulation between Volynov and incineration, and he was not wearing a protective space suit—not that it would have been very effective. He quickly reported his situation to the controllers on a Soviet tracking ship, but there was nothing anyone could do. Gaskets sealing the entry hatch began to melt and smolder, filling the cabin with acrid fumes. Meanwhile, the g-forces were building up, but instead of being pressed against his couch, Volynov was being pulled outwards against his harness straps in the back-to-front descent.

Voice contact was quickly lost with the controllers, and the situation became more disturbing with each passing second. Volynov quickly tore all the pages detailing the rendezvous from his flight log, rolled them up as tightly and possible and stowed them where he hoped they might somehow survive, even if he did not. There was also a small onboard tape recorder, and knowing his chances of survival were slim, Volynov calmly began to record his impressions of events as *Soyuz 5* tore a fiery path back to Earth.

"I looked out of the window of the capsule and saw the flames," he later recalled, "and I said to myself, 'This is it, in a few minutes I'm going to die.' It's hard to describe my feelings; there was no fear but a deep-cutting and very clear desire to live on when there was no chance left." According to journalist Aleksandr Milkus, in a chilling account he wrote of the unfolding drama for *Komsomolskaya Pravda* in 1998, "In the Mission Control Center one of the officers took off his cap, put three rubles in it and sent it further. Little by little it was filling up with money. This had already happened once, on the day when Vladimir Komarov died." Milkus wrote that many controllers, unable to do anything more, buried their faces in their hands as they waited for the inevitable bad news.

Volynov heard dull thuds from outside as propellant tanks located in the instrument module of *Soyuz 5* began exploding in the heat, and he watched in alarm as pressure waves resulting from the explosions caused the hatch to buckle inward toward his face and then outward, somehow holding against the lethal pressure. Now Volynov's body was straining against his harness at more than nine g's, and he could feel the heat steadily rising within the cabin.

Eventually, thermal and aerodynamic stresses built up to such a point that the thin metal struts between the two modules began to bend and disintegrate in the ferocious heat and friction of reentry. The exploding hydrogen peroxide tanks had probably been a major contributor in this process, for soon after the instrument module finally tore away from its mounts. The descent module, which had taken enormous aerodynamic and heat loads around the highly vulnerable area of the top hatch, now swung around to the normal reentry attitude with a welcoming lurch and made a purely ballistic reentry, with the blunt heat shield finally facing into the terrible furnace of heat, then nearing five thousand degrees Celsius.

Volynov's relief would be short lived. Though the module was still rotating wildly he realized he would probably now survive the plunge through the atmosphere. However, he had another major concern—the integrity of the parachute system. The failure of this system eighteen months earlier had been the prime factor in the death of his colleague, Vladimir Komarov. He could not know if the heat and pressure had somehow damaged the automatic parachute release system, causing a similar malfunction. Once again, the only thing he could do was sit back and ride it out—and hope. He would only know his likelihood of survival when he passed through ten thousand feet, at which point the parachute was scheduled to deploy.

Just as he had feared, the main parachute only partially deployed. The prospect that he might plow into the ground at high speed remained a very grim reality. Then, almost miraculously, the parachute lines untangled with a reassuring thump, and Volynov felt a noticeable deceleration in the cabin. For the rest of the descent, the charred spacecraft mostly operated according to procedure, and the soft-landing rockets fired as planned, moments before touchdown. Nevertheless, Volynov had been approaching the landing faster than planned; his spacecraft was still rotating and oscillating under the parachute, and the rockets were inclined away from the ground when they fired. To make matters worse, the snow-covered ground below *Soyuz 6* was frozen solid. When the spacecraft hit, it hit hard.

Volynov was thrown violently against his harness, which gave way, and his head and shoulders were slammed into the instrument panel. All the teeth in his upper jaw were smashed or broken off on impact. It could have been a lot worse, but fortunately his shock-absorbing couch protected him from much of the force of impact. The spacecraft had endured an uncontrolled ballistic reentry and was nearly four hundred miles from the planned landing area. As he sat contemplating the incredible fact that he was still alive, Volynov could hear his hot spacecraft hissing and popping in the extreme cold outside as it cooled.

Later published reports would suggest that Volynov decided he could not stay in his spacecraft and sought help after seeing chimney smoke rising from a distant farmhouse. Had he left the relative protection of the descent module as these reports suggest, he would likely have perished in the bitter cold, as he was badly bruised and bloody, and dressed in little more than a lightweight woolen track suit. Volynov indeed denies that he ever left the safety of his spacecraft until help reached him, as he explained to Bert Vis. "There was nobody to help. It was two hundred kilometers from Kustanay [Kazakhstan], and it was deserted all around. No one, and no settlements at all . . . no farm, no nothing. For maybe sixty or eighty kilometers there was nothing. I couldn't go anywhere as it was minus 38 degrees [Celsius] and I was wearing that suit. Minus 38 degrees . . . cold! I waited inside the capsule."

Vis asked Volynov how long it took for rescuers to reach him. "Not so long . . . about an hour, maybe," he said. "I noticed a rescue jet after about, maybe, forty minutes. They looked for me in different directions;

. . . then they saw the parachute. They started descending and then parachutists came. Three soldiers and a senior lieutenant. Then there was a radio link and helicopters came. It turned out that I was alive, so all were pleased to see me and they were all surprised, because all of them were sure . . . Mission Control was sure that this was a repeat of Komarov's flight. But it didn't turn out that way." At the later medal ceremony in Moscow, Volynov was in considerable agony with his smashed teeth and jaw, and subsequently spent several months in hospital receiving treatment and recuperating. In fact his injuries were so severe that he was grounded from further spaceflights for two years.

Meanwhile, postflight investigations revealed that the most likely cause of the instrument module's failure to separate was underpowered explosive charges—a problem that was easily remedied. Had Volynov been incinerated along with his tumbling spacecraft, it certainly would have brought another tragic halt to the Soviet space program, with an immediate, incalculable effect on the future of Soviet space exploration. Western analysts did not learn the full, amazing facts of Boris Volynov's perilous return to Earth for nearly three decades.

After the flight, Soviet designers and planners shrugged off any prevailing rumors concerning the complex mission, instead referring to it as an outstanding success that had clearly demonstrated the Soviet Union's space rescue capability. By docking in orbit, they said, *Soyuz 4* and *5* had unquestionably opened the way for the imminent creation of the first Soviet space stations.

The story of the *Soyuz 4* and *5* joint mission received a bizarre ending on a bone-cold day in Moscow on 22 January 1969, just four days after Volynov's near-fatal reentry. Soviet citizens everywhere had been celebrating the docking mission, ostensibly an impressive display of technology and human precision. It was certainly one of the Soviet Union's finest moments in space, and a hero's welcome had been organized for the arrival of the four triumphant cosmonauts.

Once they had undergone postflight debriefing and medical examinations—particularly Volynov—the cosmonauts flew into Moscow's Vnukova Airport. Here they were saluted by a band playing the Soviet anthem and were handed bouquets of flowers by children whose teeth chattered from the freezing cold. A three-hundred-foot-long red carpet was rolled out for the nation's newest space heroes, at the far end of which stood a small cluster of government leaders. Each of the cosmonauts was then bear-hugged and kissed by General Secretary Leonid Brezhnev.

Following the airport parade, Vladimir Shatalov, Boris Volynov, Alexei Yeliseyev, and Yevgeny Khrunov were escorted to a waiting open-top Zil limousine, the lead vehicle of twenty forming an impressive motorcade into Moscow. A well-publicized street parade was to take place, followed by a televised ceremony at the Kremlin Palace of Congresses in front of six hundred political guests, space scientists, and academicians. The second, closed Zil sedan in the convoy carried four veteran cosmonauts: Georgi Beregovoi, Alexei Leonov, Andrian Nikolayev, and Valentina Nikolayeva-Tereshkova. Moscow was ready to celebrate.

As the gleaming black cars swept toward the Kremlin's Great Borovitsky Gate, the four cosmonauts in the lead vehicle stood up and waved to acknowledge the cheering crowds lining the streets. Then, without warning, a Red Army lieutenant wearing a hat, dark glasses, and a military-style uniform burst through the crowd and began firing two automatic pistols at the occupants of the second vehicle. The driver, who had earlier expressed his pride at being given this prestigious chauffeuring assignment, caught the full force of the assassin's bullets. Struck three times, he was fatally wounded, and the Zil swerved out of control, away from the gunman. A police motorcycle outrider was also hit in the fusillade of shots and fell to the street, badly wounded.

In the rear of their car the four cosmonauts ducked as the windscreen suddenly exploded into thousands of fragments, spraying shards of glass everywhere. Then a side window also shattered as the sound of gunfire continued to resound. Terrified Muscovites flung themselves to the pavement in panic. Brezhnev's driver reacted instantly, accelerating away from the area, and did not stop again until the limousine was safely within the confines of the Kremlin. Tereshkova had thrown herself onto the floor of the limousine in disbelief. Nikolayev flung himself on top of her, hoping to protect his wife from the bullets and flying glass. By this time, Beregovoi and Leonov were also seeking cover on the crowded floor, as more glass fragments showered over them.

Within seconds it was all over. The four cosmonauts remained where they were, pressed to the floor. Beregovoi was the first to move, cautiously clambering back onto the seat and cursing from the pain of glass fragments embedded in his knees. His face was streaked with blood. Outside the limousine there was pandemonium. Tereshkova choked back a scream as the passenger door was wrenched open. But it was just someone yelling that it was all over—the gunman had been captured. Leonov, shocked to see bullet holes in his clothing, gripped the top of the front passenger seat and began to haul himself from the floor.

Nikolayev and Tereshkova, pale and badly shaken, clutched each other for comfort as they were helped from the bullet-pocked limousine. Through the shattered, bloody front windows they could see that their driver had been severely wounded, and nearby the downed motorcycle policeman was receiving urgent attention. White-faced spectators were also clambering to their feet, trembling and sobbing. Just moments earlier they had been cheering wildly as the entourage swept by, the open, leading vehicle carrying four smiling, waving cosmonauts, just returned from the latest Soviet space feat. Many had cheered loudest when they glimpsed the occupants of the second, closed limousine. Their happy cries of "Valya, Valya!" had abruptly changed to screams of terror when the assassin, named Viktor Ilyin, had surged through the crowd, pointed the handguns, and begun firing into the vehicles.

When the chambers in Ilyin's automatic pistols were finally empty, the crowd had moved in and grappled him to the ground, seizing his weapons as security guards arrived on the scene. Then Ilyin was hauled to his feet and dragged away. A deserter from a Red Army unit stationed near Leningrad, he had stolen the weapons from his barracks, then borrowed the dark blue uniform of a militia captain from his unsuspecting brother-in-law, ostensibly to give him a better vantage point as the cosmonauts drove by. It would later turn out that they were not his intended victims. He had mistakenly believed the enclosed limousine following the lead car was carrying Leonid Brezhnev and Soviet president Nikolai Podgorny. Georgi Beregovoi was not unlike Brezhnev in appearance, with the same swarthy looks and bushy eyebrows, which could have led to the gunman's error.

Alexei Leonov would later recall the bizarre incident in graphic detail, saying, "I looked down and saw two bullet holes on each side of my coat where the bullets passed through. A fifth bullet passed so close to my face

I could feel it go by. This man was shooting at me, thinking that I was Brezhnev. He was angry because he had been conscripted into the army. When it was over, Brezhnev took me aside and told me: 'Those bullets were not meant for you, Alexei. They were meant for me, and for that, I apologize.'" The cosmonaut occupants were incredibly lucky—a total of fourteen bullets had penetrated their limousine, but the spacefarers' wounds were only superficial.

Despite the drama and the death of the driver, the motorcade eventually resumed its journey into the Kremlin Palace. Tereshkova and the other cosmonauts had escaped serious injury, although a furious Beregovoi was still tending to some minor cuts on his face and knees. The official ceremony honoring the shaken Soyuz cosmonauts was completed only a few minutes behind schedule. Each was presented with the Medal of the Order of Lenin and declared a Hero of the Soviet Union. According to newspaper reports, guests at the function were unaware of the recent drama. In fact, the whole incident was so effectively hushed up that Western journalists chasing news of the shooting were unable to find a single Russian who would admit to being an eyewitness. Nevertheless, it would be the last official motorcade of this kind permitted for fear of a similar attack.

After he had been arrested, Ilyin—not surprisingly for those times—vanished without a trace. He was declared insane and locked up. The motivation for his attack was believed to have been his enforced recruitment into the army, but this was known only to Ilyin and, presumably, his KGB inquisitors after their interrogation. The officer in overall charge of security for the procession was later sacked for incompetence. According to an entry in the autobiographical papers of cosmonaut Yeliseyev, Ilyin served a twenty-year sentence before his release. He was by then no longer regarded as a threat to society.

Georgi Beregovoi would never make another spaceflight. In June 1972, having been promoted to major general, he became director of the Cosmonaut Training Center (TsPK), a position he held for the next fifteen years. During this period he continued his academic studies, earning his candidate of psychological sciences degree in 1975. He received a further promotion to lieutenant general in 1977, and stood down as director of the center in 1986. He officially resigned from the cosmonaut team on 3 January 1987, retired from the air force, and took on a position at the Lidar

All-Union Information Service of the Soviet Academy of Sciences. On 30 June 1995, at the age of seventy-four, Georgi Beregovoi passed away from natural causes. He is buried in the Novodevichi cemetery in Moscow.

Following his flight, Boris Volynov would direct the training of a cosmonaut group selected in May 1970 and serve on ground-support teams for several Soyuz missions. He had been in line for another assignment following his convalescence, but once again became the subject of undisguised bigotry in the Soviet political and space hierarchy. On 23 April 1970, Chief Designer Vasili Mishin submitted crew manifests for the first four flights of the new Salyut Long-Duration Orbital Station, or DOS, to Colonel-General Kamanin, who wielded almost dictatorial authority. The general took one look at Mishin's list and immediately rejected several of the recommendations. He wanted Konstantin Feoktistov's name removed from the list because he had medical problems and had recently divorced his second wife, while Yevgeny Khrunov was ruled out because he had recently been involved in a hit-and-run accident and had failed to render aid to the victim. Boris Volynov's name, however, was said to have been removed simply because he was Jewish, which Kamanin found unacceptable. According to Asif Siddiqi, "Kamanin had been instructed by Ivan D. Serbin, the chief of the Defense Industries Department in the Central Committee, not to allow Volynov to fly again."

Volynov mentioned to Bert Vis that doctors at the training center told him he would never fly again after his perilous and injury-sustaining mission, that there was "a psychological barrier." He then added,

The doctors said, "Everything you saw will be an obstacle for you, for your next flight. You'll never fly again, not even on board a passenger jet. You will be afraid of that." That is why I was grounded. Kamanin suggested to me a few times to take an administrative job; . . . he wanted to use my experience as a cosmonaut. The level of training was high, and they decided to use me as an administrator. That's why they wanted to appoint me commander of the cosmonaut detachment. I refused: I wanted to fly. But in the end they made it an order, and I was appointed instead of asking if I wanted to do it.

Ironically, it was the death of the three cosmonauts on the *Soyuz 11* mission in June 1971 that revived Volynov's faltering cosmonaut career and brought Kamanin's to a jarring halt. Prior to the *Soyuz 11* mission, Kamanin had already been informed that he was going to be automatically moved onto reserve status when he reached sixty, so it is fair to assume that a combination of this and the fallout from the tragedy probably hastened his exit from the scene. Vladimir Shatalov, who had successfully flown the *Soyuz 10* mission two months earlier, was named to replace Kamanin, and he in turn appointed Georgi Beregovoi as director of the Cosmonaut Training Center. The duo had no prejudice against Volynov, and he was reinstated to full flight status. "If you really love the sky," Volynov explains, "you do not retreat before difficulties. You cross all barriers, and if you fall, you pick yourself up and walk on."

Volynov then resumed his backup role once again, as second backup commander for Pavel Popovich on *Soyuz 14*, then as principal backup for Gennadi Sarafanov, the commander of *Soyuz 15* in August 1974. Eventually, Volynov would achieve a second spaceflight, as commander of *Soyuz 21*, together with engineer Vitaly Zholobov. This military reconnaissance flight, scheduled to last for two months, was launched on 6 July 1976. The following day they successfully linked up with *Salyut-Almaz 5*, the Soviet Union's last dedicated military space station. The two men would use special cameras to record strategically important military sites, and observe a massive Soviet military exercise known as Operation Sevier, then taking place in Siberia. They would also conduct numerous medical, biological, and technical experiments. Their workload was so heavy, however, that Volynov and Zholobov began to show signs of stress after weeks of working sixteen-hour days, twice what they should have been. It also caused them to neglect their planned exercise regime.

Several informed reports, including those from other cosmonauts, suggest that an overwhelming workload was not the only problem. Volynov and Zholobov were constantly bickering with each other and controllers on the ground, many times resulting in explosive confrontations between the demanding commander and his distressed rookie crewmate. Zholobov was not only exhausted and feeling the strain of minimal privacy aboard the cramped station, but was also exhibiting classic symptoms of claustrophobia and disorientation. In the last two weeks of their mission Zholobov became increasingly ill, and contemporary reports suggested this was perhaps because pungent nitric acid fumes from the station's propellant tanks may have permeated Salyut's air supply.

Medication did little to improve Zholobov's steadily worsening illness, and the tension between the two men is said to have made the continuation of the flight untenable. The situation became so bad that Volynov asked for permission to shorten their mission on medical grounds, and this was given on 23 August. "There were a few meetings by the doctors after analyzing the data through the radio connection," Volynov told Bert Vis. "And there was the decision not to complete the flight, because that would be dangerous to Zholobov's health. The deputy chairman of the State Commission who was in charge of the flight of that station was Gherman Stepanovich Titov. . . . Zholobov was a great friend of his. He talked to me and they decided—the State Commission decided—to stop."

By now, Zholobov was so weak that Volynov virtually had to prepare for the landing by himself. "I was working alone; he couldn't help me. I was to load the transportation vehicle; I was to prepare the station to fly in automatic mode. There were lots of jobs to be done; gather all the results of the mission, place them on board the transportation vehicle, and make everything ready for the return. That why before landing I practically didn't have any sleep." When the time finally came to leave the Salyut station, Volynov had to guide the engineer into their Soyuz capsule and then strap him in.

The undocking of *Soyuz 21* did not go quite as planned, and the two craft remained locked together. It took nearly an entire orbit for the problem to be rectified, but to everyone's relief the next separation attempt was successful. "We opened the passive part of the docking port using a radio command," Volynov explained.

Soyuz 21 returned to Earth around midnight local time, landing hard in strong gusts of wind 125 miles southwest of Kokchetav, Kazakhstan, and literally bouncing to a standstill on its side in a farming collective field. Volynov, still considerably weakened from his seven weeks in orbit with limited exercise, barely managed to unbuckle his harness and open the spacecraft's hatch before crawling out onto the soft ground. He then found he was unable to stand unaided. After a great deal of exertion, he managed to drag Zholobov out of the charred spacecraft, and both men fell to the ground, totally exhausted. Fortunately, they did not have to spend the rest of that night out in the open; forty minutes after landing the recovery team moved in. At first, there was a little confusion, as the rescue teams mistook the farm's harvesting equipment for the descent module and other equipment, but then Volynov sent up a flare and the helicopters found them.

"For three days and nights we practically didn't walk at all," reflected Volynov. "Although I used a running track [in space] my muscles were 'out of order.' I lost seven kilos compared to my preflight status, even though I had no excess weight. We worked a lot on the station and we slept too little."

Interestingly, and supporting reports that the troublesome, so-called acrid odor may actually been used by the crew as an excuse for shortening the mission by sixteen days, the next crew, in *Soyuz 24*, was said to have found absolutely no evidence of any odor or contamination of the air supply aboard Salyut. They entered the station wearing breathing apparatus, but after exhaustive tests they reported that the air quality inside the station and in its oxygen tanks was both pure and untainted.

Following his *Soyuz 21* mission, Volynov worked as a flight controller for *Salyut 6* and *7*, headed the *Soyuz 28* recovery team in 1978, and helped train two Mongolian cosmonaut candidates for the *Soyuz 39* mission in March 1981. In November 1982, now a senior cosmonaut, he was given overall command of the cosmonaut detachment. It was a post he would hold until his retirement on 17 March 1990, as a colonel in the air force reserves. Since then Volynov has traveled the world, representing the Russian Federation and the cosmonaut corps at official and commercial functions.

Nine months after his command mission on *Soyuz 4*, and after serving backup duties on *Soyuz 6* and *7*, Vladimir Shatalov flew into space a second time as commander of *Soyuz 8*, together with Alexei Yeliseyev, on a joint rendezvous flight with *Soyuz 6* and *7* known as the "troika" mission. As the only cosmonaut to have made a successful docking, he was a natural choice for a mission that hoped to practice the maneuver again. His third space mission came in April 1971 as commander of the three-person *Soyuz 10* mission, again with Alexei Yeliseyev and also with "rookie" research engineer Nikolai Rukavishnikov, in which their spacecraft docked with the *Salyut 1* space station. They were not able to enter the station, however, due to a faulty docking apparatus.

Shatalov was nevertheless promoted to major general after this flight, retired from the cosmonaut team, and replaced the sacked Nikolai Kamanin as cosmonaut training chief at air force headquarters following the *Soyuz 11* tragedy. Shatalov had risen to a position of great authority very quickly; in fact, he had one of the most impressive cosmonaut careers ever. It would not be without its frustrations. "The most complex spaceflight," he once stated, "is simpler than all this terrestrial red tape." In 1975 he was again promoted to lieutenant general, and from 1987 held the post of director of the Cosmonaut Training Center (TsPK) until his enforced retirement in September 1991, following the collapse of the Soviet Union. Since that time Shatalov has mostly taken on a roving ambassadorial role, and has been honored for his work by the president of the Russian Federation, Vladimir Putin.

Alexei Yeliseyev would make a total of three spaceflights, all involving Vladimir Shatalov. Together, they became the first cosmonauts to make three spaceflights. Yeliseyev would also train with Khrunov again, this time for a flight to the moon, which never materialized. "I believe that a cosmonaut should fly continuously," he remarked at the time, "until the doctors say 'enough.'" Yeliseyev received a doctorate in technology in 1973, and in that same year became flight director for the Soviet sector of the Apollo-Soyuz mission in 1975. Until 1981 he served as flight director for every Soviet manned spaceflight with the exception of the *Soyuz 18/Salyut 4* flight that took place simultaneously with the Apollo-Soyuz mission. In 1981 he was appointed a professor at the Bauman Higher Technical School and five years later was elevated to the role of rector (president) of the school. In 1991 he resigned from this post and spent five years as project coordinator for IBM's Russian branch; he then became president of the Russian branch for the Festo group of companies, suppliers of automation technology, and told the authors that since 2002 he has been the president of the Education Support Fund in Moscow.

The joint flight of *Soyuz 4* and *5* would prove to be the only spaceflight for Yevgeny Khrunov, although in 1966 he had involved himself in the failed Spiral manned orbital space plane project that was under development for several years. From 1972 to 1974, he served as a training director for the *Salyut 3* cosmonaut teams. In September 1979 Khrunov was teamed with Cuban cosmonaut José López-Falcón for a possible flight on *Soyuz 38*, but they ended up as the nonflying backup crew. Khrunov was then offered command of another flight, this time teamed with a Romanian cosmonaut, but he is said to have declined the assignment. Khrunov resigned from the cosmonaut team on Christmas Day 1980, and later became chief of administration for the USSR State Committee for Foreign-Economic Relations. He died of a heart attack in Moscow on 19 May 2000.

Without doubt, the achievement of a successful rendezvous and crew transfer from one Soyuz to another provided an enormous boost to the Soviet space effort at a time when the Apollo program was in the ascendancy. In fact, with just that one flight, almost all of the achievements of America's Gemini program had been equaled. Although such advances came too late to allow Soviet cosmonauts the distinction of being first to reach the moon, it can be argued that the early Soyuz flights have been more meaningful and beneficial to the long-term use of space than those of Apollo. Unlike its Apollo counterpart, the reliable Soyuz spacecraft is still providing a continuous link to orbit today, routinely carrying spacefarers to and from Earth orbit and docking with manned space stations. It has been greatly modified and improved many times over five different decades, yet it essentially fulfills the same function. With Soyuz, the Soviets created the enduring workhorse of space travel.

For Boris Volynov, flying in space took on a deeper significance than he could ever have imagined as he contemplated the home planet from orbit, and today he is just one of many space explorers who believes that spaceflight reshapes the psyche of space travelers.

"Having seen the sun, the stars, and our planet, you become more full of life, softer," he once wrote. "You begin to look at all living things with greater trepidation and you begin to be more kind and patient with the people around you.

"At any rate, that is what happened to me."

7. Leaving the Good Earth

'Tis likely enough that there may be means invented
of journeying to the moon and how happy they shall be
that are first successful in this attempt.

Dr. John Wilkins, 1640

The house looks simple and modest from the street—a thoughtfully designed but not overly ornate front garden, an American flag, and a wooden deck with an exquisite view over the Pacific Ocean. Though the neighborhood is a desirable one—you have to possess a good deal of money to live there—it is not snobbish. Instead, in typical laid-back, Southern California style, it makes the most of its location and beauty, while retaining a relaxed and unpretentious ambience.

When the owner arrives home he looks as laid back as the house; his jeans and hands are smeared with engine oil as a result of ongoing work on a beloved P-51 Mustang suffering from an engine problem. Yet Bill Anders isn't completely at ease. As well as the frustrations of losing a few more hours of flying time in the World War II–era fighter, he and his wife are concerned for the safety of their son. It is the first day of a new war in the Middle East, and they are anxiously waiting to hear if he has returned safely from a sortie in his A-10 Warthog attack aircraft.

Historically, the Anders family is no stranger to wartime and the unrelenting dangers that combat holds. When Valerie Anders finally gets a call later that day bringing news that their son had flown a successful and safe mission, it's yet another chapter in three or more successive generations of Anders men going face to face with the enemy and making it back alive. The fact that Bill Anders was a member of the first crew to journey to the moon, with all its inherent risks, seems almost tame by comparison.

William Alison Anders was born on 17 October 1933 in Hong Kong, at that time part of the British empire. His father, Arthur Ferdinand ("Tex") Anders, had joined the U.S. Navy in 1922 and received his Annapolis commission as an ensign in 1927. He was stationed in Hong Kong as part of the neutral Yangtze River Patrol, designed to protect American nationals and shipping along the river, and to support its economic interests in China. In typical navy fashion this was just one of many postings for the family, who moved around the world with each new set of orders. Bill Anders remembers his father being stationed next as an instructor in the Naval Academy postgraduate school, then on a battleship while the family lived in Long Beach, then San Diego, Norfolk, and back to China again, where he still vividly remembers rides in sedan chairs and huge crowds of people. His father, having been promoted to lieutenant, returned to the Yangtze River Patrol as executive officer of the USS *Panay*. The *Panay* was an antiquated navy river gunboat that was to play a historic role in the worsening relations between Japan and the United States, as Japan made increasingly bold forays into Chinese territory.

The gunboat had been increasingly involved in escorting neutral vessels carrying Americans and other civilian passengers who were fleeing the Japanese invasion of the city of Nanking in east central China. By Sunday, 12 December 1937, tensions had risen ominously in the area. On that day the *Panay*, under the command of Lt. Cmdr. James Joseph Hughes, was on escort duty for three Standard Oil ships. The gunboat was clearly marked as an American vessel and was flying large American flags. Under the Treaty of 1858 the United States had the right to provide safe passage for all American merchant ships and to ensure the safety of American citizens living in China. Suddenly, bomber pilots from a renegade section of the Japanese Army Air Force, angered that British and American ships on the river might give the Chinese more resolve to defend themselves, deliberately and callously attacked the *Panay*. On a day when he anxiously awaited news of his own son, Bill Anders recounted to the authors the story of his own father's actions that day many decades before:

The first bomb struck the bridge where the captain was, putting him out of commission. My father was back somewhere in the ship; his station was at the guns amidships, so he took over and ordered that they open fire. We believe that he was the first one in the American Navy to order "open fire" on the Japanese. He was pretty severely wounded, and couldn't talk—he had shrapnel in his throat, and he dipped his finger in his blood to write orders on the navigational chart. They fought back, and finally had to abandon the ship when it sank from underneath them. They were in a desperate situation. Extremely sick, my father was carried in a litter by the Chinese, hidden in the brush by the river for about a week, until they were rescued. Any longer and he would have died—by then he weighed only ninety pounds.

Neither the United States nor Japan wanted the incident to result in war, so the matter was quickly resolved through compensation and openly gratuitous apologies, with Japan stating that the vicious attack was simply the result of "bad visibility." As it turned out, war between the two nations was not all that far away. The wounds that Arthur Anders suffered in the heroic action, for which he was awarded the Navy Cross (the highest honor allowable in peacetime) and the Purple Heart, were enough to retire him from active duty in early 1941, by which time the family had returned to San Diego. Bill Anders and his mother had also managed to escape from China in a hurry—with the Japanese bombing them, they had the dubious honor of taking the first boat to freedom down the Pearl River after it had been mined.

However, the Anders family would not remain unaffected by events in the Far East for long. No sooner had Arthur Anders retired from the navy to convalesce than full hostilities broke out with the Japanese following the surprise attack on Pearl Harbor. Still fighting off infections he had picked up from the delayed treatment of his wounds, he was recalled to duty with the navy and stationed as a personnel officer at the Naval Training Center in San Diego. He was finally well enough to take his place on the battleship USS *Mississippi*, but the ship did not manage to leave the coast of the United States before Japan capitulated, and the war finally came to an end. Once again released from the navy, Arthur Anders and his family had a chance to settle down at last. Instead, they moved to Texas, where he dreamed of raising cattle. Fortunately, he was persuaded into the more practical option of buying and operating a drug store in La Grange, but it was a job he hated. Matters were not helped by the fact that the central Texas climate was vastly different from the perfect weather of San Diego.

Muriel Anders, Arthur's wife, had moved around all her life. She was born in Panama, where her father had worked on the canal, but in her son Bill's words she was still at heart "a California girl." The day came when she announced to her husband that she was taking their son back to San Diego—with or without him. Happily, he had few regrets at leaving his Texas dreams behind and returning to San Diego, where he became an inspector in the motor vehicles department before finally retiring for good to spend many enjoyable years growing avocados.

As Bill Anders recalled, given his family background "it seemed like it was never a question that I was going to go to the Naval Academy." The young man planned on becoming a ship's captain, just like his father. In 1951 he entered the United States Naval Academy at Annapolis, and was eagerly anticipating the prospect of joining a ship at sea. But if his dream had been to follow in his father's footsteps, he was soon profoundly disillusioned. He found the machinery and the engine room of the ship he was assigned to relatively interesting—but not much else, as he recalls:

I spent two months on an aircraft carrier, an old one. After they killed about eight pilots, bashed up a bunch of airplanes, and got caught in a hurricane, I decided that maybe a nice ten thousand feet of concrete runway would be good! So I elected to get commissioned in the Air Force. Prior to graduation, each service sent down a representative to talk to the upcoming graduates—Marines, Air Force, and Navy. The Navy had a guy who looked like he was about seventy, and the Air Force sent down a guy who looked like he was about twenty-five, and they were both one-star admirals and generals. So a lot of us said hey—we want to do that!

Upon graduating from Annapolis in 1955, Anders was commissioned into the air force. On 26 June of that year, he also made another commitment in life when he married his sweetheart, Valerie Hoard. The two had met briefly a few years before, and with the constant family moves it was somewhat miraculous that their paths had crossed again. Anders recalled:

When I was in the fifth grade—she must have been in the second—we lived in La Mesa, San Diego, about a block apart. I'd ride my bike home, and my only recollection is that her dad, who was a California Highway Patrolman, had a motorcycle parked out in front of the house. And they had a Doberman pinscher named Sally, who was a mean dog. So all the boys riding by the Hoard house carried squirt guns with ammonia water in. Then when Sally came out, we'd squirt her in the face—except one day I had forgotten to reload. So Sally grabbed my pants and brought me down. I remember Valerie's mother rescuing me, and I probably met Valerie at that time. But we both moved away. I went to Texas; she went to Barstow. And then it was later, through a mutual friend—when she had returned to San Diego for high school and I was in the beginning of the Naval Academy—that we got together. And that was it for me!

The young couple had dated in some of the more romantic venues in San Diego, such as the Starlight Opera and the Globe Theatre in Balboa Park, where they still go on occasion. But marriage to an air force pilot was not always going to be so glamorous. At the end of 1956, after flight training in Georgia and Texas, Anders received his wings. His first assignment was to the 84th Fighter Squadron at Hamilton Air Force Base, near San Francisco, where his first son, Alan, was born. The family soon had to contend with tough decisions and long absences, as he recalls:

I was sent to Keflavik, Iceland, and I was to report and leave on the 5th July. Valerie was pregnant with our second son, Glen, and was expecting about the 1st July. I could have waived that assignment because of her pregnancy, but I knew I was going to get sent to Greenland if I didn't go to Iceland. And Iceland in my view was a much more favorable place—that was almost everybody's view. Since the baby was late, the doctors induced labor, and I left the day mother and son came home. I was only able to return home fourteen months later.

For more than a year Anders was an F-89 Scorpion interceptor jet pilot in Keflavik, trained and ready to shoot down Soviet heavy bombers if they displayed any overtly hostile intentions. These bombers would constantly test the alertness of the Iceland base, and the Americans would scramble to intercept them, demonstrating their preparedness to engage if necessary. "We chased Russians around," Anders recollects. "I think I was the first guy to make the famous Tom Cruise gesture to them and they waved back—it took them about three months to figure out what it meant! We were armed, we'd come in, and their tail gun would be tracking us all the way. We always had two airplanes; one of them would come in, the other would be out where they couldn't shoot. Had they done anything to the first plane, the second would have turned and fired, shot them down."

It was a dangerous time, but Bill Anders found the flying conditions more hazardous than any posturing standoff with the Russians. Flying over the gray and white surface of lava and ice features, it was hard to gauge distances. With no trees and few signs of habitation, Iceland resembled the lunar surface in places. Constantly operating under icy conditions, in rain and very low cloud, he made six landings in which he was not able to see the runway until his wheels actually touched down. There was even one occasion where he landed on an icy runway, spun the plane completely around, and then continued down the runway as if nothing had happened.

With Valerie unable to join him, Anders filled his spare time by volunteering for the role of assistant maintenance officer, then taking on the duties of General's aide when the incumbent aide broke his leg. It kept him busy, but working with the general had its own perks, and he recalls he "got to go on a few extra fishing trips, which was nice."

After Iceland, the next air force posting took Anders to Hamilton Field, flying F-101 Voodoos. With his piloting career progressing well, Anders applied to the test pilot school at Edwards AFB in 1959, partly because he saw it as a way of getting involved in the space program. To his chagrin he was rejected, as he did not have an advanced degree, so the following year he decided to enroll instead in the Air Force Institute of Technology in Dayton, Ohio, where he would study nuclear engineering and aeronautics. He had hoped to take the astronautical engineering course, but all of the places were taken; he was only able to study aeronautics as a night course.

In 1962 Anders graduated with a master's degree with distinction, and he soon reapplied to Edwards, assuming that it would now be very easy for him to get in. However, more frustration was in store when he discovered that the regulations had changed in the interim, with far less emphasis on educational qualifications and new requirements specifying a greater number of logged flying hours. Edwards was now under the command of Col. Chuck Yeager, who valued flying far more than academic achievements.

Swallowing his disappointment, Anders accepted a posting as a nuclear engineer at the Air Force Weapons Laboratory in New Mexico, working on the effects of radiation and methods of shielding against it. He also arranged a "night job" as a T-33 jet trainer instructor pilot. He had spent the last few years moving in precisely the wrong direction to fulfill his ambitions to get into Edwards, but unknowingly he was becoming exactly what NASA would look for when choosing its next set of astronauts.

Bill Anders was driving home from work at the weapons laboratory one day when he heard a radio announcement that NASA was seeking applications for a third group of astronauts. He had been inspired when he heard Kennedy's speech about journeying to the moon and dearly wanted to be part of the enterprise, but had assumed it would never be possible without test-pilot experience. He was therefore surprised to hear that test pilot training was no longer a mandatory requirement for NASA, and in fact he fit the bill of astronaut qualifications. To make sure, he pulled his van over to the side of the road and waited for the announcement to be repeated. When he got home that night he told Valerie he was going to apply. Already in his mind was the fantastic idea of voyaging to the moon, and Valerie, for her part, was happy at the prospect that he would not be flying into combat anytime soon if selected. It was a Friday afternoon. By Sunday, his application to NASA was in the mail.

A few months later Anders was pleasantly surprised when he was called in for the required tests, and even more surprised when, after more testing, the list of candidates not only grew smaller but also still included his name. He believes that his degree specialization in space radiation shielding must have been a big factor, as that was "a big worry for NASA" at the time. Thinking creatively, Anders did find some interesting methods for passing the many tests astronaut candidates went through. One aptitude test consisted of a machine that required him to play ten video-like games at once. When he realized that there was no way he could win them all, he worked out that some games were harder than others, but that points could be gained in all of them not just by selecting the correct answers, but by also choosing any answer fast. By playing the easier games while instantly hitting any answer for the harder games, he pushed his scores past other candidates.

All the time Anders was taking the tests, he was aware that psychologists were watching him and assessing his stress levels. During one test he heard a distracting whirring noise coming from the other side of the supposedly soundproof room. He indicated to the psychologists, watching from behind mirrored glass, that he could hear the noise, but it wasn't going to break his concentration. Since no distracting noises were actually part of the tests, the psychologists entered the room, wondering if he was cracking under the strain. All of a sudden, the end of a drill bit came tearing through the wall. What Anders had imagined to be part of the tests was in fact a worker installing fire extinguishers in the corridor.

On the day of his thirtieth birthday, Anders received a call from Deke Slayton, asking him if he'd like to become an astronaut. He had absolutely no reservations about accepting, and quickly. During the final screening, while he was answering questions from space agency astronauts and doctors, the head flight surgeon asked him to talk about the concussion he had supposedly sustained five years before. Anders was a little puzzled by this; he'd never sustained a head injury, but he did not want to lie. So he answered straight-facedly that he had never had any problems from concussion. He has wondered ever since whether his medical records might have been accidentally switched with those of someone else, who may have been rejected as a result. For all he knows, he may never have been meant to make it that far in the selection process.

Two days later, Anders received another phone call. It was from Chuck Yeager, who was calling to let him know that his latest application to the test pilot school had been rejected. Anders could not resist the temptation to tell Yeager that he'd had a better offer, but it nearly backfired on him. Yeager, who headed the aerospace research pilot school at Edwards that had been set up specifically to groom potential astronauts for the air force and NASA, was most annoyed to hear that Anders had slipped past him and had still been selected. He even went as far as to try and get him unselected, but it was too late. Bill Anders was now a NASA astronaut.

Joining NASA in October 1963, Bill Anders was interested to discover that Walt Cunningham and Rusty Schweickart were part of his selection group. The two men had also been pilots, but not test pilots, and they had also engaged in physics research relevant to NASA's plans. "I think that was probably the reason we were selected," he states with conviction. "They wanted at least a few there."

But if the three new candidates were hoping that their science expertise would give them credibility within the astronaut office, they quickly found the opposite was true. Anders would come to discover that though it was no longer necessary to be a test pilot to become an astronaut, those outside the piloting brotherhood would have a hard time being noticed when it came to early crew selections. By his own admission, Deke Slayton viewed the selection criterion for this early group as those who were test pilots versus those who were not. In his view only those in the former category were up to the most demanding tasks. Many of the astronauts from the first two groups considered the new guys who had more scientific qualifications than flying skills to be "eggheads." Though nearly everyone was very pleasant to them, Anders claims it was obvious they were at the bottom of the pile. "It was pretty clear that not being a test pilot was considered kind of second rate. You know, I think it made some sense. I just figured we had to work harder. Deke was pretty fair, and eventually you got to fly. But the fact of the matter is that the non-test pilots flew last, or almost last."

Taking a very pragmatic approach, and understanding why test pilots were important in these early days of spaceflight, Anders made the best of it—but on his own terms. Instead of trying to become something he wasn't, and attempting like others to ingratiate himself with the powerbrokers, he did what he thought was best for his personal development as an astronaut, such as volunteering to go on all of the geology training trips. A few people found Anders very intense and grim-faced in that he seemed to focus solely on the job at hand; he was not easy to get to know. But Anders believed that doing a good job was more important than putting his time into currying favor at the top end. "Maybe if I had figured out the networking I would have tried," he admits,

but I couldn't figure it out. It did seem like sometimes the personal connection paid off, but I did what I was told to do and hoped to do it well. I didn't want to screw up. It wasn't clear what the measures of merit were. So I'd say, well, I'm going to do what I think I ought to do, and I thought learning geology would be useful, and I found it interesting. That probably worked against me a little bit, but I was interested in it, so I figured what the hell—I might as well enjoy it!

During tropical survival training in Panama, one aspect of Anders's background did impress the others. Anders was very comfortable with the outdoor life, having grown up that way. "One thing my parents did was to always try to live out in the country," he later recalled. "Dad was a country boy, and so I always had a horse, or was living out in the sagebrush, or in the forest. And so I had a more outdoor attitude." Anders was paired with Michael Collins for the training, and his new partner quickly grew to appreciate being with someone who had years of experience as a fisherman and general outdoorsman. Anders, for his part, was amused to watch Collins "thrash around and burn himself out, while I decided to take it easy!"

Despite the celebrity status now thrust upon him, and its many temptations, Anders never paid more than a passing interest in the hordes of attractive women who found it a distinct challenge to get up close and personal with the astronauts. Not only is he one of the few astronauts still in his original marriage, but he was also one of the few who never thought to take advantage of these distractions. After nearly half a century of marriage, Anders can afford to joke about it: "I really wasn't into the night life. Al Bean and I were probably the two straight arrows and hard workers of the group. Borman, Lovell, and Collins—those guys didn't carouse around either. But I must have missed something—I didn't really realize there were all those extra other attractions around! Had I, maybe my head would have been turned!"

On the far more serious task at hand, Anders soon found himself assigned to work on the Apollo spacecraft's environmental control systems, or ECS, which controlled and measured important components like spacecraft pressure, CO_2 levels, water, temperature, and humidity. These components had to work perfectly to ensure the crew's survival. Anders's earlier research into radiation effects was put to good use as he studied how much radiation astronauts might endure on a return journey to the moon. He also had to ensure, either by studying flight procedures and maneuvers or by proposing design changes, that the Apollo spacecraft would not get too hot or cold as it moved through the space environment.

With a confidence typical of a fighter pilot, Anders felt in his heart that he could fly as well as any of the test pilots, in fact, better than many of them. He had been practicing his philosophy of working as hard as he could and assuming that would get him noticed, but it did not seem to be working. Everyone in the office was working hard, so doing the same didn't make him stand out. A personal assessment of his astronaut colleagues also did not seem to match what was going on around him. Men Anders really would not have wanted to fly with for a variety of reasons were getting early flight assignments, while others he considered to be the best were not getting any assignments at all. He was not at all sure how the decision-makers would even notice his hard work. Alan Shepard possessed an enigmatic but changeable personality that made him difficult to know, and Anders hardly ever saw Deke Slayton. Unlike some assignments that other astronauts were given, his work on the ECS did not give him much opportunity to interact with them. He actually learned a great deal more about how the crews were selected from Slayton's posthumous autobiography than he ever did from the man himself.

Because Anders was not one of the test-pilot brotherhood, it took him two and a half years to receive his first formal crew assignment. It seems he was given his first real opportunity to impress a number of people when he served as a CapCom for the *Gemini 8* mission in March 1966. He doesn't believe he contributed much to the mission and was simply concentrating on the job at hand, like he did every other day. Yet his cool-headedness during a life-threatening emergency on the rendezvous flight was noticed. "That was a very close call for those guys," he recollects, and adds in all modesty: "I had just taken over the capsule communicator's site; Lovell had just left. I'd like to think that I contributed to their solution, but I didn't. I mean, we just kind of did our job at the bottom, but didn't have much to offer. They solved the problem basically on their own."

Perhaps it was because that mission's commander, Neil Armstrong, was so impressed with his work that Anders soon found himself with his first placement on a crew—as the backup pilot for *Gemini 11*, with Armstrong as the backup commander. As far as the prospects of actually flying, the assignment was almost certainly too late in the sequence to translate into a prime crew role in the Gemini program. Anders also knew it might keep him out of the early Apollo missions; but it was an encouraging sign.

As the *Gemini 11* flight was a high-altitude flight, the amount of radiation the astronauts would be exposed to was of understandable concern. It was a subject in which Anders had considerable familiarity, and he proved to be extremely helpful during the mission planning. The flight included experiments that would measure the amount of radiation the spacecraft absorbed, and he took part in reviewing these. With Anders's assignment to the backup role on *Gemini 11* his hard work seemed to have paid off at last, but it had taken a space emergency for him to really get noticed.

Within a couple of months of *Gemini 11*, and a little to his surprise, Bill Anders was assigned to an Apollo crew prime mission, with Frank Borman as commander and his old survival training partner Mike Collins as command module pilot (CMP). Anders was the lunar module pilot (LMP) for what was planned to be the third manned Apollo flight and the first manned launch of the mighty Saturn V rocket. Though the Atlas and Titan boosters had undergone several unmanned tests before their respective use in the Mercury and Gemini programs, the Saturn V was only scheduled to experience two unmanned tests before carrying this crew. The mission plan called for insertion into an orbit that would take the spacecraft four thousand miles above Earth's surface, affording the crew a pole-to-pole view of Earth, while they tested the LM for the second time in raw space.

According to Deke Slayton, the original crew for this flight was to have been Frank Borman, Charlie Bassett, and Bill Anders, then slated as the backup crew to the second manned Apollo mission. "I thought they would be a good lunar landing crew," he wrote in the autobiographical *Deke*. When Bassett was killed in a T-38 accident together with fellow astronaut Elliot See during training for their *Gemini 9* mission, his place in the Apollo crew fell to Mike Collins.

Anders was one of the best-prepared astronauts to fly the LM. He had spent thousands of hours training in LM simulators as well as flying the Lunar Landing Research Vehicle (LLRV), a weird-looking, jet-propelled contraption designed to give a close approximation to flying and landing the LM. In fact, Anders and Armstrong were the first two astronauts chosen to fly this ungainly vehicle, and Anders had enjoyed the brief but challenging flights, which reminded him of flying helicopters. He takes pleasure in noting that, while Neil Armstrong crashed the LLRV, he never did. Not only was the training preparing Anders to test the LM in the first high-orbit Apollo mission, it was also making him a likely contender for a future lunar landing mission. He felt at the time that there was a very good chance he'd be walking on the moon in a couple of years. Being intensely interested in geology, Anders felt that this suited him perfectly.

The *Apollo 1* fire, of course, changed everything. While the program was on hiatus, Anders continued to shepherd the LM he planned to fly through the manufacturing process. He recalls that it was frustrating at times, but he did what work could be done while the Apollo program was getting back on track: "When the fire came along, everything got shuffled. I was sort of watchdog for the second lunar module, but things got screwed up pretty fast, and there really wasn't a lot of lunar module testing and evaluation. I went to Grumman a few times, but I don't ever remember training in the lunar module simulator. Al Bean and I were assigned to do the lunar module checklist, and we conspired to make sure that the lunar module pilot got out first before the commander. But Neil [Armstrong] got in there later and changed it!"

Mike Collins felt that the procedural, hardware, and safety changes resulting from the *Apollo 1* fire had made the command and service module (CSM) assembly a spacecraft system in which he had complete confidence. He was more concerned with how the Saturn V rocket would fare. The final unmanned Saturn V launch prior to *Apollo 8* had fallen victim to several technical problems, including severe oscillations in the first stage that exceeded the design criteria, engines shutting down in the second stage, and an orbit a hundred miles too high due to incorrect guidance.

As it turned out, Michael Collins soon encountered problems of a more personal nature that would shape his destiny. During the first half of 1968, he had begun feeling something was wrong with him physically—the worst fear of an astronaut with an upcoming flight. A tingling in his leg and an occasional tendency for his left knee to buckle without warning led him to seek medical advice. X-rays revealed that he had a bone spur between two vertebrae in his back, which would require surgery. Recuperation after surgery would take at least two months, and the Apollo program could not afford the luxury of leaving him on a training crew.

In July 1968, Deke Slayton officially announced that Jim Lovell would replace Collins on the *Apollo 8* crew. Slayton felt that Lovell and Borman would work well together, as they had on *Gemini 7*. To Anders, Lovell would seem like a last-minute replacement, someone who never really had time to become a full crew member. As a consequence, he would later wryly suggest that they actually operated the mission with a two-and-a-half person crew. Anders and Borman would now take turns monitoring the CSM, with Lovell working almost independently, focusing his attention on the spacecraft's computer navigation system. "By that time," Anders states, "I was really command module pilot, and Lovell's job was navigation. He generally did good at that, I suppose, but he hadn't really learned the spacecraft. Borman and I were more of a pair than Borman and Lovell. Frank basically considered me the command module pilot, and didn't want Lovell to mess around with any of the switches. He told me, don't let him touch anything!"

Changes in crew personnel were less worrying than the delays in the hardware. Setbacks and frustrating, ongoing delays in the production of the Apollo LM were soon threatening to put the lunar-landing program well behind schedule. Not only was the delivery of the module assigned to Anders's flight delayed until the spring of 1969 at the earliest, but the one that was supposed to fly on the preceding mission was overweight and even more rudimentary. The flight's command module (CM) was also months behind schedule. If the first flight in the program, now renamed *Apollo 7*, was launched on schedule in late 1968, it would then be six months before an Apollo mission with an LM would be ready to fly.

During the latter part of 1968 there was growing support within NASA management for an audacious concept that would not only make the wait for the LM meaningful, but would also change the destination of Anders's flight. Instead of going 4,000 miles out with two spacecraft, why not go 230,000 miles out with just one—all the way to the moon? In one bold move, the United States would almost certainly beat Russia to the moon, even if it did not yet fully accomplish President Kennedy's commitment to land and return before the end of the decade. The navigation system would also get a thorough test and workout. Achieving this with one spacecraft in lunar orbit would present far less of a risk than trying it with two the first time. It would also give the command and service modules a second testing, without simply repeating what *Apollo 7* had already done in Earth orbit. It all made perfect sense.

It was still quite a risk, and there was no easy way back if something went wrong. The command and service module had only one engine capable of getting them in and out of lunar orbit. Unlike missions in Earth orbit, where there were usually multiple options for getting back down quickly if something went wrong, reentry would be days away if the crew suffered a mishap around the moon. Borman's crew would also be attempting this pioneering mission using a spacecraft type that had only flown manned once before.

When the decision was made, there were also only four months left in which to write the necessary computer programs; procedures for reentering Earth's atmosphere at the much higher speed generated by returning from the moon had yet to be devised. NASA Administrator James Webb thought the idea was crazy at first and did not even want to consider it. However, other external pressures lay on NASA that made the lunar orbit option an attractive one. The Soviet Union, notorious for upstaging any American space first, had been showing increasing signs that it was capable of flying a person around the moon. Even if they did not land a cosmonaut on the surface, a simple manned loop around the far side would be enough for the Soviets to boast that they had beaten America in the race to Earth's celestial neighbor. Anders feels that this is the principal reason why his Apollo flight became a lunar orbital mission.

When he was informed of the change to their flight, Anders was not only surprised at the bold decision to send *Apollo 8* to the moon, but crestfallen over losing the chance to fly the LM. Having worked so hard with a craft similar to one that would someday carry humans to the surface of the moon, the resolution to fly *Apollo 8* without an LM was hard to accept. Even if theirs was the first manned mission to the moon, admittedly an exciting prospect, the revised manifest meant that much of the training Anders had been undergoing would become instantly redundant. He also felt, with good reason, that it could mean he'd never get to land on the moon.

Bill Anders knew the score. He realized that if you were known as an LM person, somewhere down the line you would get to make a landing. Once he became aware that he was going to have to become expert in the CM's systems, he knew he would be stuck with that spacecraft and would consequently have little chance of ever flying an LM to the moon's surface. Even if he happened to make a second flight as CMP and then graduated to the role of mission commander, he would probably not be in line for such a command until the last two planned missions—*Apollo 19* or *20*. At that time, Anders didn't think that Apollo would last beyond five lunar landings. Even in 1968 he felt that public interest, however strong during the initial landings, would not sustain the program that far.

Despite his keen disappointment and misgivings, there was little time for speculation; he now had just a few months in which to become expert in the systems of the CM. He was philosophical about his new duties: "I became a lunar module pilot without a lunar module. I was disappointed, yeah. And on the other hand, you rationalize it. I didn't dwell on what I didn't have as much as just getting the mission done. We had to train hard on the command module. Frank gave me the responsibilities of making sure the spacecraft worked, and he took on the responsibilities of the dynamics, the rocket, and the trajectory. I was bound and determined to learn the spacecraft, and learned it probably better than anybody." With little time left for this dramatic shift in focus, Anders dove into his new responsibilities with his customary attention to detail. As his knowledge and proficiency quickly grew, Borman began to growl that his one-time LMP suddenly seemed to know more about some CM systems than he did. Despite being the only rookie on the crew, Anders felt that when it came to flying the Saturn V rocket and flying to the moon for the first time, everyone had as much prior experience as each other—none.

For his part, Frank Borman had absolutely no hesitation when it came to accepting the new assignment. He really wanted this mission and saw it as an opportunity for NASA to really prove that it was going to the moon; all the doubts about whether it was possible would be erased. For the first time in history, the moon would be seen as a place where people had ventured and not merely as a distant, future destination. He accepted the change without even consulting Anders or Lovell to get their opinions.

There was a monumental amount of work to be done in regard to mission planning. With the moon's presence as an obstacle for radio transmissions a factor for the first time, the spacecraft would be in and out of radio contact with Earth. It would also travel in direct sunlight, in the shadow of the moon, and in the shadow of the Earth at different points in the mission. It was not just a simple matter of rotating the spacecraft to distribute the heat or to get in radio contact, as doing so for one function could easily disrupt the other. Planning too many maneuvers would use too much fuel. Anders could also see that this was turning out to be an important mission for many reasons; not just as a test of the engineering of a command module on a flight to the moon, but also the question, filled with many unknown factors, of long-distance communications between the astronauts and Mission Control.

While engineering and communications challenges were the primary focus of the revised mission, the flight of *Apollo 8* would also be the first time humans could make observations and record close-up images of the lunar surface. NASA was already planning where the first manned landing might take place, and it was hoped that the crew would be able to return as much valuable information as possible about the moon's surface features.

Borman, however, remained mostly uninterested in anything but piloting the spacecraft. "The main objective was to go to the moon," he later stated, "do enough orbits that they could do the tracking, be the pathfinders for *Apollo 11*, and get your ass home." Bill Anders was enough of a pilot and engineer to know that this was an appropriate attitude for the mission's commander. However, Anders was also a scientist and therefore more open than his crewmates to the scientific potential of the flight. With this in mind, he took the time to study more intensively what the crew might see while looking down at this alien, largely unknown world.

Two months before launch, Collins had recovered sufficiently from surgery to be back on full flying status, and personally felt he could have rejoined the crew without any problems. Although Lovell had replaced him, and despite the lost training time, Collins still felt he knew more about the mission overall. But the decision had been made, and it was too late to disrupt crew training again, so all he could hope for was the chance to be assigned to a future Apollo mission. As it turned out, he would not have too long to wait. Jim Lovell, for his part, was impressed by the intensity that Anders put into his work, as he told the authors: "There's a good medium between being serious, and taking life easy, enjoying life. Bill's very, very intensive, but he later became chairman of General Dynamics, so I guess that proves he was pretty good."

The crew was placed in quarantine in the weeks before the mission; NASA was hoping to avoid the head colds that had disrupted *Apollo 7*. However, on 9 December, President Johnson broke the quarantine by bringing the crew and their families to Washington and dinner at the White House. It wasn't the greatest time to do so—the Hong Kong influenza pandemic, which would eventually claim over thirty-three thousand American lives, was reaching its peak. Valerie Anders remembers the dinner as "a complete fiasco." As she looked into a side room where an operetta was being per-

The *Apollo 7* and *8* crews at the White House in December 1968. Front row (*from left*): Walt Cunningham, Donn Eisele, Wally Schirra, Bill Anders, Jim Lovell, Frank Borman. At rear: Charles Lindbergh, Lady Bird Johnson, President Lyndon B. Johnson, former NASA Administrator James Webb, and Vice President Hubert Humphrey. Courtesy NASA.

formed, she listened to the audience vigorously coughing and sneezing. Almost everyone she knew had symptoms of the virus that month, and most of her close friends ended up following the flight on television while confined to their beds. Political showmanship had won out over a sensible medical precaution, and Valerie could only hope that Bill and his two colleagues would not get sick during the mission.

As if to underscore the historic nature of the journey they were about to undertake, Charles Lindbergh visited the crew two days before the launch. Anders greatly enjoyed the meeting, but with the mission so close there was little time to appreciate it. "He was really a fine guy, one of my heroes. Now that I am into old airplanes I would have enjoyed it even more. But we were so busy training, getting ready, that I won't say Lindbergh was a distraction, but it wasn't the event that it could have been if it had been later."

In the hours before launch, at sunset, Anders caught a glimpse of the crescent moon, a view that he would not forget. He was not nervous about flying a spacecraft far beyond where any human had previously ventured. He was a pilot with a job to do, and he felt that much of the piloting he had undertaken in the air force had been far more dangerous than this mission. His confidence came from a thorough knowledge of the machines and systems in which he trusted his life. In fact, Anders felt that his engineering background gave him far more insight into the spacecraft and booster than many of his test-pilot colleagues. He had implicit faith in these engineering marvels.

Despite his self-assurances and trust in the *Apollo 8* hardware and systems, Anders realized it would be foolhardy to imagine that the flight would be without risk, and there was an undeniable chance that the crew might not return safely to Earth. He therefore made two audio messages for Valerie before he left—one that she should play on Christmas Day, but also one to play if he did not make it back to Earth.

The countdown for the first manned Saturn V launch began five and a half days before scheduled liftoff, with final preparations carried out on the morning of 21 December 1968. Lying strapped into a spacecraft on top of the biggest and most powerful booster ever used for spaceflight, and about to become one of the first three humans to put it to the test on humankind's first trip to another world, Anders was surprised at how calm he actually felt. That all changed when the moment of launch came.

Frank Borman and Jim Lovell had ridden a Titan booster before, and they found the first stage of a Saturn V launch an amazingly smooth ride by comparison. Anders, however, was surprised at the noise and sheer violence of the initial moments of launch, especially when the rocket engines corrected their direction and threw the astronauts around despite their ultratight harnesses. He felt like he was on the end of a car radio antenna, being whipped around during a high-speed drive. No amount of simulation had prepared him for the disconcertingly jerky sideways movements, and a noise level so loud he couldn't hear his crewmates. Feeling like he was being shaken around in an animal's jaws, Anders said he even imagined that the base of the booster might have hit the launch pad as they rose.

Observers at the Cape also felt the difference in this launch, compared to all preceding manned flights. At first, the incredibly bright light of the engines dazzled them as the Saturn V slowly cleared the launch tower.

Then the sound caught up to them—a low, crackling, rumbling sound that they could feel beneath their feet and in their chests. To those who had witnessed earlier manned launches, it was obvious that this was a wholly different kind of flying machine—a new level of fierce power, capable of taking humans beyond the confines of Earth orbit.

When the first stage engines cut off and the 6-g load was suddenly gone, Anders had the uneasy feeling that he was being hurled straight through the instrument panel. The severe g's had made breathing and lifting his arms difficult, and to suddenly go to no g-load at all made him instinctively throw his arms up to protect himself from the impact with the instrument panel he felt was sure to come. Just as he did this the second stage engines ignited, throwing him back again, and smacking the wrist ring of his spacesuit into his helmet. Anders was certain that his crewmates would make fun of this mark of a space rookie. Later on, when he was stowing their helmets away, he was amused and gratified to note that they both had exactly the same scratch on their helmets.

The Saturn V took its first human payload into orbit precisely, with none of the major problems encountered on the booster's prior launches. Lovell, however, didn't feel quite as good about being back in orbit again. The moment he unbuckled and began to move he felt ill. He managed to keep the feeling under control by moving very slowly, keeping his eyes fixed on one particular spot, and soon the queasiness passed. The first experiences of weightlessness were not particularly enjoyable for Bill Anders either. Feeling fine at first, he'd overconfidently tried a backward spin to get back into his couch after removing his spacesuit. He said the feeling it gave him in his throat told him he should not do any more acrobatics until he'd acclimatized more. "My guess is everybody has somewhat of a reaction—some of them just won't admit it. Initially, you could feel the blood in your head kind of pumping, because it is like hanging by your heels. But after a couple of days it became quite comfortable; it was like lying in a swimming pool."

The crew spent the mission's first two orbits fully checking out all of the spacecraft's systems. Anders did not have much of an opportunity to look at the Earth in the brief time he spent in Earth orbit, while Borman, eager to ensure all was ready for the rocket burn that would take them to the moon, told him not to look and just to tend to his scheduled duties.

Anders recalled: "We were so busy testing the spacecraft that he—I think more than jokingly—said 'I don't want to see you looking out the window.' I peeked a couple of times, but he never caught me! Once we had got going and off to the moon, then we could sit there and stare. But we never did get a very good close-up view."

The spacecraft systems checked out, and everything seemed to be in perfect order, so the next major step after launch was to reignite the third stage for TLI, the trans-lunar injection burn that would carry them to the moon. Mike Collins, working in Mission Control as flight CapCom for this phase of the mission, gave the call. "*Apollo 8*, you are go for TLI!" These seemingly innocuous words were a key moment in the space program. For the first time, humans were exceeding the escape velocity of Earth orbit, truly leaving the confines of the planet in a way perhaps just as momentous as Gagarin's flight. The furthest that humans had ever traveled away from Earth before this moment was 850 miles, during the Gemini program. That figure would become a mere fraction of the distance *Apollo 8* was about to travel. In fact, it would soon be meaningless to speak of how high *Apollo 8* was above Earth's surface. The spacecraft and its crew were leaving Earth's influence altogether, relegating it to a remote place to be left behind and, if all went well, returned to at the end of the mission. Gene Kranz, observing in Mission Control, was so overwhelmed by the significance of the moment that he had to leave the building in order to settle down and wipe away a few tears.

A five-minute rocket burn accelerated the spacecraft and crew to over twenty-four thousand miles an hour. Less than twenty-five minutes later came the separation of the third stage, giving the astronauts yet another tremendous jolt. At this point, there came the first opportunity for the crew members to look back at the Earth. It had been just over forty minutes since they had left Earth's orbit, yet the Earth was already a spectacular blue sphere over twenty thousand miles away, diminishing in size even as they watched. The crew had felt no sensation of speed, and the view caused them to pause momentarily just to let the occasion sink in, and realize with a stunning clarity just what it was they were doing.

Anders had not expected the Earth to look so beautiful and delicate. At first he found it hard to recognize anything, until he realized why he was confused. Compared to the familiar north-south orientation he'd seen on maps and globes, the planet seemed upside-down. In zero gravity, it only took a simple 180-degree turn of the body to make the view look far more familiar. He photographed the incredibly glorious sight of the Earth as it shrank against the blackness of space. Soon, it became surprisingly difficult to locate the brilliant blue sphere in the spacecraft's windows.

For the first rest period, and for about ten other sleep times throughout the mission, Anders found it hard to get to sleep. He had the unshakeable feeling that he was constantly falling, and he recalls that disruptive noises within the spacecraft were also conspiring to keep him awake. "That happened a lot. Plus the sleeping restraint was sized for a much bigger guy, so I was like one pea floating around in a pea pod. The spacecraft makes whines and clicks, and Borman and Lovell were on watch together while I was on watch by myself. So they were right there, talking back and forth. It wasn't a very good sleep environment."

Borman was supposed to rest after the first midcourse correction was made, but he also found it difficult to relax and decided to take a sleeping pill. Two hours later he was still awake, and now feeling distinctly unwell. Then, without warning, his throat constricted and he threw up. Another vomiting spell soon after was followed by a bout of diarrhea and a wrung-out, feverish feeling. In space everything floats, so Anders quickly donned an emergency oxygen mask and set to work with a paper towel, catching any disagreeable stray particles floating around the cabin. "When Frank both threw up and had diarrhea, that didn't help my own space queasiness," Anders says of that particularly unpleasant time.

I mean, I could probably get queasy in one g if somebody's puking up next to me. He was quite ill, and I was just worried he was going to get worse. He claims it was the sleeping pill. But we'd all tested that on the ground. I would think that the commander, whether on a spacecraft or an aircraft, is always under a lot more stress. Even if he is not flying the vehicle, he is responsible for it. I don't think that it's a sign of cowardice or weakness. Frank, being a very thoughtful and conscientious guy, and I think this being a very hairy mission, was under a lot more, well-deserved stress and was just feeling it. That all built up and triggered something in his system. The sleeping pill probably didn't help.

This sickness was the kind of personal information that Borman had no wish to share with anyone, and his contrariness initially extended to Mission Control. When Anders suggested that he should inform them he had been sick, Borman replied that he had no intention of letting the whole world know he couldn't control what he thought was a simple illness. Anders finally persuaded him that the controllers really needed to know, and that there was a way of doing it more privately than the space-to-ground transmissions they knew were being constantly monitored by the media, who were on the lookout for anything out of the ordinary.

"I thought we ought to have some medical input here," was his reflection on this incident. "But you know, no fighter pilot wants to admit he threw up. We put it on a special loop of the tape recorder, and tried to give them enough hints that they would pick it up. They didn't for eight hours, and then of course they went bananas. But by that time, it was too late to do anything, and he'd recovered." The subject was debated privately on the ground for much of the next day. One of the most vital parts of the mission, insertion into lunar orbit, would be taking place soon, and if the commander was incapacitated by illness this could present grave risks. The flight's medical director cautiously but incorrectly ventured that it might have happened because the crew had flown through the Earth's radiation belts. On the other hand, if it was indeed a simple virus, the controllers were concerned that the entire crew might come down with it. Some serious thought was given to aborting the mission. When neither of the other crew members showed signs of illness, and Borman seemed to have recovered, the incident was dismissed. Borman certainly didn't want to discuss it any more; illness wasn't the kind of thing that test pilots liked talking about.

Mike Collins later wrote that it was the first severe example of space adaptation syndrome, or SAS, in America's manned spaceflight program. He theorized that before Apollo, the spacecraft had been far too small for crew members to move around in, if at all. With the advent of the roomier Apollo spacecraft, it was possible to float, spin, and twist—all of which were enough to bring on feelings of nausea in those more susceptible to the mysterious, so-called space sickness. Years later, Borman also concluded that he was probably suffering from SAS, which seems to strike randomly and even today can still affect any space traveler, no matter what precautions might be taken.

It would take them three days to get to the moon. In communicating with Mike Collins in Mission Control later in the mission, Anders pointed out that it was Isaac Newton, not the crew, who was driving the spacecraft. As they arced between the gravity fields of Earth and moon, the influence that each had on the spacecraft was constantly changing. The tug of Earth's gravity slowed the spacecraft to the point where it was only speeding away from the Earth at two thousand miles per hour, some twelve times slower than their speed when leaving Earth orbit. Yet it was enough to propel them into the influence of the lunar gravity, which would relentlessly pull them in. For the first time in history, humans were in a gravity field other than Earth's.

With their course precisely plotted in advance, there was little left for Anders to do but continue monitoring the spacecraft's systems; nevertheless, proper rest still eluded him. Forty-eight hours into the mission, he had barely slept and could not relax. Having left the navy for the air force, he found himself on a spaceflight that seemed to him more like operating a submarine. With no sensation of movement, and no places to stop on the way, it was hard at times for him to believe that they were actually going anywhere. It wasn't even possible to see the approaching moon, which was lost in the glare of the sun. In fact he never saw it at all on the outbound journey. He did, however, take a good look at the Earth from two hundred thousand miles out.

Now it looked to him like a Christmas tree ornament.

Back on Earth, Hong Kong was celebrating the fact that a person born in the British colony was soon to venture into lunar orbit. Anders remembers that after the flight "somebody sent me a newspaper from Hong Kong or Kowloon that said 'Hometown Boy on Way to the Moon.' Then they claimed they found my *amah*, my nursemaid there, and interviewed her. But I don't know how much truth there was in that."

Jerry Carr, their CapCom at this stage of the mission, gave the call of "safe journey home" as their spacecraft slipped behind the moon and radio communications were cut off by the moon's mass. Humans had studied the side of the moon that faces Earth for eons, either with the naked eye or through telescopes, mapping every detail they could. Now Borman, Lovell, and Anders would become the first humans ever to see the far side of the moon with their own eyes. There had been photos returned by unmanned probes before, but looking on it with human eyes was a moment long anticipated, both in history and in science fiction.

At first, there was nothing to see. As the spacecraft flew over the unlit side of the moon, with even the reflected sunlight from Earth blocked out, the long-awaited lunar unknowns remained hidden from view. Borman preferred that Anders not even look yet; with lunar orbit insertion, or LOI, approaching, he felt there would be plenty of time to look outside once the engine burn had been made. However, as Anders continued to look, he saw something that chilled him. Stars, not easily seen before in the glare of the sun, were now so numerous it was hard to find the constellations among them. But there was also a huge black area where no stars shone. It was as if someone had taken a bite out of the galaxy. The dark mass of the moon loomed beneath them, and the hairs stood up on the back of Anders's neck.

As he continued to peer out the window, Anders saw what appeared to be oil streaks moving slowly across the window. "We were mentally programmed for oil," he remembers, "because there were some kind of smears on the window. They were very smeared up, very disappointing." Yet, as he focused his eyes, he realized that the "streaks" were in fact far below them. "It looked to me like oil," he says of that unforgettable and indelible moment. "But then I refocused and realized that we were looking at the lunar surface, the lunar mountains, with very long shadows." What he was seeing was sunlight streaming low through the lunar valleys, and he quickly called out to his crewmates, telling them of the fantastic sight below.

Unfortunately, there was little time for slipping into the role of lunar observers. It was time to fire the engines for LOI, which would slow the spacecraft enough to be captured by the moon's gravity. If for any reason their engine malfunctioned and did not fire, their trajectory had been designed to loop the spacecraft around the moon and make a safe return to Earth. Alternatively, if something went wrong during the actual burn, the spacecraft might either crash into the moon or spin off into deep space, to enter an eternal solar orbit.

As LOI happened on the far side, Mission Control would not know the outcome until the spacecraft had circled around the other side of the moon. The spacecraft's engine was turned in the direction they were headed and fired for four and a half minutes, slowing them into lunar orbit. The crew could feel the rumble behind them as the engine fired, and the reassuring sensation of being gently pushed back in their couches. Checking the instrument panel, Lovell found that their speed had decreased by the precise amount needed.

When Anders was finally able to gaze out at the moon again, the view had changed dramatically. Now the lunar expanses the spacecraft was moving across looked to him like a big, dirty beach, stirred up by many footprints. Though Borman was the mission's commander, Anders was in charge of specifying which photos were to be taken. As Borman maneuvered the spacecraft to ensure the windows stayed pointing down for the best view, Anders moved from window to window, taking photos and movie images of views no human had ever seen before. Unfortunately, three of the five windows had purple smears on them, which, he states, made things very difficult. "I had the only decent window, over on my side. Borman's was almost unusable, the big one for Lovell was bad, the one right over me was smeared but usable. They think it was the outgassing of some of the sealant around the windows."

To aid him in finding and photographing the most important far side features, Anders used maps compiled from photos taken on Russian and American unmanned missions. Many of the landmarks he saw had never been named, so he chose to perform this little task himself—including naming a trio of craters after each of the crew members. When the spacecraft rounded the moon, Mission Control was at last able to learn whether *Apollo 8* had successfully achieved lunar orbit. Time had seemed to crawl by for the controllers as they watched and waited, each of them hoping that the burn had gone well. After all, they had sent a spacecraft a quarter of a million miles from Earth, firing it into a narrow flight corridor just a few dozen miles above the moon's rocky surface. There was little room for error. The news that the spacecraft was safely in lunar orbit brought them to their feet, cheering and waving. The crew of *Apollo 8* had reestablished contact to within a second of the estimated time.

That first rounding of the moon also brought the first opportunity for a human being to see the Earth rise above the lunar horizon, a classic image that has come to symbolize the Apollo program. Unfortunately, no one saw it; the engine had been pointed in the direction of travel to put the spacecraft in lunar orbit, and the windows were facing away from the incredible view. Although it seems an obvious move now, the mission planners had also never really considered taking a photo of Earth from lunar orbit. Not only did the *Apollo 8* crew miss seeing this sight on their first lunar revolution, they didn't see it the next time, either. For his part, Anders was not even giving the Earth much thought, as his attention was focused on studying the moon. It was only when the crew made a smaller burn, trimming their elliptical orbit to a near-circle and bringing them close to sixty nautical miles above the lunar surface, that the spacecraft was rotated "heads up" and facing forward. The windows were now positioned to face the Earthrise.

As the spacecraft rounded the far side of the moon, Earth began to peek over the horizon like a brilliant, pale blue beacon. Borman, awed by the spectacular sight, snapped a quick black and white photo. Not realizing what his commander was photographing, Anders grew annoyed. He did not want any film wasted on objects that were not part of the mission objectives. Then he got to see the incredible view for himself. He was no longer annoyed—in fact, he found the view just as riveting and beautiful. Grabbing the color camera, Anders pointed it out of the window and took a photo that would become one of the most famous images ever taken.

"There was no thought of taking pictures of the Earth that I can remember in our photography training," Anders points out when discussing that awe-inspiring moment. "The photography was so canned that they had the f-stops calculated from the lunar albedo, based on our longitude around the moon, which I could basically do with my watch. I had it worked out that for every revolution I knew what time we were, at such and such a time I'd change the f-stop without even looking. We didn't even have a light meter. Then suddenly the Earth popped up." Until that moment, according to Anders, the three of them had simply been pilots and engineers doing their jobs. The moment they saw the Earthrise was an almost transcendental experience, and one that pulled them away from their assigned tasks. Their eyes irresistibly drawn to the sight, each of them came to the realization that Earth, barely the size of a fist at the end of an arm, was fragile and finite. Bill Anders was struck by how delicate and colorful the Earth looked, compared to the starkness of the lunar landscape below. None of them would ever imagine Earth in the same way again. "I hadn't thought about it that way before, but I've thought about it that way ever since," Anders told the authors. "That lunar horizon was so ugly, so beat up—and here was this one colorful thing up there on that stark horizon. And then to realize that it was so small. You don't have to be much of a mathematician—if it's that big at one lunar distance, at one hundred lunar distances you would hardly see it. Yet that's hardly anywhere in space. At best, it's just like a little dust mote out there. Apollo was designed to go to the moon, but really the biggest thing for all of us was not the moon, which was kind of disappointing. It was being able to look back at the Earth."

Jim Lovell was also fascinated by what he saw, and it later led him to think more profoundly about his experience: "That first earthrise was really amazing; an amazing view when we saw for the first time the Earth coming over the lunar horizon. And we all wanted to take a picture. But more than that, we enjoyed seeing the Earth as it really was. We learned more about the Earth, how fragile it is. How it is cloaked in the atmosphere, which really protects us. That the Earth could be easily overcome if we are not careful. You know, resources are limited; they can be depleted, we have global warming, the ozone layer—all sorts of things."

As Anders transmitted his descriptions of the moon back to Mission Control, the director of flight operations, Chris Kraft, heard him use him use the word *earthshine*—an evocative term that made him shiver at what they must be seeing. It was a word that had never been needed before during manned spaceflight, and it brought home to the controllers just how much they were venturing into new territory on this mission.

The *Apollo 8* spacecraft would spend the next twenty hours making ten lunar orbits. Mission Control continued to monitor the spacecraft, deciding with each orbit whether to allow the crew to make another, or to fire their engine and return to Earth. Always thinking as a practical test pilot, Borman did not wish to spend a moment longer in lunar orbit than was necessary to meet the mission objectives. In fact, he'd pushed to have *Apollo 8* spend even less time in lunar orbit once it got there. Others had wanted them to orbit as long as possible, simply to maximize the science return. Ten orbits had seemed a reasonable compromise. Borman's reasoning was that the first voyage to the moon was achievement enough, and any additions to the flight plan could prove to be unnecessary distractions.

As he continued to photograph the lunar surface, frustrated by the poor view through the small, smeared windows, Anders realized that the moon was a lot less interesting than he had expected. He'd imagined the sharp crags of science fiction, but what he was actually seeing was a rounded landscape covered in impact craters, with only hints of volcanic activity. The holes and bumps he could see to the horizon reminded him of a child's sand pile, and he could not see any smooth parts at all. Everything was pitted from meteorite bombardment. The far side was even rougher and more pitted than the near side, something the crew had not expected. To his disappointment Anders found the moon unfriendly, monotonous, and even ugly, although he was greatly impressed by the lunar sunrises and sunsets, when the moon looked its most stark and desolate. Otherwise, the endless procession of meteorite craters interested him only in a technical sense. He'd hoped for the moon of the movie *2001: A Space Odyssey*, released just two months before the flight. He says there was no comparison: "The moon, when we saw it up close, was quite different from the *2001* view, which was where the moon had a lot of sharp edges. This moon was well rounded, sandblasted. It was kind of disappointing. *2001* had everything so sharp. I mean, it was still interesting, and I would have been very pleased to walk on it, to see it up close. But at a distance you looked down, it's like a rug, with every hair. The closer you look, there are just more hairs. You know, more craters—no real definition."

Jim Lovell agreed with Anders's assessment. In the crew's first postflight news conference, he stated that "after seeing that rather cold and lifeless world, it is rather hard to see how songwriters can say all those romantic things about it." Later missions to the moon detected some color on the surface, depending on the angle of the sun, but this crew did not. It was a fact that Anders attributes to the annoying window smears they were trying to look and photograph through:

We were looking through these gooey windows, and to me it looked gray, though the photography we brought back initially had a green tint. Some of the photo lab people called me early the morning after we got back, and said, "We think we've found life on the moon!" I thought they were kidding, but they were serious. They'd found what looked like green mold. I said, "Well, I didn't see any green mold up there!" It turned out that somehow the processing chemicals had not been properly adjusted for the wavelength of reflected lunar light.

Despite his reduced enthusiasm for lunar observation, Anders still wanted to make a definitive and accurate scientific record of what they passed over. Neither he nor Lovell wanted to sleep while they orbited the moon; they felt every second was precious in making new discoveries. Finally, Borman had to order them to sleep, but only after promising to keep filming the lunar surface. Anders understands why his commander insisted on this, but still wishes that they had taken more photographs: "I think he was not feeling well and was tired. Both Lovell and I were feeling pretty good. So we were quite prepared to stay awake. We came all the way up here; . . . sleeping while around the moon struck me as being not a good call. On the other hand he was the commander, and he said to sleep. I went down there, but I kept looking out of the window."

On Christmas Eve the crew conducted two live television broadcasts from lunar orbit. A great many people did not watch the first telecast, broadcast in the morning in America. However, the three major television networks carried the evening broadcast, and tens of millions were watching as the crew described their impressions of the moon, while sharing the view with the use of their tiny TV camera. Frank Borman had thought hard about what appropriate words to say, thinking not only of the overall symbolism of the mission, but also that it would be taking place over Christmas. He had eventually recommended that the crew read a passage from the Bible, and Bill Anders was first to begin reading from the book of Genesis as their spacecraft swept toward another lunar sunrise.

The use of Genesis was clever, since this story of creation is one shared by many faiths around the world. The vivid imagery of darkness, void, and water contrasting with light and land, had parallels with beliefs about the Earth's creation in Yoruban, Pawnee, ancient Egyptian, Norse, Inca, Japanese, Indian, and many other cultures. Most beliefs surrounding the creation center on a watery world of chaos, to which light and order are introduced. In the politically chaotic world of 1968, the *Apollo 8* astronauts were perhaps gaining the clearest view of the world as an ordered whole, and their words resonated deeply within the hearts and consciousness of millions of listeners back home.

There were also many detractors back on Earth who felt that the use of religion was inappropriate on a mission that was said to be an achievement for all humankind, in an American-taxpayer-funded program. Anders says he did not share the details of the religious convictions of his mission commander, yet he believes those few words had a power that went beyond individual faiths:

I don't ever remember reading the Bible. I mean, I am just not a Bible-reading guy. So I didn't care what they read from; I was too busy making the spacecraft work. The Old Testament—that's Judaic as well—and I thought about it, wondering whether our Muslim brothers, and Buddhists, how they'd find it. But, you know, the whole creation myth is pretty standard in all religions. Somehow you have to get here, whether it's on the back of a turtle, or however. And I personally don't believe what we read; it's allegory, if anything, but it struck me as being the proper tone. It was like being hit in the solar plexus; it really caught their attention. Particularly when they weren't expecting it.

The message ended with a wish of good luck for everyone back on "the good Earth." It is believed that almost a billion people, a quarter of the planet's population, were following the broadcast live. Earth had truly become another place to humankind. Humans had journeyed somewhere else, looked back, and begun to put our place in the universe in perspective. But now it was time for the crew to prepare for the journey back.

Once again, the CSM's engine had to fire perfectly to bring the crew home. This burn was called TEI, or trans-Earth injection, and it also had to take place behind the moon and out of radio contact. Just as before, in a feat of engineering precision, the spacecraft's systems performed superbly, and the burn came right on schedule. After the engine firing and as they headed home, the crew could see the Earth begin to grow ever larger in their windows. Anders remarked that he knew how explorers in the days of sailing ships must have felt; while he was happy that they had accomplished so much, it felt good to be returning home.

The ride back to Earth seemed to take a long time to Anders, to the point where he was actually growing a little bored with it all. "Borman and Lovell would try and sleep, and I just floated in zero-g for about two days. After a while, you look at the Earth, it's getting bigger. Still looks the same; you know, kind of cloudy; couldn't see many features. I was a bit bored on the way back." It wasn't, however, going to be a completely uneventful ride. As they grew ever closer to their home planet, Lovell accidentally erased the navigational data from the CM's computer. The spacecraft attempted to align itself to what it thought was the correct orientation and began to fire its thrusters, which Anders tried to counteract. The control panel indicator he used in the attempt was a ball about the size of a grapefruit, on which was marked longitudes and latitudes, known to the crew as the "eight ball." Struggling to realign the spacecraft, the situation immediately reminded him of the out-of-control thrusters on the *Gemini 8* mission, and he wondered if something similar might be happening:

As I'd been the CapCom for Gemini 8 *I was attuned for that, and coincidentally about the time the alarm went off and the ball was moving, a thruster fired. I heard the thruster, saw the ball moving, heard the alarm, and I thought "Uh oh!" Frank was sleeping; I was on watch; Jim was down there frenetically punching buttons. I came in with the hand controller, to counteract the rotation. The ball kept going, so I came in with some more, and then of course the spacecraft was rotating.*

I realized shortly what the real problem was, and by this time Borman was really getting involved in it, but I was still flying the spacecraft. Now, though, the ball was frozen and the spacecraft was still going around. So I basically stopped it, based on sunlight coming through the windows and picking up different dust motes. Then Frank took over and spent some time trying to find out where we were.

We weren't that far from reentry, and had no idea what our attitude was. We were on the reentry path, but we didn't know which way we were coming in, and we had to come in backwards. So that was a pretty tense period. By then there was enough light shining on urine crystals or whatever else was floating along with the spacecraft that you couldn't see stars. There would be just one or two little bright spots, so what we had to do was assume that one was Polaris, or another of the navigation stars, and another bright spot was another one, measure and take a shot. The computer would say no—it wouldn't tell you what it was, but it would tell you it wasn't those two. Finally, on the ground they figured out which bright spots could be which stars. So in about an hour we were able to get the platform stabilized, or oriented.

Returning to Earth from lunar orbit meant that *Apollo 8*'s reentry speed would be much faster than any previous flight. The spacecraft would be tearing into the atmosphere's upper reaches at the almost unimaginable speed of 24,500 miles per hour. They were aiming for the edge of the atmosphere, and would slice through at an angle that would cause sufficient drag to slow them down enough to continue the long drop to the ocean. It was a very precise point they had to aim for, considering the speed and distances traveled: a corridor only ten miles wide. Too shallow an angle and they would skip off the outer layers of the atmosphere and miss the Earth altogether, too steep and they would be incinerated.

Borman and Lovell told Anders to prepare for the ride of his life. The two Gemini commanders had experienced reentry before, but even they were surprised at what happened next. Having floated in weightlessness for six days, the reentry g-forces felt stronger to the astronauts; they had become acclimatized to space, and now they were undergoing an uncomfortable reintroduction to gravity. This time they were also going eight thousand miles per hour faster than on their Gemini reentries, with a grueling 7-g deceleration force applied over a particularly uncomfortable six minutes. A glow slowly began to envelop the outside of the spacecraft, and Anders began to wonder if they were going to fail at this final hurdle:

We were making a night reentry, so we were in darkness. And we were looking at the accelerometers to measure five-hundredths of a G, which would tell us we'd started entry. But before then a small, pink glow started developing around the spacecraft. I commented on it and Borman said, "Oh, that's sunrise." I was the rookie, what did I know? But he wasn't looking out, I guess, or something, and then pretty soon it got to where I said if this is sunrise, we are going right into the sun! Reentry had started, and as it progressed I thought I could feel heat on my back. I could see all these little pieces, which looked like baseball-sized blobs coming off. I was thinking, you know, if we have that big a chunk missing, we're going to be in trouble. But I realized that it was just a tiny little grain that had been ionized.

The reentry was in fact so precise that the spacecraft passed over the recovery carrier USS *Yorktown* waiting for them in the waters southwest of Hawaii. The Apollo spacecraft splashed down less than three miles away, so close that it was later decided to keep waiting ships further away on future missions, in case a descending spacecraft happened to hit one of them. *Apollo 8* hit the ocean hard, and the astronauts felt a sudden surge of water spray across the interior of the craft. It was later determined that this moisture was simply condensation that had formed during the mission and pooled during the reentry. At the time, Anders thought it was a sign of something worse.

"Coming in at night, we couldn't see the parachutes," Anders reflected,

> *but we could feel them. We must have hit a wave coming up, because we really hit hard. There was enough water in the spacecraft that it splashed up, so I was thinking, what the hell, have we split open? But we'd trained for that. Frank Borman had his hand on the switch that was to blow off the parachutes. It had a guard on it. But we hit so hard it jerked his hand away, and I think maybe even stunned him a little bit, because by the time he got back there the wind had caught the spacecraft and flipped us upside down.*

Frank Borman's motion sickness returned with the violent oscillations. Now that their commander was nearing the end of his duties, his crewmates chose to have some fun with him. "He threw up again, once we got down on the water," Anders recalls with a chuckle. "By then the mission was over, so Lovell and I were unmerciful. We reminded him that he was an Army guy, not a Navy guy like the two of us."

Whatever illness had plagued Borman during the flight, it does not seem to have been investigated too thoroughly afterward. He went through the usual postflight medical tests on board the recovery aircraft carrier, but soon after he was thrust into the maelstrom of a rapid world tour. He would shortly resign from NASA to take up a political posting under President Richard Nixon. Anders still speculates that the space illness Borman suffered might be the reason his former commander chose never to fly in space again. "It could be other things, but it may be that's why Frank Borman didn't want to fly again. You know, it probably embarrassed him more than anything else."

Unbeknown to him, when Bill Anders returned from the *Apollo 8* mission he had vaulted from the ranks of the relative unknowns to one of the contenders for the first crew to land on the moon. Deke Slayton gave a lot of consideration to immediately recycling the *Apollo 8* crew for use as the prime crew on the first lunar landing mission. They were, after all, the only crew with experience in lunar navigation. Borman, on the other hand, did not believe that they would have enough time to train for such a fast-approaching flight. They did not have any experience with the LM, he told Deke, and besides, he had decided to leave the astronaut corps. On behalf of the whole crew, and without consulting them, he turned down the idea.

On reflection, Anders is not surprised that Slayton informally offered the flight to his *Apollo 8* commander. "Borman was considered a pretty top guy. I don't think it was the number of missions he had flown; it was the work he had done on the Apollo fire, and [also] that Slayton had a lot of respect for him." Anders is, however, still a little irked by the fact that Borman turned it down. He didn't even know that it had been a possibility until decades later, when he was reading Deke Slayton's posthumously published autobiography. "I wish we had flown that mission," he states with a wistful air. "I was disappointed to read that the crew was offered the first lunar landing flight, and Frank turned it down. I wouldn't have passed up *Apollo 11*. I would have traded landing on the moon on any mission for flying *Apollo 8*, and I still would. But we were the first ones who went, so I mean it's not that it wasn't something that was historic. Oh well, it's too late to cry over spilled milk. I was disappointed—but on the other hand, you rationalize it."

Despite his disappointment, and the fact that Borman and Anders see each other regularly and go flying together, Anders says he has never asked his former commander about this contentious point. "No, I never have. Borman is just a different kind of guy to me, but he and I always have been mutually respectful, I think, and we are now good friends. I mean, I have always thought, 'Well, okay, he didn't want to do it, but he could have lined up the rest of us who did!' But that didn't happen either. We just don't talk about it."

Instead of flying on *Apollo 11*, Anders was assigned to the mission's backup crew with *Apollo 8* crewmate Jim Lovell as his commander. It was an unfortunate irony for Anders that, because he was now flight-experienced and regarded highly as an astronaut, he was promoted from LMP to the more demanding role of CMP, the person who would have to be able to fly solo around the moon awaiting his crewmates. This meant, however, that he would not get to walk on the moon on his next flight. He dearly wanted to set foot on the lunar surface, and felt he'd already done enough orbiting the moon on *Apollo 8*. Going all the way to the moon just to watch others land on it held no appeal for him at all:

When Borman turned down the lunar landing flight, that kind of set me off on a command module track. It was clear to me that as a command module backup guy, my next flight would be with Lovell, and then another backup, and then maybe I would have gotten a command. So I'd have to go to the moon three times before I could be a commander. I didn't think there would be that many flights. I figured if there were about twenty Apollo landings I might get a flight, but you could work out the program and it didn't look good. So that's why I bailed out; that's when I went to Washington. I hadn't really been looking for a way out, but when an offer to go to Washington came along I thought "Okay, I'll do that."

Deke Slayton and Thomas Paine, the NASA administrator, had asked Anders if he wanted to take a posting in Washington DC on the National Space Council. Before accepting the posting Anders discussed the implications with Slayton. He wanted to turn it down if he could be promised either of the two crew positions that would allow him to land on the moon, or even if there was a 70 percent chance of such an opportunity. Deke would not commit himself to make this promise, and Anders thinks it may have been due to the number of well-qualified Gemini astronauts then in line for a very limited number of Apollo seats. He finally agreed to take the Washington assignment, but only after he had received a commitment from the administrator that he could remain on active astronaut status, and therefore be eligible for any assignment to land on the moon that came along.

Although Anders still hoped to walk on the moon, and feels that he would probably have received command of a Skylab mission if he had stayed in the loop, he soon saw a different career path opening for him once he had begun working in Washington. Proving adept as an administrator, he was offered the position of executive secretary of the National Aeronautics and Space Council by President Nixon. It was a position he felt he had to take. By then, Anders had decided he wasn't going to get the call to come back, and he was quite content with that; there were more important things he could do for the space program.

Anders had realized that NASA and the American taxpayers perceived the space program in very different ways. As far as he could ascertain, public support was certainly there to beat the Soviets, to prove American technological superiority, and to put the Stars and Stripes on the moon. He understood and believed in NASA's vision of lunar geology and further exploration of the solar system, but saw that public support would drop off the moment the first manned lunar landing was completed. He thought that it was more important to push for programs like Skylab, the Viking Mars landers, and the space shuttle than to seek another flight for himself.

Although Anders would not undertake any geological work on the moon, he would become instrumental in ensuring that a geologist, a scientist instead of a test pilot, would figuratively go in his place. Although he had been relegated to the bottom of the ladder when he had joined NASA as an astronaut, he was now in a political position where he could exert a little pressure and make sure the same did not happen to others:

I was the guy who pushed the powers that be to put Jack Schmitt on the lunar surface. I thought it was really stupid if we were to go to the moon and not have an expert up there, nothing but a bunch of test pilots. We were working with the satellite called Landsat, and there was a top-level policy issue as to who would control the imagery. Would it be NASA or the Department of the Interior? If it was the Department of the Interior, would it be USGS? NASA really wanted to control the imagery, and my recommendation was that NASA do it, but the submerged geologist in me had caused me to gravitate a bit towards the then-head of the USGS, Bill Pecora. Bill and I were on a trip together, and I told him, "Don't just give up—make a deal. Tell them that you'll do it if they put a real geologist on the moon before the program's over."

I think that the Apollo 17 *crew had already been picked by this time, but not announced. Jack Schmitt was a friend of mine, and had been a very helpful trainer—but it wasn't just because he was a friend; he was a good geologist. So Pecora made a strong pitch, and Deke was told to put Jack Schmitt on there.*

Despite moving into administrative positions, Bill Anders never lost the urge to fly airplanes. He only took government jobs on the proviso that he still be allowed to fly NASA aircraft, and he enjoyed this privilege for five years. When NASA pilots flew into Washington DC, Anders would be first in line to take the aircraft up for a spin.

After serving on the Atomic Energy Commission and as the first chairman of the Nuclear Regulatory Commission, Anders was appointed by President Gerald Ford as America's ambassador to Norway. Even then, Anders stayed on active flight status, often flying army helicopters and F-4s from visiting U.S. Navy ships, as well as NASA's T-38 jets on quick trips back to Houston. He remained in the air force as a reservist until 1988, when he retired as a major general:

One of the things I made sure of when I left NASA was that I would keep flying. When Tom Paine asked if I wanted to take the Washington job, I said, well, on one condition—that I could keep my union card current, I could keep flying T-38s. So he wrote me a letter saying, "Anders can fly T-38s as long as he is working for the government." And he didn't put a termination clause in there! So every time I changed jobs, they'd say, "Okay, well I guess that's it." I'd say "No, I'm still working for the government." So I was the only ambassador on active flight flying jet airplanes!

After many years working in government positions, Anders proved similarly skilful in the private sector. He'd grown a little disillusioned with government politics after working for Vice President Spiro Agnew in the Nixon White House. Having worked with many of the key players who became entangled in the Watergate scandal as it unfolded, he says that the experience was "a pretty big baptism of fire for an idealist. I'm not much of an idealist anymore!" From 1977 onward, he worked for General Electric as vice president, once again specializing in nuclear energy, then later in aircraft equipment. He then moved on to the Textron Corporation in 1984 where, among his duties as executive vice president for operations, he stayed proficient by flying the helicopters that came with their Bell Helicopter subsidiary. "I couldn't quite land an airplane at General Electric," he reflects. "In fact, I was interviewing also with Boeing, and the guy who was trying to recruit me knew I wanted to fly, and he said, 'We'll have this helicopter here, you can fly that.' But I did spend some time in GE without flying, which was stressful!"

After six years with Textron, Anders became chairman and CEO of General Dynamics, winning awards for ensuring the company's survival in the rapidly changing business environment after the end of the Cold War. He still found time to fly. General Dynamics was working on night attack systems for the F-16, and Anders assisted as a test pilot when he could. These days Frank Borman considers Bill Anders to have had one of the most successful post-NASA careers of any astronaut, and it is hard to disagree.

The fact that he could continue flying was still a major drawing card for Anders, and he was shameless in listing it as a mandatory requirement in his post-NASA career: "At General Dynamics, I decided that one of the prerequisites of being chairman was to be able to appoint myself as assistant test pilot for the F-16 doing night attack, because you had to fly a day flight before you could fly the night flights. So I got two flights! I probably flew the F-16 every other month, and got on the average one flight a month."

One fact Anders understood when he left NASA was that his *Apollo 8* fame would quickly diminish. He didn't feel that his astronaut career had given him any particularly revelatory insights into life, and yet he encountered people who expected him to know the answers to all sorts of unrelated questions, simply because he'd been to the moon. He also noticed with disapproval that some of his former astronaut colleagues half believed some of their own publicity, and were accepting all sorts of impressive-sounding post-NASA positions with companies they were not necessarily useful to, except in the area of high-profile publicity. Generally, these relationships were short lived.

In the harsh realities of the business world that Bill Anders entered, no one cared much if he had been to the moon, and in fact it could count against him if people felt he was using it to his advantage. It seems that no one ever did, because Anders worked hard. He may have faded from the public eye, but he was making a big impression in the corporate world. He almost feels there is something awry in the fact that people do still recognize him from time to time and consider him a hero. To his way of thinking, there have been many people in the past few decades who should be considered heroes, and the achievements of his generation should by now have been overshadowed. Anders watched those in the astronaut corps who expected to make big money on their name alone, and was not surprised when most did not succeed.

Now retired, Anders still flies aircraft, but purely for his own pleasure. It is somewhat ironic, considering that he was initially looked down upon by some of the test-pilot astronauts, that he is probably the most active of all the Apollo astronauts today when it comes to flying aircraft. He still logs between three hundred and four hundred hours a year, and not long ago added solo gliding to the many other ratings in which he stays proficient.

A contemplative Bill Anders in 2003. Courtesy Francis French.

In 1996, along with his family, he founded the Heritage Flight Museum, dedicated to preserving and flying historic military aircraft:

Frank Borman got me started in this. He got a Mustang and I admired it, and I bought a DeHavilland Beaver from him. I never really thought about flying the Mustang, but I told him, "Well, if you ever find another one let me know." So I was flying another airplane right over his house on the way to Houston and called him on his cell phone, and he said, "Oh, I found a Mustang, do you want it or not? You've got to decide right now!" So I quickly landed and bought it! Then I thought, "Well, why don't I start a little museum?" I have several sons and a son-in-law who like to fly. So we started the museum with one airplane, then we added to it, and now we have got quite a collection. I do the air shows with Heritage Flight like Frank does, and I also do the Navy shows with the Legacy Flight, in the F-8F Bearcat.

Believing strongly in educating the public about the role these aircraft played in his country's history, Anders tries to keep his museum on the move. More people will see the aircraft flying in their full glory at air shows across the country than if they were simply in some static display. Frank Borman, who occasionally joins his old crewmate in the air, restored one of the aircraft he owns. The aircraft names reflect Anders's life: his wife's name, echoes of his time in Iceland, and aircraft he flew with in the 1950s. He also enters the air races at Reno as number 68, commemorating the year he flew to the moon.

Though now a civilian, Anders has clearance to fly in formation with navy and air force airplanes in air shows; one of the very few civilians to do so. "There are only ten of us approved for Heritage Flight, and I think there are only eight of us approved for Legacy Flight. I'm one that overlaps them both. It's a great way to spend a summer!"

It's an expensive pursuit, but one that the Anders family can luckily afford to maintain. As well as Heritage Flight they have also created The Anders Foundation, which is a big contributor to worthy causes, many of them strongly environmental. Bill and Valerie Anders remain very active, hands-on contributors to the cause of preserving America's last pieces of wilderness and ensuring that young people are educated about their ecological value. "I've been fortunate, particularly with General Dynamics but a little bit with prior companies, to get a little more wealth than I need," he says in talking of his current involvements. "If you don't need it all, it's good to share it with projects that do. Valerie has been very heavily involved in the Yosemite Institute, and she's the chairperson of the Olympic Park Institute. I was one of the original founding members of the Yosemite Institute. Yosemite and the other national parks need help, so we try to help them a little bit. But," he adds with a sudden grin, "it's really terrible when I have to do that instead of buying an airplane!"

Bill and Valerie Anders have stayed married throughout his ever-changing career. Bill never loses sight of the fact that, whatever he was doing, Valerie was also undertaking an important mission of her own in bringing up their children. Their sixth and last child, a girl, was born after he returned from the moon, and was appropriately named Diana, after the Greek and Roman goddess of the moon.

The incredible and audacious success of *Apollo 8* seemed almost forgotten after the glory of *Apollo 11*. But perhaps, in the longer sweep of things, it will prove to have been the more important mission: the first view of Earth as a small blue planet, and the first time humans escaped Earth's gravity and traveled to a new destination. Jim Lovell was glad he flew *8* rather than *11*, and Neil Armstrong considers *8* to have embodied the true spirit of Apollo.

Today Anders sees both flights as about the same in importance; one being the first to leave our world, the other the first to truly visit a new one. In terms of what he achieved personally, he sees the photo he took of the Earthrise as perhaps the most important thing he did on that mission. It has been called the most influential photo ever taken, in terms of how humankind thinks of the environment, and it is the largest photo on the wall of his living room. Together with other photos and impressions from the mission, it helped create an environmental awareness of our fragile planet that was sorely needed at the time.

Anders is under no illusions as to why he went to the moon. Though he then felt himself to be more of an explorer than a pilot, he does not believe the flight was done for exploration, nor for technological advances on Earth. With decades to reflect on his nation's spaceflight accomplishments, he is adamant that it was purely and simply another vast battleground in the Cold War. With that rationale for NASA's existence now gone, he questions what much of NASA's purpose is today, particularly as it has retained much of the same infrastructure it had during Apollo.

Although he is sure that the American public likes the concept of flights to the moon for the purpose of pure exploration and is sure they'd do the same thing for a mission to Mars, Anders does not believe that they did or ever will again wish to commit to the high costs unless there is a rationale such as Cold War competition. Simply put, exploration is not enough. He hopes that, rather than pushing for a grand Mars mission, NASA will quietly and carefully put the pieces in place for a gradual, thoughtful return to the moon. Mostly, however, when it comes to the politics of spaceflight, Anders would rather spend time flying than worrying about such things anymore.

When Bill Anders looks at the moon, he does not see what other people see. Though he feels it broadened his perspective on life, he also sees it as an uninviting place that only looks attractive from Earth because of the distance. In fact, if he had the choice of an Earth orbit or a moon orbit flight now, he'd prefer to stay in Earth orbit, looking at the ever-changing colors and landforms. Instead of spaceflight, he is indulging his earthbound loves and the beauty of Earth close up.

A vivid memory from the age of four, when his father was stationed in Washington state, meant that he remembered the area as a sunny playground of wildlife and blackberries. Returning decades later, he discovered that his memory had not been correct about the weather, but correct about the stunning natural beauty. A former astronaut, ambassador, and CEO never really retires, and Anders has yet to slow down much. His energies these days, however, are directed more toward his beautiful home life and his lovingly restored aircraft. After a lifetime of moving around the globe he has yet to stop, but this constant roving and the experience of seeing the Earth as a whole has given him an appreciative overview of our planet: "I think Valerie and I have lived in twenty-three different places in our marriage. You don't really get to know every last neighbor on your block, because you don't live there long enough. There are pluses and minuses, but we have always made the best of it. Our son, who's over in Kuwait right now, was asked that question. And he said, you know, none of our roots are deep—but they are really, really broad."

That first photo of Earth taken by Anders from lunar orbit has had a profound effect on millions of people ever since it was developed and published. In 1970 the image was used on Earth Day as a symbol for the fragile ecosystem in which we live, and it has been used many times since to aid environmental causes. For the first time in history, there was no remaining excuse for humans to see the Earth as a vast, seemingly flat and endless expanse of land and water, in which animals and plants are inexhaustible and pollution can have no real effects. Anders's photo brought home to everyone the undeniable truth that the Earth is in fact small, finite, and capable of being affected by the negative effects of human activities. The Anders Foundation does much to support environmental causes, yet all the work it does will probably never have as much impact as that single moving photo of a distant, remote Earth. This historic image captured by Bill Anders has done much to keep our planet an inviting place for future generations.

8. A Test Pilot's Dream

When we contemplate the whole globe as one great dewdrop,
striped and dotted with continents and islands,
flying through space with all other stars,
all singing and shining together as one,
the whole universe appears as an infinite storm of beauty.

John Muir

Frank Borman was a deservedly satisfied man. The *Apollo 8* mission had been an outstanding achievement, successful beyond anyone's wildest expectations. As he set off on a postmission world tour, Borman knew they had done something of tremendous importance in the Apollo program. Even as they were fulfilling the global public relations duties, the crew's mission photos were undergoing detailed scientific scrutiny, with much interest centering on a relatively flat, unexciting area. The Sea of Tranquility lacked much of the drama and majesty of photos of the more rugged areas, but its uniformity was important—it would make for a safe landing zone.

Apollo 8 had proved conclusively that NASA could send astronauts to the moon and return them safely to Earth. It had given the program renewed vigor, and enthralled the public with its audacity. Nevertheless, it had also required a hasty shuffle of the Apollo flight schedule, causing Borman's crew to jump ahead in line and fly the second Apollo mission. That mission slot had originally been promised to Jim McDivitt and his crew. Instead, they sat out *Apollo 8* and flew the next mission, one that never left Earth orbit.

McDivitt has stated in recent years that Deke Slayton would probably have granted a request to keep his crew's place in the original order—thus making him the commander of the first flight to the moon—if he had really pleaded for it. Jim McDivitt did not do so, and his crew members, Dave Scott and Russell "Rusty" Schweickart, were in complete accord with their commander. They recently told the authors they were actually pleased with his decision to pass over the attention-grabbing *Apollo 8* in order to fly their own, historically lesser-known mission. On the face of it, it seems an odd reaction. Why would they not prefer to fly to the moon?

To understand the reason, it is important to remember that while the astronauts were not averse to being in the history books, they were on the whole far more interested in being pilots. *Apollo 9* would be a test pilot's dream mission, the first to test all the elements needed to land on the moon. It would be the first flight of the lunar module, a bold new step in spacecraft design. An incredibly delicate craft, it had been pared down to the absolute bare essentials needed to descend to the moon's surface, serve as a habitat, and then return to lunar orbit. It was a craft designed for space, and space alone. *Apollo 9* would also be the only opportunity to test the Apollo spacesuit in the environment for which it was created before an astronaut used it to step onto the lunar surface.

Apollo 9 was not going to fly to the moon, as it was considered wise to test these elements in Earth orbit first. Yet this mission would be just as risky as the preceding *Apollo 8* mission—perhaps even riskier, since the LM had no heat shield. Once the two astronauts testing it had flown out of sight of the command module, they would need to fly with absolute precision in the untried spacecraft in order to rendezvous with the CM and return safely to Earth. Testing the LM alone would be challenge enough, but this mission would test the entire Apollo system—the Saturn V rocket, the command and service module (CSM), the LM, and the Apollo EVA suit. It was a truly appealing mission for a test pilot.

Nevertheless, in the wake of the *Apollo 8* mission, any flight that did not venture beyond Earth orbit, however crucial, was not going to fire the public's imagination. In his autobiography, Walt Cunningham imagined "an Apollo family portrait would picture *9* as an engineering student standing between *8* and *10*, two strapping glamour-boys who traveled to the moon." *Apollo 9*, however, was the payoff for the many years of hard engineering preparation. The machines all had to work together on this first Apollo mission of 1969, because if anything went seriously wrong there would probably be no moon landing before the end of the decade.

The command module pilot (CMP) for this mission needed to be extremely reliable and efficient, able to work under intense pressure in unforeseeable circumstances. If there was a malfunction in the LM, he would need to fly the CM solo, rendezvousing to allow his crewmates to transfer into the only spacecraft capable of reentry. Slayton had no reservations about choosing Dave Scott for this task. Scott had demonstrated his skills and quick-thinking initiative on the troubled *Gemini 8* mission.

As the launch date grew near, McDivitt's crew was proving to be one of the best trained ever. Originally the backup crew for the ill-fated *Apollo 1*, they had been working together since January 1966, and despite post-tragedy changes to flight assignments had always retained the duty of flying the first LM. Flight Director Gene Kranz believes that they were the best prepared of any Apollo crew and Dave Scott an extremely knowledgeable CMP. Scott was also considered a rendezvous expert by his astronaut colleagues.

Rusty Schweickart knew that not only was he on the best crew, but he also had the best commander: "Jim was a good leader, a good guy to work with. He was very fair, thorough, trustworthy, and he had good judgment. Not overbearing, not overly demanding, but he expected you to do your job and do it well. Both Dave and I were essentially the same way, so we were a good crew in that sense, and worked well together." It had been a long and interesting road for Schweickart. Like Bill Anders, he had come into NASA as a pilot, but not a test pilot, and with an impressive background in science. Born in Neptune, New Jersey, in 1935, Schweickart had attended the venerable Massachusetts Institute of Technology (MIT), graduating in 1956 with a degree in engineering. He followed this with a master's degree in aeronautics and astronautics in 1963. In the interim he served as an air force pilot, and remained in the National Guard as a pilot during his second stint at college. This allowed him to fly on weekends and evenings. His studies at MIT included research at the institute's experimental astronomy laboratory, where he became expert in star tracking and atmospheric physics.

Schweickart's upbringing on an isolated farm in New Jersey gave little indication that he would one day work on the leading edge of space engineering. When he was old enough, his parents would often take him and his sister on long walks at night, talking about nothing in particular. Strolling down the unpaved road leading from their farmhouse, they would pick and eat blackberries while gazing at the moon and stars. One evening a brilliant full moon was slowly rising over the horizon. He was no more than five years old, but Schweickart clearly remembers declaring that he planned to go to the moon some day. His parents chuckled at his youthful resolve.

After school it was time for Schweickart to listen to the family radio, enjoying his favorite action serials. He loved the adventures of Tom Mix and other cowboy stars, but was also captivated by the exploits of Superman. Like most little boys he tried to imagine what it would be like to fly through the air like his caped hero, and he often dreamed of growing up to be either a cowboy or a fighter pilot. Although Superman was an influence, Schweickart's youthful flying ambitions mostly came from the farm's proximity to the Lakehurst Naval Air Station. In the early days of World War II, aircraft were always flying over the farm, practicing aerial dogfights. Captivated by the spectacle, Schweickart would lie on the ground and try to imagine how it would be to fly through the clouds. Soon he was able to identify by name every airplane type that flew over.

One year, Schweickart was given a chemistry set for Christmas. He set up a small laboratory in one corner of the farmhouse and happily carried out experiments. It led to an emerging desire to be a chemical engineer. In high school he would avidly read the popular monthly science magazines, with a special interest in stories about the test firings of V-2 rockets in New Mexico. He found himself fascinated by blurry photos taken during these rockets' flights that revealed the curvature of the Earth. When issues featured the evocative space artwork of Chesley Bonestell, Schweickart would read and reread the magazines until they almost fell apart. Later, pursuing his earlier dream, he took up chemistry at MIT. After persisting with the subject for a year he realized chemistry no longer held the same fascination for him, and he switched to aeronautical engineering.

Commissioned into the U.S. Air Force after college graduation, Schweickart trained across the southern United States, mostly in propeller airplanes. Flying came very naturally to him, and he loved it right from the start. Now he was able to fulfill his childhood dreams, free to move through the skies with very little effort. His first overseas assignment was the Philippines, flying F-100s. He found he particularly enjoyed the freedom of night flying. Some nights the Milky Way and zodiacal light would surround him with incredible brightness, and he could easily spot "shooting stars," tiny meteors burning up in Earth's atmosphere. At times, alone in a star-filled sky or streaking across the land, he'd ponder the miracle of being able to fly thousands of feet above the ground, witnessing sights inaccessible to previous generations.

After four years as a fighter pilot Schweickart, known to everyone as "Rusty," left the air force and returned to MIT, this time as a graduate student in their Department of Aeronautics and Astronautics. However, as the building of the Berlin Wall heightened global tensions, Schweickart's reserve status in the Massachusetts National Guard caused him to be called up again in August 1961 and deployed as a pilot to Europe.

He was seated in a coffee shop in eastern France one morning the following February when something in the newspaper caught his attention. "It was the morning after John Glenn's flight, and I was reading the newspaper report of it when I went off into a reverie," he recalls. "Coming back, I knew what I wanted to do." The report had struck an unexpectedly deep chord within Schweickart; as he read, he came to a determination that would change his life forever. If flying in a fighter plane meant freedom to him, he could imagine that the experience of weightlessness would be the ultimate realization of that dream.

There was one serious obstacle in his path: the mandatory test-pilot experience he knew NASA required of astronaut applicants. But when applications for the next intake were announced he discovered to his delight that this was no longer a requirement. He was, however, about five hours short of the required one thousand hours of jet experience. Mailing in his application, he stated that he had the full requirement, then went to the airport and flew for seven hours. By the time his application was opened, the flight-time information was correct.

In October 1963, a week short of his twenty-eighth birthday, Rusty Schweickart was selected in the third group of astronauts, along with Walt Cunningham and Bill Anders.

Like Anders, Schweickart was never going to play the political game in order to win favor within the astronaut office. At his selection group's first press conference, he was told that the reporters would ask about his religious preference. He was not a religious person, and NASA's public affairs office gently suggested that he just allude to his parents' religion. However, he preferred to be completely honest, as he later recalled:

Walt Cunningham and I both considered ourselves to be agnostics, and I guess Walt may have even considered himself an atheist. One thing if you know me, whether it sits smoothly or not, you get the truth. I don't play games. I'm a very straightforward person, and I probably see the world a bit more simply than it is. Other people sometimes see subtleties that I just overlook, but it isn't my nature to play games. I said something to the effect that traditional religious beliefs were not something that I felt strongly about. I didn't want to just fake it by saying what my parents' religion was.

Schweickart knew that he would be somewhat of an outsider, selected in part to keep the scientific community happy. He was thick-skinned enough to be able to deal with the fact that the test-pilot astronauts were a little disparaging of their newer, more scientifically minded colleagues. There was no way he was going to downplay his scientific credentials and emphasize his piloting skills just to win favor. After all, he saw his scientific skills as one of the reasons he had been selected:

I came in when there was no such thing as a scientist-astronaut. Ours was the last of the groups where everyone was the same; after that you were either a pilot or you were a scientist-astronaut. It was very clear in the selection for the third group of astronauts that the prior requirement that one be a test pilot had been dropped under pressure from the scientific community, and in order to get people who had more educational, and in particular, science backgrounds into the program. Deke was certainly one of those people for whom science was not something that was really interesting. It was something that he sort of acquiesced in; he realized it had to be done, but it wasn't something that he ever really felt strongly about or appreciated that much. So it's fairly logical, since we were the first ones to not be test pilots, that he saw our group as test pilots and then the other pilots.

Despite the somewhat depreciative attitude of the more senior astronauts, Schweickart didn't consider himself to be any less of a pilot than the others. He still had a great amount of experience flying high-performance jet aircraft and believed himself to be on an equal level with his new colleagues:

I sure didn't consider myself second class to anybody in terms of piloting skills. I think I could have beaten any one of them in a dogfight, frankly. Generally speaking, if you've got real fighter pilots, every one of them thinks that they are the best. I think other people were more sensitive to it than I was; there were certainly people in the program who were fairly narrow-minded and had strong biases. I felt some bias, but to me it was a natural thing—you get any kind of a club, and you bring new people in, they're the newbies. If they are a little bit different for some kind of reason, then that simply adds to it. But you know, it's nature. I didn't consider it to be a great offence or anything, just the way life is.

There was another area where Schweickart differed from most of his colleagues. Those in the astronaut office were, on the whole, a group of fairly conservative military officers. Schweickart, conversely, was more attuned to what was happening in the wider world of the 1960s outside of NASA. Left wing on most social issues, he also maintained an interest in the arts, occasionally leading a literary discussion group. His wife Clare was an active campaigner for causes that were seen as surprisingly liberal for the formerly staid world of the astronaut families. The couple's wider awareness of the quickly changing country was seen as a little too idealist, intellectual, and at times irreverent by some of Schweickart's more traditional colleagues. But he was resolute—he was not about to change for anyone: "I don't think I was the only person in the astronaut office who had sensitivity to left-of-center issues and what was going on in the world at the time, but I was probably the furthest over on it. The whole social upheaval was happening, and fellow astronauts like Ed Mitchell, I think, were somewhat sensitive to it also. Ed kept his mouth closed about it; I was probably more open about it."

Schweickart does feel that recent characterizations of him as something of a "hippie astronaut" take the point too far. The astronauts who see him that way today, he believes, are remembering later events in his career, not what happened at NASA: "While I was in the office, I didn't think politically at all. To be honest with you, I don't think I started having strong political feelings until later in life. It wasn't that I didn't have politics, but it wasn't something I spent time thinking about; I don't think that was something that was evident to us until years later. We had different opinions on things, but it was only afterward that I thought of myself as more of a liberal."

There was not much time to think about the social issues facing 1960s America when NASA was so focused on landing humans on the moon. Schweickart had to quickly find his place in the team, and at times it was not easy. At first, some of his colleagues were suspicious; they felt that astronauts with scientific backgrounds would try to load up each mission with endless experiments. They did not realize that Schweickart and his science-qualified cohorts were fully supportive of what should happen first in the program:

We were certainly not advocates of adding more and more science to the missions. Early on, when we first came into the office and we were flying Gemini spacecraft, there was no question that the primary goals of the program, justifiable goals, were to prove that human beings could fly in space, the equipment was performing well, that we could develop techniques of rendezvous and docking, maneuvering during landing: all of the things that clearly had to be proven, developed, tested, and checked out. All those things were primary; there was no question about that. The goal was to land on the moon and come back before the end of the decade, and that took precedent, and we totally agreed with that; there was no conflict.

Schweickart felt, however, that there would be time on some of the missions to do a little more than the test piloting. If that time was to be spent doing science experiments, they should be done to the same exacting standards as the piloting:

In some cases, doing them correctly did not mean they were more difficult for the crew—it meant they were easier. But given someone like Gus Grissom or Alan Shepard, their basic orientation was as a test pilot. They were certainly dismissive of some of the interests that Walt, Bill Anders, and I had of ensuring that the science assigned to the flights was done well. It wasn't that we were being disruptive, but there was some suspicion that somehow we had a different set of values, that there was going to be a conflict. But we really didn't: there was a difference perhaps in viewpoint, but our priorities were the same.

This misunderstanding between the test pilots and their new colleagues caused some friction early on. Yet Schweickart remembers that he spent a lot of time defending the interests of the astronauts against those of the scientists, and in particular the medical community. They wanted the astronauts to undertake experiments that Schweickart seriously doubts they'd ever have conducted on themselves. It took a while for some of his astronaut colleagues to realize that he was on their side:

There was a big misunderstanding between Gus and I in the beginning. He clearly wrote me off as not being really supportive of his real interests, that I was being dismissive of his opinions, and represented the scientific community. Later on, when we worked together on Apollo he and I became really good friends, and we talked about it. It was interesting: Gus had misunderstood something I said, and boy, he just wrote me off! It was a real tension, but he gradually came to understand that that wasn't the case, that I was trying to get both things done: to look out for his interests, and see that the science got done well. It took a while for him and some of the other people to appreciate that.

Despite these initial misunderstandings, Schweickart was not particularly concerned about when he would get his first flight:

I just waited my turn; I figured when people saw it fit me it was my right time. As somebody said when we first got there—I don't remember whether it was Wally Schirra or Al Shepard—you know, everybody's going to fly. There's not a question you're going to fly; maybe a question of how soon or when, but everybody's going to fly. After you do fly, then there's a sort of rotation that goes on, which is a logical rotation. As far as I was concerned, that was Shepard and Slayton's job.

Other people tried to puzzle out what made the difference. Everyone wondered what were the criteria for picking people; nobody ever knew, nobody ever said. There was certainly competition: everybody tried to do their best, including me. Others probably tried to put themselves up front more. But I'd do my best and other people could judge it. I didn't go out of my way to show it off or whatever—I just did my job the best I could. If people weren't bright enough to see it, well that's the way it went.

Rusty Schweickart was not assigned a Gemini flight, unlike some of his colleagues. In March 1966, however, he was selected to join Jim McDivitt's crew, which would be backing up the first manned Apollo mission. McDivitt, Schweickart, and Dave Scott were then given the assignment to fly the second Apollo flight. The *Apollo 1* fire changed priorities and expectations once again, but the crew still expected to fly the second Apollo mission, with the first LM. It all seemed settled, until the idea of flying the second mission around the moon was introduced. Which crew was going to fly it—Borman's or McDivitt's?

Just as Frank Borman saw no need to consult with his crewmates when accepting the *Apollo 8* flight, Jim McDivitt did likewise when deciding to stick with the first LM flight. McDivitt saw *Apollo 8* as being a risky big step, but a logical one; the entire space program up to that point had been taken in big steps, and if they didn't take this one NASA would never make their self-imposed deadline. It wasn't the flight he wanted to make, however.

When McDivitt later informed Dave Scott of his choice, Scott agreed with him 100 percent. If Scott had been offered the choice between the two missions himself, he is adamant he would have chosen *Apollo 9* without a second thought. He saw *Apollo 8* as a mission with a lot of public relations glamour, but it was a relatively unchallenging one for the crew. *Apollo 9*, on the other hand, was full of opportunity and challenge, without doubt one of the prize missions of the program. Scott had not become an astronaut to be a celebrity; he simply wanted to do a good job. If he wanted recognition for what he was about to do, he hoped it would come from the people he was working closely with, who understood how crucial *Apollo 9* was to the entire program. After the disappointing curtailment of his *Gemini 8* mission he was also looking forward to flying a complete mission, one that successfully fulfilled all of its objectives.

Perhaps because the LM was his specialty, Schweickart is even more outspoken than Scott about his mission preference. In fact, he is a little disparaging about the first flight to the moon:

On Apollo 8, *you were a passenger. Eight was a kind of a quasi-political flight; it was to get to the moon before the Russians. Not many of us believed that they were really going to get there before us, but Borman was right, in the fact that there was an opportunity to get a flight in between 7 and 9, given the slow development of the lunar module. Frank wanted it, so—great, Frank—take it! We were happy to see him go and pick up that flight. As far as we were concerned, it was not a challenge. There were interesting things that were being done on the flight—you were going to see the backside of the moon for the first time—but you weren't really* doing *anything. Navigation and communication was a technical thing. Do the communications systems work out as far as the moon? Well, all you're doing is turning a switch on and talking! To us, it was a ho-hum mission.* Apollo 9 *was a challenging flight; a major test flight. Plus we had the lunar module, and we wanted to stick with it. That's what pilots want to do, fly that first flight; that's being a test pilot.*

Though he would gladly have traveled to the moon on a later flight with an LM, Schweickart was not interested in that first flight out. In the meantime, *Apollo 9* was perfect for him. He would visit the manufacturers and work in the two spacecraft well beyond midnight, as he and the crew tested systems and subsystems. During hold times when a problem had to be fixed, he would often reflect on his days as a boy growing up on that farm in New Jersey. It was quite a contrast with sitting in a spacecraft in the middle of the night—one that he would soon be flying in Earth orbit. After the years of work and effort that had gone into the design and engineering of the two different Apollo spacecraft, it was wonderful to be able to see the real things, to rap his knuckles on them and know they were a palpable reality. Training with the onboard computers, he could feel the information flowing back and forth between the spacecraft and himself. It seemed as if he was part of the spacecraft, interacting as one organism. He couldn't wait to test them in all their complexity, as he remembers: "It was a premier test flight, probably the premier test flight of all of the flights in the Apollo sequence. There were more things being tested and checked than any other mission. There were the combinations and permutations of things that you could do with two spacecraft, with two totally different control systems and all of the different configurations. It was a very complex flight, and we knew the importance of it."

There was one downside for Dave Scott, caused by the change in flight order. He had worked hard on getting Command Module 103 into a flight-ready condition, following it from its creation as a bare metal shell through all of the factory testing. He was not happy that the change in mission order meant he would now be flying CM 104, while the crew of *Apollo 8* flew "his" CM 103 to the moon. His only consolation came from the fact that CM 103 performed magnificently on the *Apollo 8* mission.

For his part, Schweickart would be flying a different LM than originally planned, but for him this was a welcome change. According to Schweickart, while he and McDivitt were training on the spacecraft designated LM 2, it had become increasingly obvious to them that this spacecraft was "really marginal." Like many other engineering inventions, such as high-performance cars, the first LM off the assembly line had a number of problems. Schweickart likened it to a patchwork quilt: "It was evident to us very shortly after we started through the testing and checkout procedure that it was basically the Grumman engineering and testing team learning how to do things, and this was reflected in the spacecraft. It was just not up to snuff, and we ended up saying that; we made it very clear. It was a learning curve issue; it's understandable."

With more time before their flight because of the *Apollo 1* fire and the ensuing switch in flight order, the crew was now given LM 3, which Schweickart says was a far better spacecraft.

Whichever LM they flew, it was obvious to the crew that this vehicle was a whole new concept in spacecraft design. McDivitt recalls that the first time he and Schweickart climbed inside the spacecraft to undertake a storage review, they broke something in the fragile interior with every turn. The interior was later strengthened, but weight considerations meant the vehicle would always be a delicate thing. Schweickart, however, grew to love the vehicle:

The lunar module was really different from anything any of us had ever seen before as a flying vehicle. It appeared to be, and to some extent was, very flimsy, with layers of cellophane on the outside. But we fully understood it. To me it was a unique vehicle designed for a unique set of requirements. It matched those requirements very, very well, including things like standing up; we had no seats. It was an interesting, out-of-the-box logical design where form fit function. It was certainly different, and there was no denying that, but I wasn't in any way concerned about it; it was a great design.

As Schweickart recalls, the flight of *Apollo 9* introduced a host of complexities into the mission planning. For the first time, a single rocket would be launching more than one spacecraft. When the two spacecraft talked back and forth, they would need to have clearly distinct call signs. As a result, an old political issue would need to be sidestepped.

After Gus Grissom and John Young named their Gemini 3 *spacecraft* Molly Brown, *NASA decided that there would be no more naming of spacecraft. Humor was seen as inappropriate—period. And that's how it was, all through the rest of Gemini and into Apollo. Then came* Apollo 9, *and we had some special, unique problems, namely three "spacecraft," which were all part of* Apollo 9: *the CSM, the LM, and me, the EVA guy. We needed to talk to each other, so we needed unique radio call signs. What would we choose*—Apollo 9 Alpha, Apollo 9 Beta*? No way. Using just CSM, LM, and EVA would also be pretty dull. We decided to think on it.*

We were sitting at the bar one evening at a steak and lobster joint in Downey not far from the Rockwell plant where we were checking out the spacecraft, bouncing around ideas about what we wanted to do. Given that my EVA helmet was bright red, and that I was to cross over from the LM to the CSM during my EVA, it became pretty obvious that I should be "Red Rover," just like the kids' game. Once that was decided, we got on a roll. We figured that we'd be as direct and unimaginative as possible. The obvious analog for the LM was Spider, *and it wasn't much more of a stretch for the gumdrop-shaped CSM to be named* Gumdrop. *How could NASA headquarters bitch about that?*

The NASA Public Relations department was not overly impressed with the crew's selection—they weren't as inspiring as those chosen for the Mercury spacecraft—but they could not argue that the choices were inappropriate. Not only did the call signs perfectly describe the shapes of the spacecraft, they were crisp—important when making radio transmissions. The call signs were not going to stir the public's emotions, but they were the perfect choice for a complex test-piloting mission. "To test it out," Schweickart recalled, "we started using those call signs during simulations. No one objected, so we just went with it. We never asked anyone about it, never asked permission, and never suggested a policy change. We just did it. After that, given that no one seemed to raise objections with us, the next missions just followed the lead, albeit with more politically correct names."

As the scheduled launch date of 28 February 1969 neared, McDivitt,

Scott, and Schweickart were acutely aware that this mission had to succeed if there was going to be a successful landing on the lunar surface that year. Though there was immense pressure on the crew to fly a perfect mission, nobody needed to push them; they knew what had to be done.

Disappointment came one day before the planned liftoff date, when medical tests revealed that the crew all displayed the symptoms of head colds. After the problems caused by similar colds on *Apollo 7* and the still-mysterious illness that had hit Borman on *Apollo 8*, the flight planners were reluctant to take any chances. The mission was delayed for a few days until the symptoms disappeared. On the new launch date of 3 March, Schweickart felt that his body and mind were precisely focused for the tasks that lay ahead: "Anytime you're going be doing something where there's a lot of tension, pressure, and even some anxiety, if you look back before the event you can hardly remember eating breakfast or what was going on. You know that when the curtain opens for the play, when they start the launch countdown, everything gets very clear, you're right there, and off you go."

Buckled up tightly inside the spacecraft and ready for launch, with the *Apollo 1* fire still fresh in their minds, Schweickart remembers that he and Scott both wanted to be able to get the hatch open in a hurry and evacuate the spacecraft if something similar happened: "Dave and I both had to make a lot of motions with our arms, and when you're lying in that spacecraft side by side, strapped in, your arms are jammed right up against each other. In fact, usually one guy's arm is on top of the others because you can't get side by side. So after Guenter Wendt strapped us in tight, closed the door, and took his head out of the window, Dave and I loosened our shoulder straps a little bit so that we could have more freedom of motion." The first manned Saturn V launch of 1969 began smoothly and steadily, carrying for the first time a full payload of lunar hardware. Unlike Scott's Gemini launch experience, where the briefings from his colleagues had given him a very good idea of what to expect, the still-novel launch of a Saturn V contained a few surprises for him. After a smooth ride on the first stage, the change between the first and second stages was an extremely violent experience for him. Schweickart also had a similar reaction, as he recalls:

If you think about those rocket engines on the back of that Saturn V, they were putting out something like seven and a half million pounds of thrust, but the vehicle now weighed much less than it did at lift-off. We were accelerating at many g's and had a long, thin metal cylinder there under tremendous compression, all the weight was at the top end now that the fuel was gone from the main tanks. So then we suddenly cut off those engines. What happened is that the vehicle suddenly grew longer! I don't know how much, but I would say several inches, maybe as much as six inches of compression. It went both ways, and we were up at the front end of it.

Just as abruptly the expansion stopped, and the loosened shoulder straps suddenly seemed like less of a good idea: "Dave and I were thrown forward right up against the instrument panel; our helmets stopped probably an inch away. I remember both of us looking at each other and going 'Whoa!' We told the guys on the next flight, don't loosen your shoulder straps! I don't remember whether Jim even noticed it, but Dave and I sure did."

The violence of staging would be the only severe motion the crew would undergo that first day, and this was a deliberate strategy. Once in orbit, they were trying a new method to ward off space sickness. They began by moving very cautiously, making each motion thoughtfully, and trying to avoid turning their heads. With their limited knowledge and experience of the illness, it seemed worth a try. Due to problems experienced by prior Apollo crews in falling asleep, it had also been decided that the crew members would all sleep at the same time. The astronauts were quick to point out in jest that this arrangement should not be phrased as "sleeping together."

One of the most critical integrity tests came at the beginning of the mission's orbital phase. If the docking probe and drogue designed to link the command and service module with the LM did not work on later flights, a lunar landing could not take place. If it did not work when the LM returned from the surface after a landing, it would severely complicate the efforts to return to Earth. Scott began the first crucial test by flying the command and service module out from the tapered adapter on the forward end of the Saturn S-IVB stage. The four sections of the adapter jettisoned "like huge flower petals," as flight controller Gene Kranz described it, looking "like something out of a James Bond movie." In an impressive display of test piloting Scott turned the CSM around, nosed it into the top of the S-IVB and extracted the LM. Somewhat ironically, given the thruster problems he had experienced on *Gemini 8*, a number of the thrusters he had planned to use did not immediately work. Nevertheless, he still did the job to perfection. According to Scott, the sequence of shocks from the launch and the separation of the CM had caused some valves to close. Cycling the switches, the problem was easily solved. The probe and drogue worked, and the mission was off to a great start.

After McDivitt and Schweickart checked the tunnel between the two docked spacecraft, they inched the assembly away from the Saturn third stage that had delivered them into orbit and began preparations for the next major task. For the first time, one manned American spacecraft would be used to push another. While the engineers were confident that the two spacecraft would stay connected, no one could know for sure until it was attempted in orbit. The first engine firing therefore lasted a very conservative five seconds, after which Scott excitedly reported that the LM was still firmly attached. For the rest of the day and well into the next, the astronauts would further test the ability to maneuver both spacecraft together. This included proving that the autopilot could handle the engine swiveling and gyrating the spacecraft. On every test, the spacecraft proved to be as tough and precise as hoped, and in some cases did even better than anticipated. The trials that the docked spacecraft were undergoing were more than the flight planners hoped they would ever have to replicate in the Apollo program. Yet, some of the techniques would be used just four missions later, at which point they saved the lives of the crew.

The first entry of astronauts into the LM was programmed for the third day, but this would also prove to be the day that space adaptation sickness became a serious threat to the progress of the Apollo program. The largely unexplained illness seemed to befall crew members at random, and it could have happened to anyone. This time, it chose to strike Rusty Schweickart, who had been assiduously following the advice given by the flight planners: "We were going totally on seat-of-the-pants, and what I did, thinking I was doing something good, was to avoid head motions after getting into orbit and avoid moving around. This did, in fact, prevent me from getting sick until the third day, when I needed to move around. Subsequently, we came to understand with all the testing I did after the flight that it would have been smarter to move around earlier."

Schweickart had a lot of moving around to do on this particular day because the crew was donning their spacesuits for the LM checkout. He would later learn that adaptation to space occurs with motion, and he should have spent the preceding days working at a level just below the threshold of nausea:

What I did was in some sense delay my adaptation until the worst possible time, thinking that I was doing well. When it came time to get suited up to go over to the lunar module, getting into the spacesuit was a major contortion. The design we had then had a zipper up the back and around to the crotch. You had to double in half to get your head into the suit, your hands into the arms; then you had to wiggle and squirm and shake to get yourself to the point where you were standing upright with the suit on. It was that extreme motion, disorientation, and contortion in getting into the suit, with eyes closed, which clearly was more of a challenge than the motion sickness was going to take. It was a huge challenge from the standpoint of the inner ear, especially when I had been avoiding motion up until that time.

The extreme motions caused Schweickart to suddenly vomit. Jim McDivitt also felt queasy due to the amount of activity, and this episode would prove to be a vital clue in understanding the illness. Once it was known what might be done to avoid the nausea, crews could work around it so a similar illness would not affect the lunar landings, now looming ever nearer on the calendar. But in Earth orbit that day, the principal concern was just getting through the day's flight plan.

Fortunately, this bout of nausea was not severe enough to prevent the astronauts from carrying out their busy schedule. First, Schweickart moved into the tunnel between the two spacecraft, opened the LM hatch, and then entered the cabin. Less than two months before, Soviet cosmonauts had spacewalked between orbiting spacecraft, the first space transfer in history. Schweickart made the first American transfer the easier way, without going outside. He felt no disorientation when what had been his "down" became his "up." The spacecraft were docked top to top, but in weightlessness these orientations had little meaning.

Working alone for an hour powering up the spacecraft, Schweickart found the systems surprisingly noisy as the spindly LM *Spider* came to life, in sharp contrast to the quieter workings of CM *Gumdrop*. Once McDivitt joined him the two men worked methodically through the day, after first ensuring that the spacecraft's life support systems functioned. While Scott remained in *Gumdrop*, the hatches were shut between the CM and *Spider*. With this, the spacecraft were effectively isolated from each other. *Spider*'s guidance and communications systems all proved to work well, as did the landing legs. On command, they sprang into the positions they would assume in order to land on the moon. Then, without warning, Rusty Schweickart vomited again: "That one was a complete surprise. It was later in the day, and I suddenly had nothing to do. I had to wait for McDivitt to get something done before we could work together on the next task. I just relaxed, and that's when I got sick the second time."

After the second incident, Schweickart felt a lot better. Perhaps he had now adapted to space, but the illness was not making his commander feel any better, and McDivitt requested a private talk with doctors back in Mission Control. He told them his LMP now seemed much healthier, but he did not want to take any chances with so much still remaining to be done on the mission and so would take each task one step at a time. Schweickart's sickness was not preventing the two men from continuing a thorough checkout of the LM's systems, however. The steering thrusters were tested, after which the large throttleable descent engine was fired for the first time. It burst into life, shifting both spacecraft in their united orbit, and proved itself to be a superb piece of engineering. The first tests of the LM had exceeded all expectations, which allowed a contented McDivitt to float back into *Gumdrop* and leave Schweickart alone to power down the new spacecraft.

The question of Schweickart's health and future well-being had still not been resolved, and that evening he was able to reflect objectively on the problem. According to the flight schedule he was supposed to perform the first EVA of the Apollo program the next day, and he was not sure if he could do it safely:

For Jim and I, it was a big question. We're supposed to do an EVA tomorrow. What's going to happen here? I was looking at it as a third party now; is this guy Schweickart going to continue to be sick? Is he going to be sick tomorrow, or is he adapting? You can't do an EVA if you have any chance of being space sick. If you are in the suit, outside in the vacuum, and you barf—you die! It's not approximately, it's not maybe, it's—you die! So it is not something you fool around with. I mean, the vomit stays right there next to your nose and mouth, and you can't breathe. You go to breathe; you drown. You can't get your hands up there—there's nothing you can do. You don't fool with that.

With such a serious concern hanging over them, the crew decided that the EVA would indeed be too hazardous. Schweickart remembers the decision being made; it was upsetting, but obviously the right thing to do. It meant, however, that he would not get his chance to walk in space, a serious blow after all his training and effort:

That night we talked about it, then we called the ground and said, we're canceling the EVA. We would go right up to the point where we would depressurize the lunar module, and then pretend to do the EVA. We'd simulate depressurizing the spacecraft and the lunar module, then come in again, close the hatch, and repressurize. This way we'd check out all the procedures. One of the major reasons for doing this test was to make sure all the procedures were right for the guys that were going to need it later on. We were going to just eliminate the actual depressurization portion of it and the EVA.

If cancellation of his EVA was not bad enough on Schweickart's morale, he had even more to think about: "It was not clear at the end of that third day, that in fact we were even going to be able to do that much. It wasn't clear whether I was going to get more and more sick. We talked about it, as in our own minds there was the question of whether we were going to have to abort the mission." Schweickart spent a troubled night, one of the most difficult of his life, restlessly pondering the consequences of his sickness. It was entirely possible that it would affect not only the *Apollo 9* mission, but also the missions to follow: "When I went to sleep that night, or tried to go to sleep that night, it was really tough emotionally. Maybe I was going to be the cause of aborting this critical mission, of not getting the tests done that had to be done. Missing the Kennedy goal of landing on the moon, because I'm *barfing*. That was not a pleasant night, let me tell you."

The crew tried to create something positive out of a difficult situation. With the EVA canceled, there would be less to do the next day. As they had not been sleeping well and were tired from the extensive work they had carried out so far, they decided to sleep in for an hour and a half. The next morning, the crew began preparing for the reduced tasks of the day. It soon became clear that while Schweickart had not fully recovered from his illness, he was feeling a great deal better than the day before. Then came the moment when he realized that an EVA might still be possible:

I don't remember exactly when, but something like an hour before we would have depressurized the spacecraft, and we had a waiting moment, Jim looked at me and said, "You look like you're feeling a lot better today." I replied, "Yeah, I am." He said, "Well, let's keep going, but let's keep that in our minds." So we went for another half hour or forty-five minutes, and had another moment when he looked at me and he asked, "How are you feeling?" I replied, "I'm feeling fine. I don't think I've got a problem." He said, "You look pretty good. Do you think we can do it?" I said, "Yeah—I think we should do it." Jim looked at me and made his decision. "Okay—let's do it!" He called the ground and told them, "We're going on the EVA."

Thus, with a minimal amount of discussion, the spacewalk was back on, and with it the chance to really test the spacecraft and spacesuit.

Originally, Schweickart was meant to spacewalk from the LM door right over to the open hatch of the CM, to prove that it could be done in an emergency situation. Given the circumstances, this now seemed a little ambitious. It was instead decided to let Schweickart venture outside as far as the front porch. At the same time, a space-suited Dave Scott made his own stand-up EVA in the open hatch of the CM, bringing in experiments that had been placed on the outside of the spacecraft, while photographing Schweickart's EVA.

Soon Schweickart was outside, in a spacesuit that provided all of his life support. He was in effect a separate spacecraft, attached to *Spider* only by a thin tether. To his relief he felt fine now, so much so that McDivitt suggested he also try the LM's handrails. Moving around turned out to be far easier than in water-tank simulations, and Schweickart was also buoyed by the fact that the spacewalk was happening at all. "That EVA was really special," he remembers with a smile, "because we went from that incredible low the night before. We went right through the EVA, and had no problem. The answer was, I had adapted. But of course the night before we didn't know that."

Within a few minutes Schweickart had concluded that it would not have been necessary to conduct the full EVA anyway. As he recalls, he was finding that controlling himself on the handrails was simple:

It was a piece of cake. The only real reason for going across the top of the lunar module to the command module on that EVA traverse was to demonstrate that we weren't going to be flopping around, poking holes in the spacesuit, and breaking off antennae. Two minutes after I was out there on the handrail it was obvious that I could control myself with one hand. There was no wind, nothing to disturb me, no problem controlling myself: it was a perfect, weightless environment. I recovered thermal samples from the front of the lunar module, took pictures, and did some experiments on the handrail. There was not a question in the world.

Dave Scott is also sure that Schweickart could have carried out the entire EVA had he been asked to do so, but agrees that it was not necessary. He believes that Schweickart proved what needed to be proved, and doing any more would have just been academic.

With the EVA objectives completed, Scott unwittingly provided Schweickart with a moment of reflection in which he had one of Apollo's most moving experiences. Up to this time, almost every second of the mission had been packed with engineering and piloting tests. Then Dave Scott's camera jammed—a camera type that Scott laughingly remembers as failing on almost every mission it flew. McDivitt asked Scott to try and fix it, leaving Schweickart to hold his position on the front porch of *Spider* in his spacesuit, his feet securely placed into restraints that had been dubbed "golden slippers." He had absolutely nothing to do for five minutes but look around, and he took full advantage of this unprecedented moment of personal luxury. He watched land masses and cloud-covered ocean slowly glide by below him, until he was called back to his tasks. He didn't overanalyze what it was he was looking at; he simply lowered his defenses and let the vivid experience flow into his mind. There would be time later to work out what it all meant. All too soon it was time for Schweickart to make his way back inside *Spider*, close it up, and return to *Gumdrop*. The next day would provide the biggest test of the spacecraft, and the crew would need to rest and prepare.

Two momentous events had taken place that day, though neither of them was apparent at the time. The answer to managing space sickness sufficiently to perform lunar landings lay in the events of the mission. The second major occurrence was connected with the awe-inspiring view that Schweickart had taken in. He did not have any time then to dwell on it, and in fact it would be some five years before the full meaning of what he had seen would burst to the surface of his consciousness and surprise him.

The fifth day of the flight brought the ultimate test of faith in the hardware—the planned separation of the LM from the CSM. Until the two spacecraft had flown separately and redocked, it would be impossible to accurately design the rendezvous trajectories and techniques necessary for lunar orbital and landing maneuvers. With McDivitt and Schweickart positioned in the LM and Scott alone in the CM, it was time to see if it all worked. Scott hit the switch that would separate the two spacecraft, hoping that everything would go as anticipated. McDivitt and Schweickart were now relying on their flimsy, untried spacecraft *Spider*. Yet all did not go to plan during the separation, as Schweickart recalls:

The only place that there was tension on the mission, the one place Dave really surprised Jim and I was right as we undocked. Dave had to reach up and push a spring-loaded switch, and that released the probe so that it would push the spring. The probe would then push out about eight inches, and it would do it with a velocity of maybe half a foot per second. At the end of that it would reach its full extension, but the lunar module would just keep going. Well, there were also three little latches in the head of the probe that would initially connect in the back of the drogue. Dave hit the switch and let it go the same way that he did in all the simulations on the ground. But this was one of the few cases where the simulators were wrong. In the simulator, what would happen is the probe would reach its maximum extension and the LM would just keep drifting away. But in flight, when Dave hit the switch and let it go, the probe extended, reached the end, and all of a sudden it went clunk! Jim looked up; we both said "What's that?"

The LM stopped, we looked out the top, and we weren't moving away from the command module. It was clear that we were still hooked onto the end of the probe. What had happened was the three little latches at the head of the probe didn't stay in. Dave had to hold the switch until it reached full extension to keep them in. When he let go of the switch, it let those latches reconnect.

After a hurried discussion, McDivitt and Schweickart agreed to redock firmly, call the ground, and then decide what to do next. They had no sooner decided to do this, Schweickart remembers, than Scott called to them that they were now free. Looking out of the windows, they saw to their surprise that they were slowly drifting away from *Gumdrop* and the attached service module. Scott had hit the switch again, releasing them.

Dave Scott watched as the ungainly *Spider* slowly diminished in size. At first it was just an irregular-looking vehicle moving away from him, and then it became just another bright star in the distance. Finally it was a hundred miles away and lost to sight. For the first time since the days of the Mercury program, an astronaut was flying a spacecraft solo. Scott kept the command and service module in a circular orbit, ready to complete a solo rendezvous and rescue if needed, while his colleagues aboard *Spider* tested the craft's ability to maneuver into different orbits. For the next six and a half hours McDivitt and Schweickart would pitch and yaw the spacecraft, while repeatedly firing its descent engine at various throttle settlings to place them into differing orbits. It was a thorough piloting checkout of this strange new type of spacecraft, as well as a test of the ability of two craft and two pilots to locate each using both radar and eyesight.

Even though he was in a spacecraft with no heat shield, Schweickart is reported to have said he felt safer than he did during the altitude chamber tests in Houston, as if this was intended as a validation of the spacecraft's integrity. What Schweickart actually meant was quite different:

What people don't realize is that I actually was *safer. I can remember standing at the bottom of the huge altitude test chamber A in Houston—this thing is something like a hundred and twenty feet high and eighty feet in diameter—and I'm the little thing at the bottom in a cage of heaters, testing and checking out the spacesuit, stepping up and down on the step to put a controlled heat input into the suit. Not only did I have all of the systems in the suit which could fail, and the back pack which could fail, but I had all of the failure modes of the test chamber, which could also kill me, and in fact could kill me even more readily than a failed suit hose could. The one thing you can count on is a vacuum in space—it ain't going to go anywhere! That vacuum is reliable, and it's simple—there are no moving parts. On the ground in those tests, I had a much higher threat to my life than I did in space.*

Flying such a relatively fragile spacecraft was no problem to him, and Schweickart also found the LM to be crisply responsive to the controls, far more so than the CM. It was somewhat like the difference between a fighter and a transport plane. He also preferred the uncluttered feel of the spacecraft, and the sensation of flying while standing up. The LM was proving to be an extremely adept piece of hardware. The LM had been designed so that its lower section, the descent stage, would serve as a launch platform on lunar landing missions. This section would remain on the moon as the upper section launched back into lunar orbit using a separate ascent engine. McDivitt's crew would need to test the separation of the two sections while in Earth orbit.

It came in a burst of pyrotechnics that sent shards of gold foil insulation flying in all directions. McDivitt mostly recalls the separation for the tremendous jolt that resulted. Schweickart, on the other hand, paid it little attention. He was too busy trying to see the descent stage out of the window as it tumbled away, and was filled with professional anxiety about what would happen once the ascent engine ignited:

Both of us were concerned about the fact that the ascent engine was right there, inches away. I could reach back and touch the top of that engine, and we were concerned that it was going to make so much noise we weren't going to be able to communicate if we had to abort. So we started making hand signals that we would give each other. We fired that thing, and looked at each other—there was no noise. Zero, we never heard a thing; we could converse normally. We looked at the gauges and we could see that the ascent engine was burning, working just right. We couldn't really feel the acceleration very much, either, it was only about a sixth of a g, and we couldn't hear it at all.

Having used up much of its fuel and relieved itself of the descent stage, the LM now had less than a fifth of its original mass. Consequently it flew even better and with greater dexterity than before. Schweickart remembers that "it did handle a lot better. There's no question that the LM was much more of a fighter plane than the CM, and with the ascent stage only it was really very snappy." Now thoroughly test-flown as an independent spacecraft, *Spider* matched orbits once again with *Gumdrop*, in the most complex space rendezvous to date. Dave Scott kept a watchful lookout as he waited for the LM to come into view, and when he spotted it, cried: "You're the biggest, friendliest, funniest-looking spider I've ever seen!"

The rendezvous duplicated as closely as possible the flying techniques needed for a crew to ascend from the moon and dock. To ensure that it could be done by either spacecraft, *Spider* was the active partner in the docking. Right at the most crucial moment, however, another problem developed. The sun was shining directly through the upper window that McDivitt needed to look out of in order to rendezvous successfully, making it extremely difficult for him to see. Adding to his problems, he also had to cross-control the inputs to the controls mentally—a difficult thing to do, and made considerably more complicated when looking straight up. Schweickart says it was a difficult time for his commander.

It was a requirement that you had to actively dock from the lunar module, and we were demonstrating that capability, like a lot of things we did on our flight. Jim was to do the docking looking out the top of the lunar module, and when you do that your controls are crossed and it's very difficult. It was simple from the command module; it was very complex looking out the top of the lunar module. It was of course complicated because the sun was shining on that top window, and it made it a mess for Jim to see the thing.

Scott believes that McDivitt could still have docked unassisted, but he made it easier by guiding him in a little. The latches worked perfectly as the two spacecraft came together in a hard dock and a satisfying thump. "That wasn't a docking," McDivitt declared. "That was an eye test!" He had encountered so much difficulty in performing the maneuver that on subsequent flights this task would fall to the CMP.

With the docking successfully completed, the last piece of the engineering puzzle was in place. The two spacecraft needed for moon landings had functioned well together, and a complex program of test piloting had been carried out with great skill. Schweickart was delighted to see Dave Scott again, but he and McDivitt were understandably fatigued after such a busy day. They would soon enjoy a well-earned rest, but first they jettisoned *Spider*, allowing ground controllers to conduct several remote engine firings using the LM's remaining fuel.

Although the *Apollo 9* crew had performed the majority of the mission tasks, the flight was less than half over. The mission's ten-day duration was designed to check how well the CM would perform over the time needed to voyage to the moon and return. During the next five days they would test-fire the command and service module engine again and make crucial navigation alignments. They would also take numerous photos of the Earth passing below them. Schweickart particularly enjoyed the time they were out of contact with the ground. Mission Control was always asking the crew to relay numbers, and when they were out of range they could get some work done, and also enjoy the view. Having already enjoyed a priceless view of the blue planet, Schweickart had time to see even more as they orbited Earth, experiencing a glorious sunrise and sunset every ninety minutes:

After we kicked the LM off the front of the command module at the end of the fifth day, the next four days were really Dave Scott's mission. Jim and I were the supporting cast while Dave was doing tests of the command module. So I had a lot of time to just look out the window. I had the map over on my side, and I would check our orbits, line up the map, and announce what we were coming over. Just before we came over the Nile Delta or something, I would say "Nile Delta coming up!" and Dave would say, "Hey, tell me when we've got the best view." He'd go on doing what he was doing, and I'd say "Come on up, the Delta's right out the window." He'd come up, look out, and say "Wow!" and I'd take a picture for him. So I was playing tour guide, and to some extent had more of a chance to look out the window and appreciate the Earth going by. It was beautiful, just spectacular.

On the tenth day, bad weather and heavy seas southwest of Bermuda caused Mission Control to advise the crew that they would be flying an additional orbit before reentry to allow conditions to settle. Just after midday EST on 13 March, the *Apollo 9* spacecraft splashed down in the Atlantic, 180 miles east of the Bahamas. They were just three miles from their recovery ship, the USS *Guadalcanal.* Rescue helicopters from the carrier had circled the spacecraft as it dropped the last few hundred feet and were hovering over it within seconds of the successful splashdown.

As a triumphant Gene Kranz delightedly said, the 151-orbit shakedown flight had been "sheer exhilaration" for both Mission Control and the crew. George Mueller, NASA's associate administrator for manned spaceflight, declared: "Apollo 9 was as successful a flight as any of us could ever wish for, as well as being as successful as any of us have ever seen." Among other accomplishments, the crew had managed to bring back more photographs than were taken during the entire Gemini program.

Everything had worked so well that there would be no need to schedule any more Apollo test flights in Earth orbit. In fact, the euphoria over *Apollo 9* even led to talk that the *Apollo 10* mission, then scheduled for May, could be bypassed and replaced by *Apollo 11*. However, most NASA officials considered that to be a remote possibility, as the *Apollo 11* spacecraft was still being assembled and was unlikely to be ready before June. Lt. Gen. Sam Phillips, the Apollo program director, said that a decision would be made before the end of March but that any alternatives to the set flight plan "stand small chance" of being accepted. In any event, whether astronauts descended to within fifty thousand feet of the lunar surface or actually touched down on the following flight, *Apollo 9*'s success had ensured that the next Apollo mission would go back to the moon.

In some histories of the Apollo program it has been stated that Jim McDivitt wanted to command another Apollo mission, but only on the proviso that he would retain his entire *Apollo 9* crew. For the record, neither Scott nor Schweickart remember it this way. McDivitt became manager of lunar landing operations in May 1969, and Schweickart recalls that "I think Jim really did opt out and wanted to go ahead into management; he felt he had flown enough." McDivitt remembers that it was clear he wouldn't be able to get back into the lineup fast enough to make the first lunar landing, unless it was a last-minute job as someone's backup. He was interested in being the first person to land on the moon, but not the second or third. In fact, these days he jokes that he should have messed up the *Apollo 9* mission to place himself in line for the first landing, well after *Apollo 11*.

There was another reason that the crew was not going to be immediately recycled. Both Schweickart and NASA were keen to thoroughly investigate why he had become so ill in space, and how he had been able to successfully overcome it: "I became a human guinea pig after that flight. I did a lot of vestibular testing, for three months or more, and that clearly took me out of the possibility to just rotate right around to a new mission. It put me into a different rotation pattern; by that time Dave had picked up a new assignment, as backup on *Apollo 12*. But firstly I wanted to know myself whether or not I was going to be a liability on a mission. It worried me, the last thing I wanted to do was to be going out on a lunar mission and screw up a lunar landing!" In retrospect, Schweickart knew that if the

Apollo 9 crew had been immediately recycled, they would have been backups on *Apollo 12*, putting them in line for a lunar landing on *Apollo 15*. At the time, *Apollo 12* could well have been the first landing on the moon, as he remembers:

This is a big misunderstanding of the public; nobody knew what the first lunar landing mission was going to be. Everybody now thinks it was Apollo 11 *from the beginning—that's baloney! Nobody knew it was going to be* Apollo 11*; that was just the first possible mission it could be. If you had taken a poll amongst the crews, it would probably have been* 12 *or* 13 *picked as most probable.* Eleven *would not be the one that people picked, because you always assume something's going to go wrong, something's going to get delayed.*

I was thinking, as a backup, I also needed to be ready to fly. If for some reason 12 *had been the first lunar landing mission, and, for whatever reason our backup crew would have taken it, because of something wrong, then the real question was, would I be jeopardizing the lunar landing mission with motion sickness? What if I was lucky and adapted on* 9, *but was less lucky on* 12*? At that time we didn't understand that adaptation comes along pretty readily if you challenge yourself right up to or short of getting sick. I knew I'd adapted on* Apollo 9, *but I didn't know what would happen if I went to the moon. I wanted to understand it better before I risked the program by saying I wanted to be in the rotation and get back on a lunar mission.*

Since all successive flights would carry LMs to the moon, the race was on to ensure that the lunar landing crews would be unaffected by the sickness. Schweickart went to the naval facilities in Pensacola, where he threw himself into a dizzying and punishing series of medical tests to try and determine what had happened. He put himself through days of nausea-inducing trials on rotating devices, balancing beams, and as many flying machines as possible. The team did not determine the causes of the illness, but they learned from the tests and the *Apollo 9* experience what could be done to help astronauts adapt, both before and during a mission. Schweickart believes there are still a lot of unknowns about motion sickness, let alone space sickness, but genuinely feels that his experiences and the tests he took provided many answers: "There are different sensitivities to motion sickness. It's all tied up with the inner ear function and the brain, and clearly in our crew I was the most sensitive to it, probably McDivitt next, and Dave least. We all adapted to it, and I don't think there was any difference between us by the end of the mission."

Thirty years after *Apollo 9*, Schweickart still gets annoyed by one aspect of the whole space sickness experience, and is not reticent about it. He says that if Frank Borman had been more open about his illness on the previous flight, they might have saved themselves a lot of trouble. He sees the difference in approach between himself and Borman as being partly due to their differences in background:

We didn't really understand about adaptation, thanks largely to Frank Borman. He got sick on Apollo 8 *but refused to acknowledge it until he had to, or take any tests postflight and give us any information. Would it have made a difference if Frank had had the testing? I don't know, that's not clear. But what is clear is that* not *doing it put me in a worse position, without question. I have never forgiven Frank for his attitude in that, for the way he behaved. I mean it was such a classic macho way to do things. It was, in my mind, so damned unprofessional as a test pilot, not to acknowledge what had happened and do what it took so that the next guys could understand it better, learn from it, and perform better. I was offended by it, and I still am.*

The reasons for Borman's illness are still undetermined decades later, so it is hard to assess whether intensive medical testing on him would have helped the *Apollo 9* crew. For Schweickart, however, an honest approach was always going to be the way he went about things. He says, for instance, that he never entertained the thought of not telling the ground when he became sick in space: "To me it's just straightforward; it's part of your responsibility. But that's the way it was, and to a lot of people the idea of getting space sick was not acceptable. In those early days, there was the Right Stuff kind of mentality, and the self-image of barfing in flight was something which somehow affected your manliness, you know. To me, that's stupid, ridiculous ego crap. I mean, professional attitude was part of the job—you just don't screw around with other people's safety, or the mission."

Rusty Schweickart may be fiercely defensive of his decision to be honest about his adaptation to weightlessness and his willingness to take as many tests as needed, but his attitude was not one shared by his fellow astronauts. Though grateful that he was trying to get to the bottom of the problem, they were pleased it was him and not them.

It can never be known how many of Schweickart's colleagues had felt sick on previous missions but chose not to disclose it. Dave Scott remembers that the astronauts had quietly discussed space sickness before *Apollo 9*, and believes that the only reason this flight became the focal point is that it impacted the mission's time line. Bill Anders remembers being pleased that Schweickart was undertaking the tests, but as he told the authors, their test-pilot colleagues did not quite see it that way: "Rusty got sick, and he probably would have got sick no matter what they did with Frank. But Rusty probably over-volunteered to research space sickness, and I don't think the general astronaut crowd appreciated that. I thought it was probably a pretty good idea. But as you know, fighter pilots never get sick, we all know that! And here was Rusty, more or less on his own, volunteering to be a space sickness guinea pig. That probably cost him any further missions."

His overt honesty put Schweickart on the losing end of astronaut office politics. As Gene Cernan has stated, Schweickart "paid the price for them all" when it came to understanding space sickness. Schweickart understands this all too well, but wouldn't have done it any differently. In fact, he's quite proud of what he accomplished:

Gene's right—in a certain sense I did pay the price for everybody because I don't hide things, I don't play games. It's part of my nature. To me it had nothing to do with being macho, or a real man, or any of that—it's physiology. I'm not unique, other people were going to suffer this thing too, so we'd better learn about it. But I did *pay the price, and it cost me. I didn't fly on Apollo again, because by the time I got done with the tests the Apollo missions were filled out, and I was reassigned to Skylab.*

Schweickart fully understood why he was assigned to Skylab. The program needed astronauts with flight experience, particularly with complex engineering and EVA specialties. He was, however, disappointed. He'd been hoping to fly an LM to the moon's surface. He knew, however, that it wasn't his choice to make. "It bothered me slightly," he reflects, "but it wasn't an appeal process; it was basically just salute and do your job. Yet I was happy to do the job I had. I'm a good soldier that way."

Dave Scott, on the other hand, hit the jackpot. His assignment as backup commander to *Apollo 12* led to him commanding the flight of

Apollo 15 to the moon in 1971. His mission turned out to be the first extended scientific exploration of the moon, and one of the high points in spaceflight history. For three days he investigated the lunar surface, spending twenty-one hours outside the LM. In many ways it was the pinnacle not only of the Apollo program, but also of space exploration to date. It was a great personal achievement for Scott too, and one he compares to going on vacation: if you study your destination thoroughly before you go, you like it all the more when you get there. Because he spent so much time studying geology before the mission, he really enjoyed the scientific exploration of the moon.

Standing on the ancient lunar surface, Scott took time when he could to look back at the home planet. He found the moon curiously beautiful, but the profound lack of life and color really brought home to him how important and unique the Earth is. Looking at the tiny blue sphere, he truly understood that it is the only place in the universe where humans know they can live. It made him realize with great clarity how the Earth's environment needed to be cared for, and sparked in him a wish to work on environmental and ecological issues in his post-NASA career. As Scott so eloquently puts it, he believes that the significance of the Apollo program was not so much to put a foot on the moon, but putting an eye on the Earth.

Moving into Apollo management had not been the only postmission option for Jim McDivitt. The air force was interested in him running a component of their proposed manned space program, called the Manned Orbiting Laboratory (MOL). After talking to some trusted colleagues, however, he was convinced that the air force would never follow through on its manned spaceflight plans and would abandon them as they had done with the earlier Dyna-Soar program. He was later proved correct; the MOL program was abruptly canceled.

McDivitt's wife Pat thought that Jim should remain an astronaut and command the *Apollo 13* lunar landing mission, which McDivitt says was being tentatively offered to him at that time. What McDivitt found more interesting, however, was the opportunity to work as a NASA manager to redesign the hardware he'd test-flown, enabling it to undertake extended scientific exploration of the moon. Having already spent a lot of time as an astronaut working closely with the contractors to shepherd the spacecraft along as they were constructed, it seemed like a natural move to him. He therefore declined the proffered flight and moved into management. He began by stripping down the bureaucracy that surrounded the position and gave himself more control over the decision-making process. This made for some memorable boardroom fights with scientists who wanted more of a role in the lunar landing process, as well as with bureaucrats who wanted more input into sensitive areas such as the *Apollo 13* accident report. Overall, however, it seems that McDivitt won people over with his management style, and in particular the science community, who welcomed the windows of opportunity they were given once the initial lunar landings had been completed. "McDivitt was near genius in the math-engineering area," according to Paul Haney. "No wonder that George Low made him head of the Apollo program in Houston after Low tired of the task and moved back to Washington."

Even when McDivitt was the Apollo spacecraft program manager, he continued to fly T-38s. Although he was not actively training as an astronaut, he wanted to keep his reflexes sharp and his adrenaline levels up. He knew that he would have some quick, crucial decisions to make, and had to be prepared. When his colleagues were flying to the moon, their lives would sometimes depend on the choices he made.

His major decisions began with his first mission as manager, *Apollo 12*. He had to decide if the flight should continue after the ascending Saturn V was hit by lightning soon after launch. During the *Apollo 16* mission, while in lunar orbit, the service module engine had a problem that could have caused an immediate abandonment and return of the flight. McDivitt was able to draw on his experience test-firing the oscillating engine on *Apollo 9*. At times like this, he was very grateful for the intensive mission he had commanded. Having personally run all the hardware through the most rigorous testing, he knew what it was capable of doing. As well as being able to interpret the telemetry, McDivitt also knew what it felt like when an engine fired or gimbaled in a certain way.

Some people, including McDivitt's wife, were surprised that he allowed some of the lunar landings to continue when such major problems arose. Yet in each case, he made the right decision. His experience was of greatest use during the *Apollo 13* mission—which he personally considers the greatest spaceflight ever flown—when the docked CM and LM were put to the

A recent photo of Gemini and Apollo commander Jim McDivitt. Courtesy Robert Pearlman.

supreme test in a frantic endeavor to bring a crew home alive. The fragile LM performed far outside of its intended parameters, and the crew lived.

Knowing the intrinsic worth of the Apollo program, McDivitt was totally opposed to curtailing the lunar landings after *Apollo 17*. He envisaged Apollo as a tremendous investment—both political and actual, considering that much of the hardware for later missions was already built. But it was an investment that was now being squandered. He believes that NASA was becoming nervous about the possibility of losing crew members on a later lunar mission and buckled under pressure just when things were getting interesting. Dying in space was a risk he had been prepared to accept, and he saw the doubts expressed by other NASA managers as an abrogation of their responsibility. As difficult as it might have been, he would have accepted the loss of another crew and moved on, just as he did following the deaths of his *Apollo 1* colleagues. A few months before the final Apollo flight to the moon, McDivitt resigned from NASA.

Looking at the moon today, he visualizes the small number of Apollo landing sites, and imagines how the lunar surface could have been dotted with them. To his growing frustration he had watched the budget earmarked for more lunar landings siphoned off into the next program—Skylab.

As the lunar program drew to a close, Rusty Schweickart was still hoping for another flight. He was originally named as backup commander for both the second and third Skylab missions, but when Walt Cunningham quit the program, he moved into the backup slot for the first mission. When the Skylab station was launched and sustained severe damage, Schweickart immediately went to work testing the repair tools in a hurried program of EVA simulations. His participation helped to save the station, allowing three crews to visit, but he wasn't on any of them. Although it was four years after *Apollo 9*, he still had not shaken the stigma of space sickness, and it is something that he says annoys him to this day. "I didn't get to fly on Skylab, which in my mind is an injustice, but then again I'm not the one who picked the crews."

With the last of the Apollo hardware used or allocated, the chances of another flight were growing ever slimmer for Schweickart. "My option was to hang around for probably six years before I would have another chance to fly on the Space Shuttle, and I didn't want to do that." In 1974 he took an assignment in the applications office at NASA headquarters, exploring ways to use space program technology for earthbound applications. "I went up to NASA headquarters, basically to learn how to become a manager," he recalls. "I saw myself as a person with a lot of potential; a lot of things were of interest to me in addition to spaceflight, so I went to learn something about budgeting, testimony before Congress—all that kind of thing."

Throughout this period, and just in case something unexpected came up, Schweickart remained an active astronaut. He gave himself a fifty-fifty chance of returning for a flight someday. Then, in 1976, he moved into a different position, where he directly influenced the planning of space shuttle payloads, and the dream of another mission faded for him: "I'm not a person who sits in one place for very long. By the time I actually got done at NASA headquarters, I found that I was a lot more interested in management and shaping where the space program was going to go, what the priorities were and what we did, than flying what other people decided on. Not just manned spaceflight, but space in general."

Not anticipating his own change in priorities, Schweickart's decision to end his astronaut career came as something of a surprise to him. By 1977, he also had the unshakeable feeling that he'd done all he could with NASA. More and more, he was convinced that the agency's administration was using him as a salesperson for their policies, and they were hawking policy directions he did not always agree with. He didn't cut his ties completely, instead taking a leave of absence that year to serve in California governor Jerry Brown's office as his science and technology assistant. "I got interested in Jerry Brown's politics, but it was also because I felt it was someplace I could shake the space program out of my hair."

In the end, Schweickart did not return to NASA. Two years later, he made his leave of absence permanent with an appointment to the California Energy Commission, which occupied him for the next six years. He chaired the commission for five of those years.

In the first five years following his spaceflight, Rusty Schweickart gave very little thought to analyzing his experiences during the five minutes he'd spent staring at the Earth during his EVA. That changed dramatically for him in the summer of 1974. He was invited to speak at a meeting on planetary culture at the Lindisfarne Association based in Southampton, Long Island. The association's mandate was a significant contrast to the technology-focused efforts of NASA; it had been founded in the early 1970s to study Eastern philosophies and Western science, and to formulate ways they could be compared and combined. This included serious studies on how to introduce meditation into the practice of science. The association had grown increasingly interested in ecological issues as it examined ways to link the spiritual with the technical. Rather than trying to promote a particular view, the association wished to be a place where new ideas could be discussed.

Always open to the new ways of thinking that had grown in popularity in the 1960s and early 1970s, Schweickart found he was becoming increasingly interested in the aims of such groups. As the time to give his Lindisfarne speech approached, however, he was growing increasingly frustrated.

"I was asked to talk about the experience of looking back at the Earth from space, and its implications for humanity," he recollects. "I never could prepare for the damn talk; I just couldn't ever get anything done on it, couldn't write even a note; I just mentally blocked." When he stepped in front of the audience Schweickart still had no real idea what he was going to say. But all astronauts are used to public speaking, and he thought he would just expound the usual rhetoric about the marvels of flying in space followed by a brief question time. As he began to speak, however, the meaning of what he had seen finally crystallized and found expression in words: "I basically listened to myself give that talk. It was a real inner expression of the experience of flying in space, and what it meant. It was related not to the immediate experience of flight, but to reflecting back on it, integrating it for three or four years afterwards. It really all came out, became conscious to me in that talk. I was almost in the audience. What came out, quite frankly, was an experience I didn't even know was in there."

To his own amazement, Schweickart found he was talking with clarity and purpose about the meaning of what he'd seen during his EVA. It was something he'd never really contemplated even as he'd gazed spellbound at the Earth below him:

It wasn't something that had happened in flight. It wasn't something that I realized during those moments when the camera failed and I was on EVA, and McDivitt said, "Stay right there, Rusty! Dave, take five minutes to try and fix the camera." But it was during those five minutes when so much of the input came in. Not the integration of it, not the working of it, not the thinking about it, but the input came in during those five minutes. I wasn't thinking about what was happening next, I just turned around basically and looked at the Earth, and just let it all flow through.

Now in tune with his previously buried impressions, Schweickart spoke of how, at first, he had looked for individual locations on the surface of his home planet. Looking down at Greece, Rome, Africa and the Middle East, he was aware he was looking at regions that had represented the apex of human civilization for centuries of history. Passing over places where he had lived in America, he felt a personal attachment. For the duration of the flight, as he continued to orbit the Earth, he began to identify with Africa,

Rusty Schweickart in 2004. Courtesy Francis French.

with India, and with other places he had never visited. His perception was shifting; looking at a world without visible borders and boundaries, he was beginning to identify with it as a whole. His relationship with the Earth was altering and blurring; yet he knew it was not an experience for him alone. He had been fortunate enough to be the person right there, sensing the silence and the view on behalf of everyone else, yet he truly felt there was nothing special about himself that would have caused him to be selected for this astonishing experience:

It became obvious to me that this was not just a case of "Look where you are. Holy smoke, how the hell did I get here?" It was not just me, as an individual—but how did we get here, and what's happening here? Again, not to me personally, and not to the program, but philosophically in terms of where human evolution is taking us all. This is clearly an evolutionary step, a major *evolutionary thing that I'm lucky enough to be taking part in, that humanity is for the first time moving away from the home planet. We're being borne out of the Earth, out of Mother Earth. I didn't think that at the time. I absorbed the experience, taking down all my defenses, allowing all of it to come in.*

What I understood in real time was that part of my responsibility was to share this experience with all the people down there; the people who paid for it, the people who are part of this evolutionary process. People have to understand; this is more than just some technological thing, a big rocket shooting people up into the air. This is a major change in the whole evolution of life on this planet. We're in the process—it may take centuries—but we're in a process of being borne into the cosmos. We're going to take all kinds of things out there with us, the whole experience of the development of life on Earth into the larger cosmos. Will we meet anybody? Who knows, but unless we blow ourselves up in the meantime, we're going to do it.

It had taken five years for everything to suddenly come together, but the *Apollo 9* mission had eventually caused a profound change in Rusty Schweickart. From a feeling of being merely a child of Earth, his thoughts evolved to those of being a parent of Earth. He felt a new responsibility to care for the planet he'd observed in all its glory. In fact, he felt like he did not have a choice. The Earth as an entity had ceased being merely an abstract, beautiful icon to him; it was now an intrinsic part of a deeply personal experience, and he wanted to act on what he felt was a precious calling. Though the flight had been an incredible demonstration of engineering precision and a huge technological leap, the success of the mission was still taking place years after the LM had been retired, in a way that had not been anticipated. Schweickart had experienced something on behalf of humanity, and he has been sharing it ever since: "I was already environmentally oriented anyway, but going around and around like that, your identity changes. It wasn't that I learned anything new—I didn't even experience anything unexpected, or unanticipated. But I experienced it; instead of knowing I would experience it, or just thinking about it intellectually, I experienced it. That's the difference; when you actually experience something, you change. And I changed; there's no question about it."

The farm boy from Neptune, New Jersey had been extraordinarily lucky; he'd been in the right place at the right time. He not only flew on one of the most important missions in the space program, but also went through a deeply moving experience, which he feels was only his on behalf of all humankind. The knowledge that Rusty Schweickart had to share only emerged once the Apollo program was almost at an end. Long after the spacecraft involved have become museum pieces, it is still a powerful message: that all humans on the Earth can continue to learn from the experiences of the early spacefarers for many decades, perhaps many centuries, yet to come.

9. The Highest Mountain

But why, some say, the moon?
Why choose this as our goal?
And they may as well ask why climb the highest mountain?

John Fitzgerald Kennedy

Once again humans stood poised to leave Earth orbit, this time on a mission of vastly increased complexity. *Apollo 8* had successfully tested the CM in lunar orbit, and the crew of *Apollo 9* had run the LM through its paces in Earth orbit. This next mission would combine the hardware and software components with the skills, experience, and techniques of hundreds of thousands of people to carry out the most complex mission to date. *Apollo 10* would be the precursor flight—the last forerunner—to the greatest technological achievement in the history of space exploration. If successful, the mission of *Apollo 10* would set the stage for a manned landing on the moon.

The astronauts chosen to fly the *Apollo 10* mission were the most experienced Apollo crew yet—arguably the most experienced there ever was. Commander Tom Stafford, CMP John Young, and LMP Gene Cernan had all flown during the Gemini program; Stafford and Cernan had even flown a mission together. As well as being spaceflight veterans, the *Apollo 10* crew members also were noted for their good humor, cool temperament, and easy-going nature. They complemented each other well, knew each others' strengths and impressed each other with their dedication during training. They were even willing to play the PR game, sending entertaining and engaging television transmissions back to Earth throughout the mission. It was one of the best space crews ever assembled.

This mission was finally doing what Apollo was meant to do: carry three capable astronauts on a voyage into lunar orbit with a lunar landing vehicle. With Kennedy's deadline rapidly approaching, the pressure to succeed was intense. Given such a time constraint, the crew has often been asked why *Apollo 10* was not the first landing mission. After all, as Tom Stafford explains, the crew was being asked to do everything but journey the last 25,000th of the distance to the surface: "Our flight was to take the first lunar module to the moon. We would take the lunar module, go down to within about ten miles above the moon, nine miles above the mountains, radar map, photo map, pick out the first landing site, do the first rendezvous around the moon, pick out some future landing sites, and come home."

There was indeed some debate within NASA management circles about whether *Apollo 10* should be the first landing mission. To the surprise of some, Stafford argued that they shouldn't land. The commander was ruling himself out of becoming the first person to walk on the moon. The reasons he did so are sound, practical ones, and show that this crew's dedication to test-piloting excellence was more important to them than personal glory. In short, there was still a lot to learn before NASA was ready to attempt a landing.

Experience had taught the mission planners that the worst problems would be the ones they had not thought of. In preparing for the unexpected, all they could do was learn to understand and overcome the major hurdles they knew existed. For example, communications from Earth with two separate spacecraft circling the moon had never been tried before. Both spacecraft, once separated and orbiting the moon, would need to be able to communicate with each other as well as be able to transmit independently to Earth. Moreover, they would frequently be around the back of the moon and out of contact with Earth, often at different times. Sometimes they would be able to contact Earth but not each other, when the moon's bulk blocked the signal between them. It was a complex problem of geometry that needed testing before an actual landing could be attempted.

A rendezvous between two craft circling the moon was a crucial maneuver that had also never been tried before. Orbital velocities around the moon were very different from those around the Earth, not only because of the weaker gravity but also because spacecraft could pass relatively close to the lunar surface while orbiting without having to worry about atmosphere. Lighting conditions around the moon would be very different from those in Earth orbit, meaning a rendezvous might be difficult even for experienced astronauts. Tracking the spacecraft from Earth presented other difficulties, as ground stations would be focusing on two spacecraft in one point of the sky, rather than an Earth-orbiting spacecraft quickly passing overhead.

Subtle wobbles in the orbit of *Apollo 8* and in the paths of unmanned lunar orbiters had also shown that there was still much to be learned about the lunar environment. The wobbling was caused by "Mascons"—areas of slightly stronger gravitational pull on the moon. Sections of the lunar seas were denser than other parts of the moon, tugging an orbiting spacecraft downward just a little as it passed overhead. This phenomenon could affect the ability of two spacecraft to find each other in lunar orbit. Twelve such areas had been detected before *Apollo 10* launched, but more investigation was needed if they were not to pose a threat to a possible lunar landing.

Gene Kranz also saw one more mission without a landing as a vital step in testing the Mission Control team. Would they be able to support two docked spacecraft around the moon, and then do the same when they were separated? It was far riskier than doing the same job in Earth orbit, where a crew could make a quick return home, Kranz told NASA historian Roy Neal. "Once you get to the moon, you don't have too many options. You have very limited wave-off options. You're either going to accomplish your mission or not; I mean, it's black and white. There's no compromise there." Experience was needed, and with this ambitious mission the flight controllers would get it.

Kranz added that there were still more reasons why another test flight was needed: "Basically, we had a lot to learn. Each one of the missions, 7, 8, 9, and 10, really contributed a major portion. The mission was designed to progressively expand our knowledge, so it would lead to a high probability of a successful landing on *Apollo 11*. One of the things that we were still fighting at the time of *Apollo 10* was the lunar landing software. It was really a question of getting it all squeezed within the very limited memory capability that existed within the machine that we built for Apollo." As well as testing the modified LM computer and the untried landing radar, the *Apollo 10* crew also needed to shake out many other parts of the LM design that weren't quite ready for a landing. "One problem as far as the landing," Stafford noted, "was that the LM was too heavy to land, unfortunately, or our crew would have been the first ones to land on the moon." The LM, still undergoing final changes to its design before a landing mission, was having trouble keeping its weight down. If it was too heavy, it wouldn't be able to carry enough propellant to both land on the moon and take off again. As Gene Cernan wryly explains, without the descent programs in the computer or sufficient fuel in the tanks, it would have been a one-way trip had they tried to land: "We came close to landing on *Apollo 10*: down within forty-seven thousand feet of the lunar surface. We thought of landing, but we didn't think of it very long, because we looked at that fuel gauge for the ascent stage, and it said half full. That wouldn't have made it. They didn't give us enough fuel to get off the surface!"

Cernan was disappointed not to be making the first landing. He wanted it as much as Stafford, as much as anyone else in the astronaut office. After all, this is what they had been hired for, and what all of their Gemini flights had been working toward. Yet he says he knew that Apollo wasn't ready; if it had been, he would have fought for *Apollo 10* to become the first landing attempt:

It was the right decision; there were a lot of other potential problems to test first. You always built on what you learned from the flights before you. Every flight to the moon is like a first flight—I don't care how many times you go to the moon; every flight is a first one. It is my understanding that at one time they were pushing for Apollo 10 *to go all the way to the moon. They were saying, if you take the lunar module and put your crew in harm's way, why not go all the way down? But the decision was made, and I think it was the right one, to take the lunar module and put it the lunar environment first, to see how it reacts.*

There would be more than enough to accomplish on this mission without a landing. In many ways, it was as much of a test-piloting leap as *Apollo 9* had been. As Rusty Schweickart told the authors: "*Ten* was much more of a pilot's flight than *Apollo 8*; it was much more challenging."

Stafford believed *Apollo 10* was going to be the most difficult mission yet. He was the perfect choice for commander. Having been a part of the first rendezvous in Earth orbit, he was a natural to command the first rendezvous and docking in lunar orbit. His crew designed its mission patch with the Roman numeral ten gently resting on the lunar surface. If they weren't going to land, Cernan explains, they were still going to make a permanent impression there. Just as on *Apollo 9*, two call signs would be needed to identify the two spacecraft. If NASA managers had disliked the last choices a crew had made—*Gumdrop* and *Spider*—they positively hated the *Apollo 10* call signs. The penultimate flight to complete the grand vision of President Kennedy would be using the names *Charlie Brown* and *Snoopy*. Even Charles Schulz, creator of the "Peanuts" comic strip that featured the cartoon duo, wasn't sure the names were such a good idea. Charlie Brown, he wryly mused, was always a failure. But Cernan thought the names were perfect—*Snoopy*, the LM, would be snooping as close as it was possible to get over the lunar surface, while his master and companion, *Charlie Brown*, was ready to call him back.

It would, however, be the last time any frivolity was allowed in spacecraft names. As Deke Slayton explained in his autobiography, "From now on the spacecraft names were going to be dignified, goddamn it."

Gene Cernan held his breath as he felt the Saturn V booster shake beneath him and build up to full power. He is not above admitting that, for him, it was an apprehensive moment. The *Apollo 10* launch on 18 May 1969 was only the second time humans had ridden the Saturn V rocket to the moon. It was still anything but routine, and this outbound journey to the moon would provide some tense, nerve-wracking moments, making the crew wonder if the flight should be aborted.

The crew had all flown the Titan booster into orbit on their Gemini missions; from the start, this ride felt different. The vehicle's first stage vibrated ominously, shaking the astronauts in their couches. When the first stage was spent and jettisoned with a particularly nasty rippling jolt, a shaken Stafford couldn't even force the words out to tell the ground. "It was quite a monster to ride," he recalls. "It had a total of seven and a half million pounds of thrust, and the first stage burned four million pounds of propellant in two minutes. When you rode that monster up, you knew you had a real tiger by the tail. It felt like a train wreck when you staged off that first stage to the second stage; it was quite a machine to ride out!" John Young, with his characteristic self-deprecating humor, later admitted he found the ride an unnerving experience. "The Saturn V launch vehicle: it's a big rocket. Three hundred and sixty-five feet long from top to bottom, it weighs six and a half million pounds. It shakes so bad that you can't tell if your knees are shaking or not! On top of it were three very nervous human beings. The space shuttle doesn't shake nearly as bad as the Saturn V. I mean, I could tell for sure that my knees were shaking when I flew that!"

Apollo 10's second stage was significantly lighter than *Apollo 9*'s, allowing more payload to be carried into space, but this also meant that the thinner structure vibrated more. "What a ride, babe, what a ride!" Cernan commented later in the flight. It was an understatement only an aviator would make: the crew was by now worrying that the spacecraft might actually shake itself to pieces. "But we got into orbit after eleven minutes," Stafford says with some relief, "then we started on the way out to the moon after one revolution and a half."

Following the pattern established on *Apollo 8*, there was a thorough check of the spacecraft systems in Earth orbit before the Saturn V's third stage was reignited. For this burn, the crew considered leaving their helmets and gloves off but, as Young confesses, they "chickened out." They were soon glad they had; the third stage would prove to be the roughest ride of all.

In the moments before dawn in the quiet, northern Australian outback township of Cloncurry, residents watched as the entire eastern sky suddenly lightened. It was not the sunrise; a hundred miles above them, the Saturn V's third stage had ignited. The glow looked like car headlights coming over a hill in the fog, they later described, and the spacecraft resembled a slightly greenish, bright comet. Most of the town's population had woken early for the occasion, and they watched the swiftly moving bright light race away from them. A meteorological officer on the scene described it as the most awe-inspiring sight he had ever witnessed.

As the burn began and the violent shaking slowly increased, John Young's unease grew. He began thinking, only half jokingly, that they might somehow lose the LM nestled below them. The rocket's increasing ignition was creating some unnerving metallic screeching noises. As he endured almost six minutes of discomforting shaking, Gene Cernan was not sure if the spacecraft would survive this punishment and thought about manually cutting the burn short. The decision to do so, or not, rested with the commander. Although the instrument panel blurred to the point where it could not be read, and Stafford's hand was ready on the abort handle, he held back. He had trusted his senses during the *Gemini 6* launch abort, and he put his faith in them again, deciding that the rocket was not yet at the point where it would shake itself apart. Once again, those test pilot instincts proved correct. The burn ended with the spacecraft and booster still intact, and the shaken crew was truly on the way to the moon.

With the Saturn V's stages completely spent, they would now be relying on the Apollo spacecraft engine. This Saturn V had been a very rough ride indeed, but it had placed them exactly where they needed to be. Now it was time to draw on the experience *Apollo 9* had given the program. Using the same techniques, except this time outside of Earth orbit, John Young separated the command and service module from the spent Saturn V third stage. As the two separated objects hurtled on a path toward the moon, Young coolly rotated the command and service module *Charlie Brown* and slowly headed back toward the booster, where the LM *Snoopy* was nestled. As the ground controllers watched the activities via the spacecraft's color television signal, Young smoothly docked the spacecraft and carefully withdrew the lunar lander from the unruly rocket stage. For the first time two spacecraft designed to be flown by humans were heading for the moon. The lessons of lunar travel and of flying the LM, gained on the previous two flights, would now be seamlessly combined for the most complex Apollo mission to date.

After the experience of leaving the Earth, the crew welcomed the much smoother ride out to the moon. Their path was in fact so precise that only one course correction burn was needed. For all the crew's experience they had never left Earth orbit before, and all three were soon marveling at the shrinking Earth. Young was greatly impressed by the planet's beauty, describing the continent of Africa as looking like velvet from far away. Tom Stafford was equally impressed with the magnificent view; the Earth seemed to be "fading away" before his eyes. "Although I had flown two missions before, space was so beautiful; it was unbelievably fantastic. We turned around and saw the Earth after we picked up the lunar module from the third stage on the way out. You could see Baja California, the snow on the Sierra Nevadas, the Rocky Mountains, Mexico, polar ice cap, and Canada. A beautiful view: you could never tell anyone inhabited the place."

Gene Cernan thought about the view on an even deeper level. Reflecting on the journey outward, he believes that orbiting the Earth and voyaging to the moon are, philosophically, two entirely different space programs. While orbiting the Earth, he was excited by the impressive view of the planet below him, but never felt he had left the familiar. As *Apollo 10* sped away from the planet and Earth shrank to a single view that he could take in all at once, his feelings changed. He felt detached from Earth, wondering where he really was in space and time. Instead of being a participant, he was now an observer of the planet receding behind him. Instead of flying through sunrises and sunsets he was watching them from afar, unaffected by them. It was a profound philosophical shift for him. As soon as he returned to Earth and was surrounded by familiar sights the feeling dissipated; yet it had touched him deeply. "It is one of the deepest, most emotional experiences I have ever had," he says today:

To fly in Earth orbit is a great, very exciting place to be, it truly is. It's a place where you can end up going 17,400 miles an hour, circling the globe every ninety minutes, flying through sixteen sunrises and sunsets in every twenty-four-hour day. But when we departed from what I consider the relative safety and comfort of our home in Earth orbit, things did become different: we were detached from our home planet and about to be captured by another.

One of the first things we saw as we headed outwards from Earth orbit was that the horizon that had been slightly curved in Earth orbit now seemingly closed in around and upon itself. All of a sudden we saw the entirety of the Earth. No longer were we speeding over oceans and continents and literally crossing the breadth of North America in fifteen minutes. Now we found ourselves literally peering across their entirety, until the time came when we could look from the giant icebergs of the north, across the deep dark blues of the Pacific to the snow-covered desolation of the pole in the south without even turning our heads. All the while the world continued to grow smaller; very quickly as we left the Earth and then much more slowly as we approached our destination, the moon. Until the time came when we could literally take the thumb of our hands and cover up the entire planet.

Three days after launch *Apollo 10* reached its target, gliding around the back of the moon in a precisely predicted path. Radio contact with Earth was blocked by the mass of the moon, and seven minutes later the command and service module engine fired to decelerate the docked spacecraft into lunar orbit. With two spacecraft now combined as one, meaning more mass to decelerate, the engine burn lasted for almost six minutes. When it was over, the spacecraft had shed nearly half of its speed. "That was a pretty good shot," Cernan recalls, still amazed at the precision of the journey. "Three days' lead time, 240,000 miles away, and you are supposed to miss the moon by sixty miles. I hoped and wondered about that; but it worked."

When the spacecraft emerged from the other side of the moon, Mission Control was delighted to hear the exultant crew reporting that the burn had been successful and precise. Young even asked the CapCom, Joe Engle, to give the tracking network controller a kiss. Engle politely declined this assignment, but promised to thank him instead. Although the crew had caught occasional glances of the moon on the journey from Earth, it had mostly been faraway glimpses of a surface lit only by earthshine. There had been the chance to have another quick look when they fired their service module engine to place the docked spacecraft in orbit, but the crew had had more pressing tasks during the burn and had instead been tightly focused on the spacecraft systems. "You fire in darkness," Cernan explains, "at least we did on *10*. Then all of a sudden you come into the daylight, and there is the moon." After one more engine burn on the second orbit, circularizing their path about seventy miles above the surface, there was time for a longer look. "We became like three monkeys in a cage," Cernan describes in his autobiography.

Almost immediately, Stafford noticed how different it was than flying in Earth orbit: "It really was amazing how slow the moon went by, compared to how fast you went by on the Earth. We were going 3,700 miles an hour, or about 5,500 feet per second. When you're in Earth orbit you do 17,400 miles an hour, or 25,700 feet per second. It was like we were in an airplane that was starting to stall." There was a purpose to looking out: the crew would soon be passing over the landing point planned for *Apollo 11*. As they radioed detailed descriptions of the geographical features, Stafford noted how majestically rugged the moon appeared. In fact it seemed quite surreal—like a gigantic plaster of Paris cast. "It was quite a sight," he recalled. "The moon had a very rough beginning; actually, so did the

Earth. But where the Earth had an atmosphere around it, and over billions of years things have smoothed out, the moon has never had that. So there is a lot of history there, which can help really unlock the secrets of how our solar system was formed. We could hardly keep ourselves from looking out: we were like three kids."

John Young also thought about the similarities between the Earth and the moon. As he told an audience in Las Vegas in 2001, it made him ponder the effect an asteroid impact would have on Earth: "Look at the backside of the moon, such as the pictures we took. What do you see? It is full of holes—man, really full of holes! We could see that the backside is very blocky, very rough. There are new craters up there we got to fly over—about thirty brand-new, bright-rayed craters on the back side. Planet Earth must have looked the same way once. Isn't that awful?" Nevertheless, Young found a strange beauty in the pitted, chaotic scenery. Although he would eventually journey to the moon twice, studying it from orbit for lengthy periods as part of his mission duties, he never tired of it or found it looked the same twice. "It was totally different," he said. "The moon is a very big place: 14,500,000 square miles of territory. You've got a lot of looking to do to see it all. We could see the Messier craters—you can see them from Earth with a pair of handheld binoculars. Those bright rays go all over the moon. We could see evidence of volcanic activity. Seventy percent of the front surface of the moon is covered with volcanoes; it is really something. But it is very old stuff, it is not very active—maybe dead."

Adding to the impressions of the *Apollo 8* crew members, the *Apollo 10* astronauts observed a variety of shades and colors in the stark lunar terrain. This seemed due to the angle of the sun as they gazed down at the alien world below. To Stafford, the peaks of the mountains and craters seemed to have a slightly red tint, and he also saw tans and yellows in the surface scenery. Young saw a variety of colors—brown, gray, black, white, and variations between them. The color differences reminded him of parts of the western United States, rugged areas without grass and trees. With the basic geological training he had, Young greatly appreciated the subtle differences. To him, it was nothing like the uniform "dirty beach sand" that *Apollo 8* had seen. Every part of the moon seemed remarkably different to him, and he was fascinated to watch it pass by below. As he told the authors, there were subtle details hidden in the lunar shadows that no camera could ever record:

The human eye is so much better than those photos we took. You can see down into the shadows of those craters in earthshine on the moon. You can see down into the shadows of the craters in the daytime. Now, you've never seen a picture come back from the moon that saw down into the shadows of the craters—not one. With the human eye, it's a piece of cake. In earthshine, it's gorgeous. You would be able to explore on the moon in earthshine. You could read a newspaper without glasses in earthshine on the moon. That's how bright it is—it's like having a sixty-watt lamp.

The beauty of Earthrise was not a surprise to the crew. After all, they had seen Bill Anders's magnificent photo and knew that the sight would be one of the highlights of their journey. Seeing the home planet slowly rise above the lunar horizon, however, was awe-inspiring even when it was expected. Looking at the whole Earth, Stafford recalls, was the only time in his entire spaceflight career when he felt odd and acutely aware of the enormous distances they were traveling. "It is quite a sight when you look out and you see that," Stafford explains. "It was a real funny feeling; really changed your view of things. You know you are a long way from home when you see that sight! Only twenty-four of us have ever seen that sight, and there are eighteen of us left alive today. I hope that number eighteen stays there for a long time."

Young was also deeply moved by the view, enough to make him think deeply about what he was seeing. "After you have flown two hundred and forty thousand miles," he explains, "planet Earth is about two and a half days away. All you can see are the white of the clouds, the blue of the ocean, and the atmosphere. Let me talk philosophy a little bit. This is the home of six and a half billion human beings right now. You don't see them—but they're out there. It is not a very successful species either. At the moment we have about twenty to forty wars going on. We've spent about forty billion dollars a year on the international arms trade so that people can kill each other—and they really do a very good job of it. There are other species on this planet, from birds to fire ants—I think I've got a trillion of those in my backyard—and we need to make sure that we all can be successful together."

After six hours of circling the moon, it was time to check the LM. Almost immediately a problem arose. As Cernan opened the hatch, a blizzard of white particles floated out toward him. It was insulation that had come loose in the tunnel between the two spacecraft. The particles had first been discovered a couple of days earlier when Cernan had checked the latches holding the spacecraft together, and they had still not successfully vented from the tunnel. "We have got a spacecraft that has beaucoup of insulation in it," Stafford told the ground. The crew set out to clean up as many of the itchy fragments as they could. Stafford laughed when he glanced at his crew member—Cernan's hair and eyelashes were flecked with the white debris. He looked as if he had been walking in a snowstorm.

With the particles finally cleared up, Cernan then set to work checking *Snoopy*'s systems, powering the LM up and testing the communications. All seemed well with the lunar craft, and the crew could finally take a well-earned rest. They had a busy and potentially hazardous day ahead. The astronauts were glad they had listened to Jim McDivitt, who had suggested they each bring along a sleeping bag. With the windows covered, nights around the moon were cold.

The next day, Stafford and Cernan spent another three and a half hours thoroughly checking out the LM. The floating Mylar, it seems, was the likely cause of other problems. By sticking to the docking ring, it had caused the LM to be a few degrees out of alignment with the CM. When the spacecraft separated there was the danger that some of the latching pins would be sheared away, making a redocking very difficult, perhaps even impossible. In addition, the pressure in the tunnel between the two spacecraft did not reduce correctly. "We did have that one little problem," Stafford explains. "Somebody had not followed instructions on how you manufacture the vent holes between the two spacecraft, and so we had to slip it. There were always little things that would come up that were not accounted for, so we had to slip the probe out until we undocked." Mission Control had decided that the misaligned spacecraft were still within safe limits to separate, and *Snoopy*'s descent to the moon could begin.

On the twelfth orbit of the moon, out of contact with Earth, the spacecraft separated. At first, they stayed just a few feet apart while they tested their radar systems. Once all the checks were complete, Young fired *Charlie Brown*'s thrusters to move the spacecraft further away from each other. Stafford and Cernan were now trusting their lives to the fragile LM in the unforgiving environment of lunar orbit. As Young watched from a safe distance aboard *Charlie Brown*, Stafford gingerly ignited *Snoopy*'s descent engine, bringing it up slowly from minimum thrust. The burn was precisely monitored; there would be little time to make corrections if it went wrong. Too long, and the spacecraft would crash into the moon below. But the engine worked perfectly, and *Snoopy* began an arcing descent that would take it down low over the cratered surface, on a long looping orbit. "John Young was there in the command module at sixty miles," Stafford recalls. "Cernan and I in the lunar module started by going way high, about two hundred miles above, and then down below to about nine miles."

Communications were not perfect. For a short time, Young was not able to talk to Mission Control, though he could talk to his colleagues aboard the LM. But it was enough to proceed; the descent continued, and Stafford and Cernan prepared to fly low over *Apollo 11*'s proposed landing site in the Sea of Tranquility. Occasionally, communications between the Earth and the LM also proved tricky, and much of the vital communication had to be passed via Young in *Charlie Brown*. Somehow, it all came together well enough.

As *Snoopy* and its occupants descended closer to the moon than any human had ever been, the curved horizon flattened dramatically and circular crater rims began to resemble looming mountains. There was no longer the sensation of being in an orbiting spacecraft; the view was now very similar to one from a high-flying aircraft. It was a speedy ride too—equivalent to over Mach 5 back on Earth. The difference in the view was striking, and as they descended to within forty-seven thousand feet—nine miles—of the surface, sharp new details sprung out at them to observe and record. They were now in fact only thirty-five thousand feet above the moon's highest mountain peaks. "We're low, babe—man, we're low!" Cernan exulted, remarking also that he felt like they could touch the top of the mountains. It was almost as if they were flying below the highest ranges, and he looked at them as an unnerving, menacing presence. To this day, Stafford still gets excited when talking about what they saw out of the windows as they zipped across the surface, continually taking photographs of the terrain: "You can absolutely see the craters, but one of the things that also amazed us so much was the awesome boulders. I said, my gosh, that boulder I saw on the east rim of the crater must be as big as a thirty- to forty-story building. It turns out that they were bigger than the Astrodome in Houston!

Just kicked up on the edge of the crater; the size of those boulders and craters was just awesome."

As *Snoopy* skimmed to its closest point over the surface, the two astronauts reached the moment where future lunar missions would begin a powered descent to landing. They were, in fact, over the spot chosen to do just this on the next mission. "It looks like we're getting so close that all we have to do is put the tail hook down and we're there," Stafford exulted. But instead of descending further, *Snoopy* swept across the lunar surface as the astronauts busily scrutinized the proposed landing site below them. They had studied photographs of the desertlike region as much as they could before the mission, but the naked-eye view was far better than any picture. To Stafford, the area looked like smooth, wet clay. The LM's landing radar worked, the site looked smooth enough to be a landing zone, and the sun angle was not a problem. Everything looked good for the next mission.

All too soon, it was time to fire *Snoopy*'s engine again. The burn arced the spacecraft up into a high loop—higher in fact than the orbiting CSM—then swooped them down low. It was a maneuver Cernan likened to a dive-bombing run, and as planned it gave them a second look at the potential landing site. When the engine fired it rattled the LM's electronic systems, setting off unexpected alarms—but the burn went well, and as they looped back around the moon Stafford and Cernan had a chance to rest and prepare for the next pass. As he told the authors, Cernan was impressed by what he saw when he was on the far side of the moon: "I was some place that not many people go. When you go behind the moon, where you are both in the shadow of the sun and the shadow of the Earth—boy, that is black. It should be light, because there are hundreds of billions of stars that just blanket the sky, but it is not: it is black."

John Young, in the meantime, became the first person to fly a spacecraft around the moon solo. He found that piloting the enormous Apollo CM alone was like "flying an aircraft carrier with an outboard motor." Looking through the sextant, he was able to watch the LM skimming across the surface from as far away as 340 miles. To him, the spidery LM appeared to be scuttling across the cratered surface like an insect.

After *Snoopy* had completed its second low pass over the landing zone, it was time to prepare for staging. This maneuver, which would take place on the moon's surface on the later landing flights, would separate the descent and ascent sections of the LM. The ascent engine would then propel the upper stage of the LM back for an eventual rendezvous and docking with the CSM. It was a critical moment in the mission, and to prepare for it Stafford turned the LM to face in the correct direction for the burn. There were ten minutes left before the separation: better to be ready, he reasoned. All too soon, he was very glad he'd been so cautious.

Without any warning, the LM began lurching and wildly gyrating, flipping end over end. "That thing just took off on us," Stafford says of the sudden drama. "The whole LM started to tumble at about sixty degrees per second and tried to rotate." Cernan swore in surprise at the totally unexpected motion, and then the two pilots tried to stabilize the bucking, swiveling spacecraft. Their immediate concern was that the twisting would lock the spacecraft's inertial measuring unit, meaning they would not know their attitude relative to the moon's surface any more. With a critical rendezvous burn fast approaching, this was a highly dangerous proposition.

Stafford quickly hit the switch that jettisoned the descent stage, while Cernan pulsed the thrusters forward to avoid colliding with it. "I blew off the descent stage because I knew we'd get a better torque–to-inertia ratio," Stafford explains with the same cool analysis he demonstrated many times throughout his piloting career. "This was because all the thrusters were on the ascent stage." Using manual control, Stafford quickly brought the errant spacecraft under control, damping out the motions before the inertial measuring unit could freeze. It had been a tense moment. "I saw the lunar horizon go by several times in fifteen seconds in different directions," Cernan remembers. "That's really six degrees of freedom of flight! That's when *Apollo 10* became X-rated. The whole world heard me; Tom mumbled a lot worse, but no one understood him!" Cernan admits that he'd been "scared to death."

It was later determined that the problem was due, in part, to Cernan being too good at his job. He prided himself on his thorough knowledge of the LMP's responsibilities, but he was also able to fly the LM as a commander and, if necessary, the CM as well. He and Stafford worked far more as a piloting duo than later LM crews, including Cernan's pairing with Jack Schmitt for *Apollo 17*. As the only LMP to later command a landing mission, Cernan was able to second-guess his commander very well. He'd flown the LM simulator as commander, and knew which switches the commander would throw before staging could take place. This time, however, the blurring of responsibilities had gone too far.

As part of the tests of the LM on this mission, the crew had planned to use the abort guidance system. The system could be switched into a position where it would hold attitude, but there was another, automatic mode in which it would try to locate the CM to help with a rendezvous. As Cernan sees it, he probably reached over at some point, thinking he was helping his commander, and correctly moved the switch to the hold mode. Stafford didn't notice. "It has probably happened to every aviator a time or two," Cernan explains. "We both had our hands on a computer switch. I switched; Tom thought it needed to be switched again, and he switched it the other way."

When the gyrations began, Stafford was troubleshooting another problem, which helps explain why neither astronaut caught the error at first. "Just before we staged off," Stafford told the authors, "I noticed we were going upside down and backwards during nighttime, down to nine or ten miles above the moon, and I noticed the thrusters started to fire. I saw the yaw go over, but I could tell we weren't yawing; I didn't see that on the eight-ball indicator. An electrical malfunction light came on, I was troubleshooting—and so we went to a wrong switch position." Because they were already looking at LM problems, Stafford and Cernan initially thought the errant movement might be due to a jammed thruster or computer failure. There was little time to decide—the countdown to their rendezvous burn was still proceeding, and there were only forty seconds remaining until it was due to fire. If they were aimed in the wrong direction they might miss their rendezvous point, or even plow into the moon at high speed. "If you make a few-feet-per-second maneuver," Stafford told interviewer William Vantine, "you want to make it in the right direction or else you'll be going the wrong way in a hurry."

Luckily—or more truthfully through sheer skill—Stafford correctly aligned the LM with seconds to spare. The burn took them on a precise path toward their planned rendezvous point with the CM and John Young. "That was the only little problem on *Apollo 10*," Cernan says with relief; "a little bit of cockpit communication. We recovered from it and made a rendezvous." Stafford talks of the piloting challenge with the same sense of pride. "Of all the emergencies we'd try to simulate, we'd never have been able to simulate that."

After almost an hour of cruising up in a high arc, Stafford used the LM's thrusters to circularize their orbit. *Snoopy* made one final lunar orbit solo, as *Charlie Brown* gradually closed the gap between them. With one last burn from the LM, the two spacecraft were at last only feet apart. It had been an impressive eight hours of test piloting, but Stafford and Cernan were ready to dock and get back into their ride home. One orbit later, the crew transfer finished, *Snoopy* was jettisoned. Now unmanned, its engines were tested one last time—firing for over four minutes until the fuel tanks ran dry, which looped the craft out of the moon's gravity field. It was a fitting sendoff for the first LM ever to visit the moon.

There was still some useful work that could be done around the moon to help the next mission. Most importantly, more landmarks could be plotted to help a landing mission on an approach course. After the demanding flying, the crew was also able to enjoy a well-earned rest. But finally, after more than sixty hours circling the moon, the CM's engine was fired for the return to Earth. It was only the second time humans had returned from the moon, but Stafford makes it sound easy: "We came up, did the first rendezvous around the moon, completed the whole mission, and then started back." Gene Cernan is equally nonchalant about what, two missions before, was an unknown frontier of flight. As he stated at the postflight press conference: "We spent most of the way home discussing what color the moon was."

On 26 May, after eight days in space, *Charlie Brown* slammed back into Earth's atmosphere on a speedy but precise course. It was the fastest trajectory possible, shaving fourteen hours off the standard return journey. No humans have ever traveled faster before, or since. "We set an all-time world speed record on the way back," Stafford claims with pride. "We were really hauling the mail when we hit the atmosphere!" Just as with liftoff, John Young was aware of the risks, though he accepted them with humor. "Your outside temperature is thousands of degrees Fahrenheit. Luckily, inside the vehicle it is about 68 degrees Fahrenheit. So everyone stayed cool." Splashdown in the Pacific took place only a couple of miles from the recovery ship.

The mission answered any lingering concerns NASA managers had: a landing on the moon was indeed achievable. Barely an hour after splashdown, NASA Administrator Tom Paine told the press: "We know we can go to the moon. We will go to the moon. Tom Stafford, John Young, and Gene Cernan have given us the final confidence to make this bold step."

Today, Gene Cernan is rightfully proud of what the crew accomplished. "We tested everything that it was humanly possible to test on the lunar module. We flew right over the landing site, we targeted, took pictures, came home, and turned it over to Neil and Buzz. I keep telling Neil and Buzz: we put the white line in the skies so they wouldn't get lost on the way to the moon—all they had to do is cover that last fifty thousand feet!" Stafford agrees: "Two months after we went down in that heavyweight lunar module and picked out the landing site, with the radar map and photo map, Neil Armstrong and Buzz Aldrin repeated the trajectory we'd flown." Other than his Gemini rendezvous mission, Stafford considers *Apollo 10* his greatest contribution to the space program. As Buzz Aldrin declares, because of the *Apollo 10* mission "the door was wide open for *Apollo 11*."

All three of the *Apollo 10* crew members would fly Apollo missions again. At first, it did not look like Stafford would get a chance. Within two weeks of splashdown, Slayton offered him the coveted position of chief of the astronaut office. The job had been Al Shepard's, but as Shepard had managed to get himself back on flight status someone was needed to fill his shoes. So, as Stafford explains, "I replaced Al Shepard as head of the astronaut group for several years, and then became deputy director of flight crew operations." Already a respected figure in the astronaut office, this new assignment gave Stafford valuable management experience that he wanted to help further his career. It did mean, however, that he was out of the running for the other moon landings. Stafford believes he could have flown one of those missions, despite Schirra's belief that you could only command an Apollo mission once. "There wasn't any hard set rule about being an Apollo commander twice. I think I could have got in the queue and gone back to the moon if I'd really pushed for it. I would like to have gone back, but by then I was also involved in management."

By 1970, Stafford was even considering leaving NASA altogether—he had an attractive offer to run for the U.S. Senate representing Oklahoma. Instead, he stayed. The success of *Apollo 11*, made possible in large part by the work Stafford had done on *Apollo 10*, created an almost immediate change in relations between the American and Russian space programs. By early 1972, it was certain that there would be one last Apollo mission to fly in 1975, one that would dock with a Russian spacecraft. Stafford wanted to command this mission. The trouble was, so did Deke Slayton, who had finally obtained medical clearance to fly. Slayton had the political clout to get on the crew, but without any spaceflight experience he was always a long shot to be named as commander. Stafford was named to command the flight, with the unenviable job of having authority over his boss. "We worked it out," Stafford says of the potentially awkward relationship.

In 1975, with Slayton and Vance Brand by his side, Stafford flew the very last Apollo CM into space, docking with a Soyuz spacecraft flown by Alexei Leonov and Valeri Kubasov. Having been a central part of the missions that got America to the moon first, Stafford ended his astronaut career by making the first steps toward cooperation with the Soviets. "He's like a brother to me now," Stafford says of his good friend Leonov. Their friendship, Leonov told the authors, was representative of a wider understanding:

In 1959, Tom Stafford was flying in West Germany. I was flying in East Germany. Our countries' airfields were only 125 kilometers apart. Every time Tom Stafford took off, I took off. Every time I took off, Tom Stafford took off. We flew, not far away from each other. We could see each other. I was a military man at that time; it was a very crazy time for people. At that time, we believed it to be a very dangerous situation between our countries. But that was 1959. In 1975, we flew together inside two spacecraft. It was truly another time, another people. I believe many people on the Earth came to understand each other much better than they did before.

Today, Stafford continues to be a very active and influential presence in the space community, called upon by presidents and NASA administrators to help foster international cooperation and space program growth. "I was deeply involved with getting the Russians to fund their portion of the space station," he explains, "and I'm still working with the Russians. It's still a lot of fun, and I still keep some dirt under my fingernails, as far as the space program goes."

Until his retirement from NASA in late 2004, John Young was still listed on NASA's books as an active astronaut. When John Glenn, America's third astronaut, flew on the space shuttle in 1998 after being away from the agency for thirty-four years, America's seventh flown astronaut was still there. He had never left, and on retirement was America's longest-serving astronaut by a wide margin. By then, John Young had made a record seven launches, including one from the surface of the moon. In an astronaut career spanning over forty years, his unbeatable résumé included two Gemini missions, two Apollo missions to the moon, and two space shuttle missions, including the very first. There could have been more, including the coveted command of the mission launching the Hubble Space Telescope in the late 1980s. But the *Challenger* disaster, it seems, put an untimely end to the career of an astronaut who would have been content to fly forever. Young had been a watchdog for crew safety both before and after the accident, and when some of his memos on the subject were leaked to the press by a third party it rubbed his bosses the wrong way. "I write memos all the time, but I don't send them to the newspapers, for cryin' out loud," Young told interviewers Douglas MacKinnon and Joseph Baldanza for their book *Footprints*. "Turn them loose to the newspapers, why, you've got to expect trouble."

The controversy was enough for NASA to promote him into a desk job in 1987. He might have hoped for more support from his colleagues, but there were those who felt that, as chief astronaut, he had not done enough to protect crew safety pre-*Challenger*. Although he never flew in space again, Young never fully acquiesced to this political move. At the time, he felt he had another two or three flights left in him, and while carrying out his administrative duties he also stayed proficient as a current astronaut for the next seventeen years. It would appear to have been an unspoken snub to those who grounded him. Young watched NASA administrators and policy directives come and go over the decades, and it is gratifying that he is finally seeing NASA return to what he believes in strongly: working toward a return to the moon. As he explained at a 2001 conference, he is convinced that humankind is still in a space race, but it is a race against time:

We need to get back and start exploring the moon. I'm glad I got to go. I think the moon is a fascinating place, and I think we should be going there every

John Young in 2001. Courtesy Francis French.

chance we get. The moon is our future, I believe. It's right there. But, right now, about the closest I can get to it is in a T-38. People say we've been there, done that. We don't know beans about it! When we really start looking around, it is likely to be a heck of a novel place. We need to go back up there, and set up a base, for a lot of reasons. It is in our future. I think the long-term benefit is the advanced technology that we will need to help survive on this planet.

Gene Cernan worked with Tom Stafford again, this time as special assistant to the Apollo-Soyuz test project manager, accompanying the crews on their training visits to the USSR. Before that, however, he had one more flight to make: *Apollo 17*, the last manned moon mission of the century. Once again, he would enter a LM in lunar orbit and descend toward the surface. This time, he was the commander, and he would land. "It would be nice to be here more often," Cernan had commented while circling the moon on *Apollo 10*. He got his chance. "Going and not landing certainly was not disappointing," Cernan told interviewers MacKinnon and Baldanza when discussing *Apollo 10*. Still, there had been a feeling of unfinished business for him. "When I came back [to Earth], that's when I wanted to go back. I wanted to cover that last 47,000 feet, and I also wanted to command my own crew; that was important to me."

Returning to the moon was an opportunity to revisit and refine the feelings he had when looking back at Earth. "I got smacked with it during *Apollo 10*," he mused, "and going back, on *Apollo 17*, I challenged my own feelings; . . . it just reinforced what I had brought back with me from *Apollo 10*." Gene Cernan was fortunate enough to travel to the moon, think about the experience for over three years—then do it again. Because he thought about the experience so deeply, and was able to do it twice, Cernan is perhaps the most important Apollo astronaut—the "Great Communicator" of the group. He strongly believes that he and the others who voyaged to the moon have a duty to share what they witnessed with all of humankind. He does it, in many ways, better than any of them:

John F. Kennedy challenged our nation and our county's commitment to the future, a commitment that led to human beings leaving this planet for the first time—not only physically, but from my experiences both emotionally and spiritually as well. We had an opportunity to go where only our dreams had taken us in the past, to literally go a quarter of a million miles out into the endless blackness of space and time, and see what has never been seen before by the human eye. We all came back with something a little different, perhaps, having a similar experience.

I had a chance to go to the moon on Apollo 10. *We came close to landing, but we didn't. It reinforced the same feeling on* Apollo 17, *I think even more than trying to accept the fact that we were literally living and calling this valley on the surface on the moon our home for the next three days, that always, over the mountains in the southwestern sky, was our security blanket—was home.*

How can you relate to what it is like to look back a quarter of a million miles at the overwhelming beauty and majesty of our planet? It's an incredible, incredible *experience. I like to think that it overwhelmed me. You could literally see from pole to pole, ocean to ocean, across continents. It is alive; you could watch the world mysteriously and very majestically turn on an unseen axis, and see that there were no strings holding it up. Three-dimensionally, within the endlessness of time, within the endlessness of space, was the Earth, a beautiful blue marble, dominated by the blue of the ocean and the white of the clouds. You could almost see beyond it, you could focus beyond, into the blackness. Into that massive three-dimensional place that I have chosen to call the infinity of space and the infinity of time, only because I don't know any other words that can describe it quite as well.*

I can't show it to you, I can't hold it or put it on a screen, but I can tell you that the endlessness of space and time does exist, because I saw it with my own eyes. Try and comprehend the spirit, the sense, the feeling of what that is like; just really let your imagination really go way out there. The time came when I had to ask myself if I truly realized where I was at that moment in space and time. I felt that I was at a point in my life where science had met its match.

The world that I looked at moved with purpose, with logic, with order. It moved in the heavens in a manner that, for me, literally defies description. And it reinforced, for me anyway, something I have believed all my life: an undeniable belief that this Earth of ours, this place called home, is just too perfect, too beautiful to have happened by accident.

Two months after Stafford and Cernan flew their LM low over the Sea of Tranquility, another LM followed the same trajectory, with a CM once again waiting in a higher orbit. Until this second LM reached the nine-mile low point of its orbit, it had done nothing that *Apollo 10* had not already thoroughly tested. But now, with the moon tantalizingly close below, it was time once again for Apollo to push into the unknown. The LM's descent engine was fired, swiftly dropping the spacecraft in a path that would take it to the surface. It was only a twelve-and-a-half-minute journey, and an insignificant distance compared to the enormous journey the astronauts had already taken. But if it was piloted successfully, Neil Armstrong and Buzz Aldrin would achieve what the entire American space program had been working toward for eight years; they would land on the moon.

The *Apollo 11* crew was originally formed to serve as backups for what became the *Apollo 8* mission. Deke Slayton had given Armstrong a limited number of choices for his crew members. When making the decisions, Armstrong looked for spaceflight experience in the areas of expertise needed for a lunar landing mission. He chose what became the *Gemini 12* crew, Jim Lovell and Buzz Aldrin, to join him. That grouping did not last long, however, because Lovell was elevated to the *Apollo 8* prime crew due to Mike Collins's health problems. Buzz Aldrin too changed crew positions, becoming the backup CMP for *Apollo 8*, with Fred Haise moving in as backup LMP. Haise was one of the best LM pilots in the entire astronaut corps and would have made a worthy permanent addition to the crew. But Mike Collins had both seniority and Deke Slayton's sympathy. He was given the first possible crew assignment after he recovered, which was *Apollo 11*. Haise was moved into a backup slot, Mike Collins became the CMP, and Buzz Aldrin went back to being the LMP.

Collins has been asked ever since whether he would have preferred to fly *Apollo 8* or *11*. It's a question he can't answer. "I think Apollo 8 was about leaving and Apollo 11 was about arriving," he told interviewer Michelle Kelly. "One hundred years from now, which is more important, the idea that people left their home planet or the idea that people arrived at their nearest satellite?" Aldrin also wasn't particularly bonded to the idea of flying with Armstrong. "If I had had a choice," he would confess in his book *Return To Earth*, "I would have preferred to go on a later lunar flight." Aldrin would have enjoyed the challenges of a longer, more exploratory lunar surface mission, but knew that making such a request would have been career suicide. At the time, there was no guarantee that this crew would be given the first lunar landing attempt, and Armstrong thought it highly unlikely that he would get the chance. It was only when *Apollo 9* was such a great success that he started to think otherwise.

As the *Apollo 11* crew began the intense training needed for the lunar landing mission, those who regularly worked with Apollo crews noticed a difference. Compared to previous crews, who "stuck together like glue" according to Guenter Wendt, the Armstrong-Collins-Aldrin crew didn't seem to operate as a tightly bonded team. They would come and go from their training separately and take lunch in different places. "They were the first crew who weren't really a crew," Wendt believed. Mike Collins mused on this in his autobiography, puzzling at the memory of a trio that only told each other essential information, never thoughts and feelings, and concluded that they would forever remain "amiable strangers; . . . we are all loners." In the book *First on the Moon*, Aldrin would admit that "Neil and I are both fairly reticent people, and we don't go in for free exchanges of sentiment. Even during our long training we didn't have many free exchanges." When asked about this period of training, Aldrin told the authors, "It was a very precise, intense time. There was more pressure on our flight, and Neil and I are more introverted people than someone like Pete Conrad, who was probably the most extroverted person! And the contrast really showed. Mike was the talkative one; Neil and I didn't say too much."

The differences in the crew personalities were also very evident to the media. Aldrin would write that "we seemed so dull that invention was sometimes necessary to attract readers and listeners." With press interest in the first moon landing at fever pitch, reporters were hoping for a commander who could meet their preconceptions of the first person to walk on another world. Instead, they got Neil Armstrong. Pulitzer Prize–winning novelist Norman Mailer, very much an outsider to the world of the space program, decided to write a book about *Apollo 11* at the time of the flight. He did his best to jump into NASA's world, and what he saw puzzled him deeply. He attended press conferences where, despite the exciting subject matter, he observed an uneasy crew and a near-bored press. He saw Armstrong as "extraordinarily remote," with no discernible personality for the press to write anything interesting about. He hilariously described the relationship between Armstrong and Aldrin as similar to "one of those dour Vermont marriages where the bride dies after sixty years and the husband sits rocking on the porch." When the husband is asked if he is sad to lose his wife, he responds "Nope . . . never did get to like her much." He concluded that Aldrin and Armstrong were such loners, it may have never even occurred to them to ponder whether they liked each other as people or not. If public interest was going to be sustained by the commander's public comments alone, this mission would be a PR disaster.

Other journalists had the same impression. Armstrong reminded *New York Times* reporter William Stevens of a Sunday school boy. "He was surrounded by protective walls of shyness, modesty and self-possession," Stevens would relate. "Armstrong comported himself formally, often pausing a long time before delivering a sparse answer to a question." For a news reporter hoping to cover the event of the century, obtaining a scarce interview only to be met with such diffidence was maddeningly frustrating. They gravitated instead to Mike Collins, who Stevens says "was as open and breezy as Neil Armstrong was reserved; . . . he sat down, threw a leg over a chair arm, and talked animatedly."

Armstrong is, of course, a far more interesting person than many of the news reports of the time would suggest. Not only was his quiet personality not suited to a media frenzy, but the press attention was also the last thing he cared about at that time. As he told historians Stephen Ambrose and Douglas Brinkley, the media focus "might have been a burden if we'd had time to notice it. But we were going full blast trying to be ready on time, and we just tried to shut anything out of our mind that wasn't focused on our principal objective. . . . I was certainly aware . . . that the nation's hopes and outward appearance largely rested on how the results came out. With those pressures, it seemed the most important thing to do was to . . . try to allow nothing to distract us from doing the very best job we could."

Armstrong was so totally focused on the flight, it surprised even those who knew him well. Dee O'Hara was amazed at how remote from surrounding events Armstrong could be, as she told the authors.

Neil is so laid back, and so quiet. A couple of days before the flight, we were over by the office at the Cape. We were walking across the walkway, and I said "Neil, you would not believe, there are just a gazillion people out there. There are even more campers, buses, trucks—I mean there's just a mob scene out there." And he kind of laughed, and said, "Well, God knows why they're making such a big deal out of this." I said "Neil! Do you realize what you just said?" I thought it was just the most bizarre statement: but that's the way he looked at things. He just couldn't get excited about things like that. He was really, really cool and laid back.

Harrison Schmitt, the only geologist among the Apollo astronauts, thinks that Armstrong's personality and skills went hand in hand. He considers Armstrong to have been the best of the lunar astronauts in his observational and sample-collecting skills—until Schmitt was able to go to the moon himself. To Schmitt, Armstrong was a quiet, intense person, with outstanding skills of perception—something he calls "observational intelligence." It is a skill that Schmitt thinks outstanding geologists and outstanding test pilots share. The media would have preferred someone who could better relate the meaning of Apollo, but the commander of *Apollo 11* was picked because he could command the mission, and other concerns were secondary. Tellingly, the *Apollo 11* crew patch was the first Apollo mission patch not to bear the crew members' names. The mission, not the people, was the important thing.

In the end, Armstrong's quiet nature has mostly worked in his favor over the decades. "Neil Armstrong today is a lovely guy, a very nice man, but pretty much a recluse," Wally Schirra told the authors. "Neil has handled the fame very well." Gene Kranz adds that "Neil was always quiet, contained, kept to himself. That became even more accentuated after *Apollo 11*—and that was smart! Neil is a guy who did not blow it—he has maintained the quality of being the first guy on the moon." Armstrong's personal life and his innermost thoughts have remained his own, allowing what he achieved, and not how he felt, to remain the focus.

Armstrong had very good reasons to distance himself from everything but the forthcoming flight: it was going to be fiendishly complicated. He had concluded that he had a very good chance of returning to Earth alive, but only a 50 percent chance of making a successful landing. He didn't even decide what his first words on the lunar surface would be until after he landed, in case he didn't get to say them. Collins thought the odds of success were about the same, and Aldrin shared his thoughts on the odds with the authors:

I remember us discussing it, and I thought we had about a 60 percent chance of landing successfully. I don't think we went into the details of, "Does everything else after that work; are we able to get out and walk on the surface?" But I think a successful mission would have been declared if we had landed, and then for some reason we couldn't go outside. We'd stay there for a while, take a bunch of pictures, and then lift off again—that would have been a success. So I remember 60 percent. But I also felt that we just bought into the high probability of survival. Now, what's the difference if somebody hops into something that's dangerous, when they say you've got 90 percent, 95 percent, or 98 percent? You are either going to do it or you are not! And once you are committed to that as a business, and there are not enough other reasons—you are not going to back out. I mean, to face the reality of it. So anyway, I think we bought into the high chance of survival.

One public relations task the crew did have to think about was coming up with names for the two spacecraft. Unsurprisingly, Mike Collins was the most active participant. Having already decided, in partnership with Jim Lovell, to portray a bald eagle on the mission patch, the call sign *Eagle* seemed a logical and fitting name for the LM. Collins had less luck thinking of a name for his CM, eventually accepting a suggestion of *Columbia* from NASA's public affairs division. To Collins it sounded a little pompous, but since he couldn't think of anything better he halfheartedly accepted it as the second call sign.

For the task ahead, Armstrong and Aldrin could count on a highly focused Mission Control team to give them every assistance on their descent to the surface. Dozens of mission controllers were supported by hundreds of backroom technical advisors, who in turn had thousands of engineers to call upon. And yet with all this energy focused on the crew, there was only one person who was allowed to talk to them during this critical time: that was the CapCom. Through long months of training with the crew, he was supposed to know exactly how they would respond to any given situation. As always, the CapCom was a fellow astronaut; the person specially chosen for this historic moment was Charlie Duke.

Duke would go on to become the tenth person to walk on the moon, copiloting a lunar module to the surface on *Apollo 16*. But at this moment he was a rookie astronaut awaiting his first mission. Five years younger than the *Apollo 11* astronauts, he'd been an astronaut for only three years, having formerly been a test-pilot instructor at Edwards Air Force Base. Like most rookie astronauts, he was eager and happy to be given any assignment. His sunny personality and disarming North Carolina accent certainly helped his chances—there wasn't anyone in the astronaut office who didn't like Charlie Duke. Three decades later, it is still impossible to dislike him. Duke has an easygoing, caring manner reminiscent of a favorite family priest—not surprising, as Duke is now a lay minister. He's still excited about and grateful for the role he had to play in the *Apollo 11* mission.

For such an important moment in the flight, it made sense to have a specialist CapCom, and Duke had been a natural choice. He was the CapCom for *Apollo 10*'s flight over the lunar landing zone at the special request of Tom Stafford. Neil Armstrong also wanted Duke as the voice of Mission Control for landing as he had more experience with the LM activation procedures than anyone, other than those already assigned to a mission. As Duke told the authors, it made sense to select him for this task: "I was a support for *Apollo 10*, and I was also part of a team that did the activation and checkout procedures for *Apollo 10*, which was basically a dress rehearsal for *Apollo 11*. And I wasn't assigned to a crew so I wasn't in training in the simulators. It was a logical sequence, Neil thought, for me to help them out also."

As a sign of their trust in him during their busy training, the *Apollo 11* crew expanded Duke's support role to where he was, as he describes it, "the bridge between the mission planners, the flight controllers, and the crew." It was not only a vital job; it was a prize assignment that Duke was very pleased to get. "I mean, it was not only fun working with the crew," he explains. "There was a pleasure and the excitement of the opportunity to be in Mission Control. Being a CapCom and support was sort of a dual role, and doing that on *Apollo 10* and *11* was very exciting. I'm sure other people could have done it, but I was very thankful that I was there, working with Gene and the team. It was really a great privilege."

Perhaps because of his positive outlook on life, Duke did not pick up on any lack of warmth between the *Apollo 11* crew members. Rather, he felt that the three astronauts were specializing in different mission areas, and he assisted them with those where possible:

I felt like each one was concentrating on more specific parts of the mission. For instance, Buzz Aldrin was extremely interested in rendezvous. He'd got a PhD at MIT, was sort of Mister Rendezvous for the astronaut office, so he did a lot of tweaking on the procedures, things like that. Neil I felt was more of an overall, as a commander should be, and Mike of course concentrated on the command module. Now, I wasn't backup there, so I didn't see them that much in training, except in the simulations that we did in Mission Control. Of course, I was in Houston; they were in Florida. But those all worked out real well, and I just felt like the team was coming together, with teamwork on their part and ours. We all fit together real well.

It had been a frenetic few months of preparation for Mission Control and for the crew. The question had even been raised of giving the crew more time to ready themselves. But in the end, Duke explains, it all worked out on time:

The schedule was really compact; compressed is a better word. We really had to work hard! But I think everybody sensed we were ready to go when the launch was set. I don't think we could have developed a better way. I think over the history of early manned flight, those techniques were refined from the early days of flight test, and so we built on every flight experience, and adaptive procedures. So it's more like an adaptive control system. You had a good foundation, so you just worked through each mission and then built more experience. After Mercury, Gemini, and now Apollo, we were in pretty good shape as far as working together, working procedures, and knowing how that was all going to fit together for the success of the mission.

Gene Kranz, flight director for the landing, described a similar sense of readiness to the authors:

You finally get to the point where your proficiency is pretty much peaked. You can accomplish the normal or the routine almost on autopilot. Since you are not tied in to the mechanics of accomplishing, you can tune yourself to pick up even the slightest deviation. All the good controllers, good flight directors, always had that capacity, an ability to accommodate the norm. It was inherent, embedded in that guy's ability. If you listen to some of the crews reporting, and the voice transcripts, you can tell. Those guys were so totally on top of their job that they were basically—this is going to sound like bullshit, but—basically in harmony with everything that they are doing, such that if anything does not fit, they immediately pick it up.

On 20 July 1969, Charlie Duke was putting the months of training to work in Mission Control. He had already given the crew the call to undock the two moon-orbiting spacecraft and sent them the data that *Eagle* would need to descend. While *Eagle* was on the far side of the moon, Gene Kranz gave a quick pep talk, to rally his team together for what was to come. A tape of the speech does not exist, but to Kranz's recollection it included the bold statements, "This is no bullshit, we're going to go land on the moon . . . and after we finish this . . . we're going to go out and have a beer and we'll say, 'Dammit, we really did something!'" Duke vividly remembers the moment: "Gene pepped everybody up, and I was focused, and I felt like everybody else was focused too. But I was, I guess, a little anxious. You know, we were going to do it for the first time. I was always concerned about communications. In the sims we'd seen some communications problems. But I thought we'd be prepared for all of that. I was anxious to see us go; we were right on schedule."

Duke's fears were borne out. Just minutes before he was due to give *Eagle* the go-ahead for powered descent initiation, the data coming from the LM suddenly disappeared. Spacecraft attitude was causing the signal to reflect off the LM's skin. The data dropped in and out right up to the moment when a decision had to be made: would they descend to the moon, or wave off? "People were a little frantic," Duke remembers. "You get a little frantic when you don't have any communications in Mission Control, and when it is dropping in and out. But Gene Kranz polled everybody, and everybody said go. We had good enough communications; everyone was satisfied with the data. We gave them a *go* for powered descent." At that moment, Chris Kraft noticed Duke taking a long, deep breath.

The communications were not the only problem. *Eagle* and *Columbia* had separated with slightly higher velocity than intended, due to a small amount of residual air in the docking tunnel. With *Eagle* moving a few feet per second faster than planned, they were already halfway to a velocity abort limit before they even began the powered descent. In addition, an electrical problem was causing problems with the landing radar, which Mission Control was trying to keep track of so that Aldrin didn't have to. Duke needed to keep the crew informed, even though it was hard for the crew to hear him, and vice versa.

Aldrin praised Duke after the mission for being able to keep on top of the communication problems, sometimes relaying details to them through Mike Collins, while remembering everything he'd said and relaying it all at one go when there was a clear moment. It was a demanding task, considering the ordered sequence in which the mission was supposed to run. As Duke explains, it was something he had learned to do during his piloting career:

I don't want to take much credit, but something just seemed to be there with the training that we had done, and all the emergencies we had in the simulations. It just programmed your mind; being a pilot, a fighter pilot, and a test pilot you tend to learn to absorb a lot of in-communication information, if you will. For instance, as you are flying along in a T-38 or any airplane, an FAA flight controller will give you a whole sequence of instructions. I remember one time my wife was flying along with me, and she was listening, and she said "What did he say? What did he say?" And I'd gotten it all, you know! But that was just experience. So I guess that the training that we had had as pilots, to listen to the right words and catch the keywords, and the key phrases, helped me immensely to adapt to the same kind of thing that I was used to in spaceflight that I was used to in aviation.

Mission Control was hoping that the communications problem would be the only anomaly. They did not know that an incorrect switch setting meant that the landing computer was being asked to constantly update radar information it did not need. As a result, the computer was becoming overloaded and having difficulty keeping on top of all the tasks it was being asked to perform—a condition called "executive overflow." A few minutes into the descent, an alarm rang out in the astronauts' headsets. For an unprepared team, such an unexpected distraction could have caused disaster, or at the least an abandonment of the landing. But Duke and the rest of the team, including a young LM computer expert named Steve Bales, were ready with senses on full alert after the communications difficulties. As Duke puts it, they were "spring-loaded" for any problems, and able to work on the problem and keep the descent going simultaneously. "That was just the beauty of our training and the thoroughness of it. Somebody would be able to handle the problem, we thought."

When Aldrin heard the master alarm in the spacecraft, he checked the computer for an alarm number, and tensely radioed back the number: "1202." He didn't know what it meant; the astronauts were relying on the ground to tell them, as well as how to fix it, and fast. "The potential for catastrophe was obvious," Aldrin would later write. It was a critical moment in the mission, when vital landing radar data needed to be acquired; a program alarm was the last distraction they needed. Duke didn't know what the alarm meant either, although it seemed familiar to him. "At first I didn't remember ever seeing anything like that. I was not familiar with the computer alarms: I reached for my guidance and navigation checklist, and was flipping through the pages." Then he remembered—it was a number they had come across in one of their last training runs, and he mused aloud on this in Mission Control:

I remembered seeing that combination of numbers, but I didn't remember what it meant. I don't remember exactly what I said, something like "That looks familiar," or "That's the same one." I didn't know what to do with it—

Steve Bales did that. Steve Bales knew it all. He voiced up right away, "We are go on that alarm." So then I was able to communicate up to the crew what it was—we were go. It was an example of the thoroughness of our training, that everybody really knew their systems. I can't say that I was real confident during the time of descent. "Oh no, here's another problem on top of communications, this and that and the other!" But when Steve said go, *I mean that gave me the confidence, "Well we're go—and let's just see what happens."*

With timely and accurate help from his backroom team, Bales had decided that as long as the alarm came up occasionally, not continuously, the computer would still be able to function well enough for *Eagle* to land. When Duke told Aldrin to continue the descent, Aldrin remembers that the tension in Duke's voice was obvious.

The alarms were coming at a busy time in the flight—*Eagle* was slowly throttling back its descent engine, preparing to pitch forward into a steeper descent path. As they did so, Armstrong and Aldrin could at last use their visual intuition, looking out of the window to see the ever-approaching landing zone. But there would only be time for a frustratingly brief glimpse. Once again, the computer sounded an alarm, a similar executive overflow warning numbered 1201. Aldrin felt a slight feeling of panic, and repressed an urge to shout in alarm. Once again, Steve Bales gave the go-ahead, Duke recalls:

When he gave those go's, I didn't even wait for Flight, Gene Kranz. It seemed like we were so time-critical. Normally, the procedure was, "We're go, Flight," and Flight says, "We're go, CapCom." But you listen to that whole internal loop: I'd heard Steve say we're go, and as Gene was about to speak I was already talking. As it went on, even though we kept having a few of those alarms, everybody else, the guidance guy, the flight dynamics guy, the propulsion, everybody was go. It was apparent that Steve's call was right, and it wasn't going to have any material effect on the descent. So he got more and more comfortable with the data, as everything was agreeing.

Duke remembers that there was little time to think about the consequences of their calls beyond the immediate technical issues at hand:

Your mind is racing, and you do start wondering, if this thing keeps up how is this going to affect the navigation, how is it going to affect the computer, the updating of the vector through the landing radar—all of those kinds of things. The computer was absolutely vital for the landing, so when we started having those problems, your mind does go to that what-if: are we going to have to abort? It was more of an, "Oh no, I hope we don't have to abort" type feeling, rather than a critical analysis. You just don't have time to do that, you have to depend on the team to make the right decisions.

The astronauts were trusting Mission Control—with no time to ask for explanations of what the alarms meant, Armstrong and Aldrin had to believe Duke's messages were accurate. "We really have to give the credit to the control center in this case," Armstrong would tell the press after the flight. "They were the people who really came through and helped us and said 'continue,' which is what we wanted to hear." Reflecting with the authors on those rapid decisions, Kranz is still amazed by the lucky coincidence of having discussed procedures for those alarm numbers in the last training simulation before the flight: "That was incredible. I don't think we would have aborted *Apollo 11* without that simulation, but the thing that would have happened . . . no matter how hard you try to run on the line, you always tend to play simulations a little more conservatively. I think that what it would have done, it could have been a distraction for the team, where another problem would have resulted in severe problems, because all of a sudden you just hit sensory overload."

The computer alarm crisis seemed to have passed, but Armstrong now had another concern. It had taken four minutes to silence the alarms completely, commanding most of his attention. This was a time when, ideally, he would have been looking out of the window, checking the landmarks *Apollo 10* had identified for their final approach, and choosing a suitable landing site. By the time Armstrong was able to look out uninterrupted, there was little time and altitude left before he had to land—or abort. "The concern was not with what landing area we were going to go into," Armstrong would relate in his NASA debriefing, "but rather could we continue at all? . . . Our attention was directed towards clearing the program alarms, keeping the machine flying and assuring ourselves that control was adequate to continue without requiring an abort. Most of our attention was directed inside the cockpit in this time period, and in my view this accounts for our inability to study the landing site and find a suitable location during final descent. It wasn't until we got below two thousand feet that we were actually able to look out and view the landing area."

Eagle was about four and a half miles off course, and Armstrong could now see that the computer was steadily flying them into a jagged boulder field that surrounded a huge crater, with rocks "as big as small motorcars." Aldrin recalls that if this had been a simulation, they probably would have aborted based on their lack of a clear landing zone. "The computer was taking us to a place which Neil didn't care for particularly," he remembers. Armstrong quickly took control to slow the descent almost to zero, while *Eagle* continued its forward motion across the crater zone. "We could have tried to land there," Armstrong told Ambrose and Brinkley, "and we might have gotten away with it. It was a fairly steep slope and it was covered with very big rocks, and it just wasn't a good place to go; . . . if I had any choice of a more promising spot, I was going to take it. There were some attractive areas far more level, far less occupied by boulders . . . a half mile ahead or so, so that's where I went." There was no time for Armstrong to explain to Aldrin what he was doing, never mind telling Duke, who was baffled by the telemetry he was now seeing: "That was a great mystery to us. We'd never seen a trajectory like that and just didn't understand what was happening. It was unusual, because we were using up precious fuel—why was he flying it like this? What's going on? And he never really told us until after touchdown. So we were just sort of sweating it out, in the dark about what was going on. That's when I think everybody began to get really anxious about the fuel state."

The descent fuel tanks only held a minute's worth of extra fuel above what was needed for an exact, flawless descent. With weight at a premium and the LM as lightweight as it could be, they could not afford to carry more. Armstrong needed to land, soon, and there was nothing Duke could do other than tell him how much fuel he had left. Armstrong changed his mind about a landing area a couple of times: he would spot what appeared to be a good area, but as he drew closer he would see that it wasn't, and press on further. There were echoes of Scott Carpenter's reentry in *Aurora 7*: a spacecraft descent with critical fuel running out.

When Duke radioed "sixty seconds," Armstrong was still two hundred feet above the surface, having only just seen a smoother area nearby. There were only two calls left to make after that, Duke remembers: "thirty seconds," then "abort." The thirty-second call was an estimate of remaining fuel, based on the throttle settings for a textbook landing: this descent had been anything but, and the fuel margins were in truth unknown. Duke has no idea if Armstrong would have aborted, even if he had told him to. It is unlikely. Armstrong stated in a postflight press conference that "in simulations . . . we are usually spring-loaded to the abort position; . . . in the real flight, we are spring-loaded to the land position. We were certainly going to continue with the descent as long as we could safely do so." It was perhaps the most nail-biting moment of the descent for Duke:

I've never talked to Neil about that—what would you have done, you know, twenty feet off the ground, and we called abort? It was tense, I tell you! When I called "sixty seconds," there was dead silence in Mission Control. You could certainly feel your heart beating. It was really getting critical. The tension was rising exponentially. I was talking a lot, just giving them updates from what we were seeing in Mission Control. Deke Slayton was sitting next to me; he kind of punched me in the side and said "Shut up and let them land." Yes sir, boss!

Duke and Armstrong did not know it, but Kranz had already decided to himself that he was not going to call for an abort. "The crew is close enough to the surface," he told interviewer Rebecca Wright, "I'm going to let them give it their best shot."

Aldrin was continuing to provide Armstrong with the vital information he needed, reading off numbers so that his commander could keep his eyes out of the window. "The landing itself obviously was very challenging," he recalls. "I was giving him all the information I could from the primary guidance, reading off the altitude, altitude rate, and velocity over the ground. So I wasn't looking out very much to see what he was trying to dodge. But as we got to the point where the sixty-second light came on, we were still a hundred feet or more above the ground. I had a lot of confidence in Neil, but I sure wanted him to get a little closer to the ground!"

By the time Duke called "thirty seconds," *Eagle* was in the grimly named "Dead Man's Zone." If they aborted now, casting off their descent stage, the ascent engine would not have time to fire before they crashed into the surface. Armstrong had decided that, if the vehicle was in the right attitude, he was prepared to run out of fuel and drop unpowered to the surface. "I could fall from a fairly good height," he'd later explain, "perhaps maybe forty feet or more in the low lunar gravity; the gear would absorb that much fall." Aldrin now felt no apprehension; instead, he felt a peculiar sense of arrogance. They were going to touch the moon, one way or another.

Clear of the boulder field, Armstrong had banked left to avoid more scattered rocks and thrusted forward once again to clear another, smaller crater. He was so close to this crater, the descent engine exhaust sent streaks of dust shooting from its rim. He'd finally seen a clear spot only a couple of hundred feet in area—"a relatively smooth area between the craters and the boulder field," as he'd later describe. The site he had finally selected was pitted with small, shallow craters and a few lumpy rocks—but it would have to do. "The terminal phase was absolutely chock-full of my eyes looking out of the window," Armstrong would relate, "and Buzz looking at the computer and information inside the cockpit and feeding it to me. That was a full-time job."

As he descended into the small, smoother area, Armstrong tried to cancel any remaining forward motion. The spacecraft legs could be damaged if they were still drifting forward when they touched down. "It's quite important not to stub your toe during the final phases of touchdown," Armstrong would later explain. "I don't think I did a very good job of flying the vehicle smoothly in that time period—it was a little bit erratic, I think." The rocket blast was now scouring the lunar surface, and the cratered ground was obscured by a rush of foggy streaks blasted away by the descent engine. "We were beginning to get a transparent sheet of moving dust, which obscured the visibility a little bit," he explained in the postflight debriefing. "As we got lower, this visibility degradation continued to increase." Armstrong focused on the few rocks he could see poking through the haze, and used them as guide points to help cancel any remaining forward motion. *Eagle* tried to bank sideways and backward, but Armstrong held her steady on the final, gentle descent. "We were essentially close to running out of fuel," he'd remember postflight; "we were hitting our abort limit."

"We were still about ten feet above the ground," Aldrin recalls, "and we were pretty much sweating it out there for the last couple of seconds, but I thought we maybe had it made at that time. I could see the shadow of the landing gear in front of us, see the dust picking up." It had taken a pair of skilled aviators to do it, and the fuel tanks were "quite close to our legal limit," as Armstrong put it, but they were going to make it. As the probes extended from the right and forward footpads gingerly touched the surface and began to bend because of the spacecraft's slight leftward drift, Aldrin called out "Contact light." They were touching the moon, but still inches from a full landing. Then the spacecraft steadied as the footpads made contact with another world. "Okay, engine stop," Aldrin called out, thus essentially speaking the first words from the moon. It had been a very near thing, Duke remembers, with telemetry indicating less than twenty seconds of fuel remaining in the tanks:

I called "thirty seconds," and we heard "Engine stop" thirteen seconds later, according to my watch. So it was real close. So we all just erupted in excitement in Mission Control. It was a great release. It took Kranz a while to get us back into the operations mode! But we got back to work real quick. Everyone cheered for just a second or two, then really got intense again, checking the spacecraft out, making sure this thing hadn't sprung any leaks, that the systems were okay. Things calmed down a little bit more as time went by: the module was stable, all systems were stable—we knew we had a good spacecraft.

The moment of landing on the moon had been so soft, Aldrin told the authors, that he and Armstrong did not feel it at all: "It was not jolting in the least. The contact light comes on when the probe hits the bottom, and that is not going to give any reaction by the spacecraft. And we were not in much of a descent then, so given the human reaction to hearing that, and then shutting the engine off, and then having the engine tail off—it is going to make for a pretty cushioned touchdown. In retrospect, I'm not surprised that we didn't feel any jolt." Although Aldrin had called out the first words from the moon, Armstrong would say the ones that everyone remembered. The commander who had been criticized by the press for being uncommunicative was not above knowing when an historic phrase was needed. When *Eagle* and *Columbia* had undocked earlier that day, he had chosen the phrase "The *Eagle* has wings." Now, using his commander's prerogative to name their landing zone, he radioed, "Houston, Tranquility Base here. The *Eagle* has landed."

Duke, back in Mission Control and still recovering from the intense elation of landing, was momentarily thrown by the change from technical language to near-poetry: "I couldn't come up with a reply, I was so excited. If you listen to the transcript, I got tongue-tied on 'Tranquility.' It was more like Tweety and the cat, you know, 'Twang-quility!' And what just came out was the way I felt. 'Roger, Twan . . . Tranquility. We copy you on the ground. You got a bunch of guys about to turn blue. We're breathing again. Thanks a lot.' Because I really believe everybody was just holding their breath. Were we going to make it?"

Eagle's reply to Duke was humble yet entirely appropriate: "Thank you." Duke was correct that the Mission Control team were about to turn blue. "For the last few seconds of landing I had been holding my breath," Kranz told the authors. "At the instant of landing and for the next two to three seconds the wave of emotion choked me up—I could not speak."

With no air on the moon for the dust to hang in, it settled fast. "I was absolutely dumbfounded," Armstrong would relate, "when I shut the rocket engine off and the particles that were going out radially from the bottom of the engine . . . just raced out over the horizon and instantly disappeared . . . That was remarkable. . . . I'd never seen anything like that." Within a few seconds, a pristine surface awaited them. Looking out of the windows, Aldrin could see rocks of "just about every variety of shape, angularity, granularity." Armstrong described the wider area—craters up to twenty feet wide, ridges up to thirty feet high, and a hill about half a mile away in what he'd later call "the cool of the early lunar morning." To him, the surface looked "warm and inviting," with "a stark beauty all its own . . . very pretty."

Humans still had not walked on the moon and yet, Aldrin would later decide, the hardest part of the mission was already over. Aldrin related in his autobiography that "From a technical standpoint, the great achievement was making the first lunar landing, and two of us would be doing that. We all expected the actual surface activities to be relatively easy." Armstrong would agree, explaining over thirty years later:

The most difficult part from my perspective, and the one that gave me the most pause, was the final descent to landing. That was far and away the most complex part of the flight. . . . The unknowns were rampant. . . . There were just a thousand things to worry about in the final descent.

Walking around on the surface, you know, on a ten scale, was one, and I thought that the lunar descent on a ten scale was probably a thirteen.

Back on Earth, Duke took a deep breath and turned in his chair to look at Deke Slayton. The pair grinned at each other—they had done it. At virtually the same moment on the moon, Armstrong and Aldrin took a second to congratulate each other. The teamwork, on the Earth and the moon, had paid off. It had only been two and a half years since the *Apollo 1* fire, yet the first landing had now been achieved. For Charlie Duke, it would be his greatest experience in the space program, until he landed on the moon himself three years later.

For Gene Kranz, it was also a moment of elation, triumph, and proof that he had the right team, as he told the authors:

The team that took America to the surface of the Moon in July 1969 had an average age of twenty-six. We literally fought every step of the way to get there. When we finally shut those engines down on the lunar surface, we had less than seventeen seconds of fuel remaining, or we would have had to execute a very difficult abort decision. When we shut those engines down, and when we got those guys back home, we had fulfilled the pledge we had made to our now-dead president, John F. Kennedy.

It had been an almost unreal moment for them all. Despite the relentless training, the actual landing was a peak moment in their lives, one where a moment formerly confined to science fiction had come true. Kranz remembers how it took some time for the achievement to sink in for some of his team:

Before we go into any critical phase in Mission Control, we always go into "Battle Short." The Flight Director makes the call, we lock the circuit breakers and bolt the doors, and nobody can get in or out. Many years after we had done the lunar landing Steve Bales, my guidance officer, came up and said, "You know, it really hit me on Apollo 11, *when we went to Battle Short—I realized then that those doors weren't going to be opened until we either landed, aborted, or we crashed. Those were the only three outcomes that day." For a guy who was twenty-six, twenty-seven years old, to have that kind of a feeling, that's pretty interesting.*

Charlie Duke, the duty CapCom when *Apollo 11* landed on the moon.
Courtesy Francis French.

Kranz also has special words of praise for Charlie Duke. Of all the astronauts, he believes Duke could have been an outstanding flight director. "He's probably the best from a standpoint of the astronauts," Kranz told Wright. "Charlie Duke was just absolutely a master of timing." Duke was shocked, but pleased and honored when he heard this: "I felt like Mission Control was my second home. I got to participate in so many Apollo flights, as backup for 13 and 17. That put us also in Mission Control, not necessarily as flight CapComs, but in the system. So I got to know those guys, most of them are still great friends, the ones still around. We got to be like brothers, it was a real wonderful experience for me."

Although Duke would be in Mission Control many times afterward, the *Apollo 11* landing is still a prized moment for him:

I don't think in many respects it was appreciated how difficult that mission was to accomplish. Being the first time we had done it, and the short time span that we had to get ready, to have it come out so successfully was almost miraculous. And it was a very, very difficult thing to accomplish, I thought. We were able to have the good fortune of good spacecraft, confident people, and a dedicated, enormously focused team. To have been able to pull this off certainly shows a commitment to training and teamwork, and I was delighted to be there.

For the public the highlight of the flight was yet to come: the first steps on the moon. But for those who were most intimately involved with the flight, the landing was the real moment of achievement. Armstrong said at a postflight press conference that "the most exciting and rewarding part of the . . . flight was the descent to the lunar surface." On the tenth anniversary of the mission, he would still be describing it as the "emotional high point." For him, the climb down the ladder and the first steps on the lunar surface were far less important. They weren't even part of Kennedy's goal, which specified a landing on the moon, not a walk. As Armstrong has said many times in recent years in public speeches, "Pilots take no special joy in walking: pilots like flying. Pilots generally take pride in a good landing, not in getting out of the vehicle." The line generally gets a big laugh, but it is accurate, and also underscores an important point. Aldrin and Armstrong landed together, working as a team, and the achievement belongs to them both. Walt Cunningham goes even further in his assessment, stating in his autobiography, "Make no mistake, the significant event was the landing, not man's footprint in the lunar dust which came several hours later. That was show biz."

Cunningham's judgment may be a little overstated. Climbing down a ladder was not difficult compared to landing a spacecraft on the moon, but it was the beginning of something else, something very different. The short time Aldrin and Armstrong spent on the lunar surface was the beginning of a whole new space program. The "Space Race" to the moon was over, and so were the complex piloting challenges needed to get there. Instead, for the next three years, there would be an outstanding program of lunar surface exploration. Flying in space was no longer the challenge. Now it was a means to a new goal. Instead of exploring the space environment, outer space was something to be passed through in order to visit a new world. As Armstrong stated at a postflight press conference: "I see it as a beginning, not just this flight, but in this program, which has really been a very short piece of human history—an instant in history—the entire program. It's a beginning of a new age."

It was a new age that Yuri Gagarin had not lived to see. Though he had not witnessed a successful manned Soyuz mission, the first human spacefarer had been hopeful that his country's space program would recover and thrive. "We hope that, before long, man's foot will step on the moon's surface," he had stated in 1963. Though Gagarin would obviously have preferred his own country to have made that first landing, it was a sentiment that Armstrong had shared and knew how to respond to. Less than a year after he landed on the moon, Neil Armstrong visited Star City, and described the *Apollo 11* landing to a packed audience in the auditorium. He told them how he had left medals commemorating Yuri Gagarin and Vladimir Komarov on the lunar surface. Valentina Tereshkova, acting as one of the social hosts, noted how much Armstrong's round face and innocent smile reminded her of the much-missed Yura. She accompanied him to Gagarin's former office, which had been preserved exactly as it was on the day Gagarin died, as a kind of shrine. When presented with the visitors' book to sign in this hallowed room, Armstrong considered what to write. Just as he had done when landing on the moon, he came up with the perfect phrase. Taking the pen, he wrote something that touched his Soviet hosts deeply and paid tribute to the person who had begun humanity's journey beyond the Earth:

He called us all into space.

Epilogue

Destiny is not a matter of chance. It is a matter of choice.
It's not a thing to be waited for; it is a thing to be achieved.

William Jennings Bryan

Over recent decades, in the Babylonian plains of Shinar, in an ancient land formerly known as Mesopotamia (now Iraq), archaeologists have discovered the remains of what is thought to be the basis of the story of the Tower of Babel. Built by Nimrod, known as the Great Hunter and recognized as the founder of Babylon, this and other clay-brick towers called ziggurats (from the Akkadian *zagaru*, "to rise") were built as high as structural methods then allowed, with circular staircases ascending to a shrine at the very top, around which were inscribed astronomical symbols. Nimrod's dream was to expose himself and his people to the mysteries of the universe. His great tower—believed to have been around 153 feet high—pointed like a rocket to the heavens, and he was one of the first to physically try to ascend into the firmament of space.

In mid-2003, one of the authors of this book spent a day assisting the Aerospace Legacy Foundation in some "urban archaeology" in Downey, California, at the plant where the Apollo spacecraft were constructed. The buildings had by then been abandoned for four years; some were in the process of being demolished, others were being refurbished as movie studios. Science fiction moviemaking was fast replacing the science reality that had taken place there.

The building's vast work hangars were pitch black and crumbling; the formerly pristine, vacuum-sealed clean rooms covered in bird droppings. Executive suites and boardrooms where crucial decisions were made that allowed humankind to travel to the moon were now dark and empty, except for scattered signs of occupation by the homeless. Engineering workshops that had once bustled with the creation of cutting-edge technology were now left dusty and undisturbed, silent as a tomb. The site had much in common with the ziggurats of Shinar; it had helped humankind reach for the stars and, like Shinar, its time had come and gone. The demolition crews were moving in, and the Foundation's team moved hurriedly through the buildings with flashlights, retrieving filing cabinets of unique and irreplaceable documents and blueprints before they could be destroyed.

The glittering age of high-technological human exploration was already corroding, disappearing, becoming lost to time like the ziggurats. Yet, just as Shinar was abandoned, so a future site will replace the abandoned Downey. The ways of reaching for the stars come and go over the centuries, but the same dream remains, and never dies.

References

Entries in this bibliography are organized in four sections: books, periodicals and online articles, interviews and personal communications, and other sources (films, personal papers, public documents, presentations, CD-ROMS, and reference sources).

Books

Aldrin, Buzz, and Malcolm McConnell. *Men from Earth.* New York: Bantam Books, 1989.

Aldrin, Buzz, and Wayne Warga. *Return to Earth.* New York: Random House, 1973.

Armstrong, Neil, Edwin Aldrin, and Michael Collins. *First on the Moon.* Toronto: Little, Brown, 1970.

Baker, David. *The History of Manned Spaceflight.* New York: Crown, 1981.

Bartos, Adam, and Svetlana Boym. *Kosmos: A Portrait of the Russian Space Age.* New York: Princeton Architectural + PHS, 2001.

Benson, Charles, and William Barnaby Faherty. *Moonport: A History of Apollo Launch Facilities and Operations.* NASA History Series Publication SP-4204. Washington DC: NASA HQ, 1978.

Beregovoi, Georgi. "Not to Be Forgotten." In *Pioneers of Space*, edited by Viktor Mitroshenkov, 298–99. Moscow: Progress, 1989.

Boomhower, Ray E. *Gus Grissom: The Lost Astronaut.* Indianapolis: Indiana Historical Society Press, 2004.

Borman, Frank, and Robert J. Serling. *Countdown: An Autobiography.* New York: Silver Arrow Books/William Morrow, 1988.

Brand, Stewart, ed. *Space Colonies.* New York: Penguin Books, 1977.

Burgess, Colin, Kate Doolan, and Bert Vis. *Fallen Astronauts: Heroes Who Died Reaching for the Moon.* Lincoln: University of Nebraska Press, 2003.

Caidin, Martin. *Rendezvous in Space*. New York: E. P. Dutton, 1962.

Carpenter, Scott, Gordon Cooper, John Glenn, Virgil Grissom, Walter Schirra, Alan Shepard, and Donald Slayton. *We Seven: By the Astronauts Themselves*. New York: Simon & Schuster, 1962.

Cassutt, Michael. *Who's Who in Space*. International Space Year edition. New York: Macmillan, 1993.

———. *Who's Who in Space*. 3d ed. New York: Macmillan, 1998.

Cernan, Eugene, and Don Davis. *The Last Man on the Moon*. New York: St. Martin's Press, 1999.

Chaikin, Andrew. *A Man on the Moon*. New York: Viking Penguin, 1994.

Chappell, Carl L. *Seven Minus One*. Madison IN: New Frontier, 1968.

Chrysler, C. Donald, and Don Chaffee. *On Course to the Stars: The Roger Chaffee Story*. Grand Rapids MI: Kregel, 1968.

Clark, Phillip. *The Soviet Manned Space Program*. New York: Salamander Books, 1988.

Collins, Michael. *Carrying the Fire: An Astronaut's Journeys*. New York: Farrar, Straus & Giroux, 1974.

———. *Liftoff: The Story of America's Adventure in Space*. New York: Grove, 1988.

Compton, David. *Where No Man Has Gone Before*. NASA History Series Publication SP-4214. Washington DC: NASA HQ, 1989.

Cooper, Gordon, with Bruce Henderson. *Leap of Faith: An Astronaut's Journey into the Unknown*. New York: Harper Collins, 2000.

Cunningham, Walter. *The All-American Boys*. New York: Macmillan, 1977.

Duke, Charlie, and Dotty Duke. *Moonwalker*. Nashville TN: Thomas Nelson, 1990.

Fairley, Peter. *Man on the Moon*. London: Arthur Barker, 1969.

Fallaci, Oriana. *If the Sun Dies*. Trans. Pamela Swinglehurst. New York: Athenaeum, 1966.

Furniss, Tim. *Manned Spaceflight Log*. London: Jane's, 1983.

———. *One Small Step*. Somerset UK: Haynes, 1989.

Gagarin, Yuri, and Vladimir Lebedev. *Survival in Space*. New York: Frederick A. Praeger, 1969.

Gainor, Chris. *Arrows to the Moon: Avro's Engineers and the Space Race*. Burlington, Ontario: Apogee Books, 2001.

Gibson, Edward G., ed. *The Greatest Adventure.* Sydney: C. Pierson, 1994.

Golovanov, Yaroslav. *Korolev: Fakty I Mify* [Korolev: Facts and myths]. Moscow: Nauka, 1994.

Grissom, Betty, and Henry Still. *Starfall.* New York: Thomas Y. Crowell, 1974.

Grissom, Virgil. *Gemini! A Personal Account of Man's Venture into Space.* London: Macmillan, 1968.

Gurney, Gene, and Clare Gurney. *The Soviet Manned Space Programme: Cosmonauts in Orbit.* New York: Franklin Watts, 1972.

Hacker, Barton C., and James M. Grimwood. *On the Shoulders of Titans: A History of Project Gemini.* NASA Special Publication 4203. Washington DC: NASA HQ, 1977.

Hall, Rex, and David J. Shayler. *The Rocket Men: Vostok & Voskhod, the First Soviet Manned Spaceflights.* Chichester UK: Praxis, 2001.

Harford, James. *Korolev: How One Man Masterminded the Soviet Drive to Beat America to the Moon.* New York: John Wiley & Sons, 1997.

Hartmann, William, Ron Miller, Andrei Sokolov, and Vitaly Myagkov, eds. *In the Stream of Stars: The Soviet/American Space Art Book.* New York: Workman, 1990.

Hawthorne, Douglas. *Men and Women of Space.* San Diego: Univelt, 1992.

Hooper, Gordon. *The Soviet Cosmonaut Team.* Vol. 2: *Cosmonaut Biographies.* Suffolk UK: GRH Publications, 1990.

Hurt, Harry. *For All Mankind.* New York: Atlantic Monthly Press, 1988.

Kamanin, Nikolai. *I Feel Sorry for Our Guys: General N. Kamanin's Space Diaries.* NASA Publication No. TT-21658. Washington DC: NASA, 1993.

———. *The Hidden Cosmos.* Moscow: Infortekst, 1995.

Kelley, Kevin W., ed. *The Home Planet.* Reading MA: Addison-Wesley; Moscow: Mir, 1988.

Kelly, Dr. Fred. *America's Astronauts and Their Indestructible Spirit.* Blue Ridge Summit PA: Aero (Division of TAB Books), 1986.

Klerkx, Greg. *Lost in Space: The Fall of NASA and the Dream of a New Space Age.* New York: Pantheon Books, 2004.

Kraft, Chris. *Flight: My Life in Mission Control.* New York: Penguin/Dutton Books, 2001.

Kranz, Gene. *Failure Is Not an Option*. New York: Simon & Schuster, 2000.

Lambright, W. Henry. *Powering Apollo: James E. Webb of* NASA. Baltimore: Johns Hopkins University Press, 1995.

Lattimer, Dick. *All We Did Was Fly to the Moon*. Gainesville FL: Whispering Eagle Press, 1985.

Lebedev, L., B. Lukyanov, and A. Romanov. *Sons of the Blue Planet*. NASA publication. New Delhi: New Delhi Amerind, 1973.

Leonov, Alexei. *The Sun's Wind*. Moscow: Progress, 1977.

Leonov, Alexei, and Vladimir Lebedev. *Space and Time Perception by the Cosmonaut*. Moscow: Mir, 1971.

Leonov, Alexei, and Andrei Sokolov. *Life Among Stars*. Moscow, 1981.

———. *The Stars Are Awaiting Us*. Moscow, 1967.

Lewis, Richard S. *Appointment on the Moon*. New York: Viking, 1968.

Lovell, James, and Jeffrey Kluger. *Lost Moon: The Perilous Voyage of Apollo 13*. Boston: Houghton Mifflin, 1994.

MacKinnon, Douglas, and Joseph Baldanza. *Footprints: The 12 Men Who Walked on the Moon Reflect on Their Flights, Their Lives, and The Future*. Washington DC: Acropolis Books, 1989.

Mailer, Norman. *A Fire on the Moon*. London: Pan Books, 1970.

Maranin, I. A., S. Shamsutdinov, and A. Glushko. *Soviet and Russian Cosmonauts 1960–2000*. Moscow: Novosti Kosmonautika Publishers, 2001.

Murray, Charles, and Catherine Bly Cox. *Apollo: The Race to the Moon*. New York: Simon & Schuster, 1989.

Oberg, James. *Red Star in Orbit*. New York: Random House, 1981.

———. *Star-Crossed Orbits: Inside the U.S.-Russian Space Alliance*. New York: McGraw-Hill, 2002.

O'Leary, Brian. *The Making of an Ex-Astronaut*. Boston: Houghton Mifflin, 1970.

Outer Space and Man. Moscow: Mir, 1967.

Pogue, William. *Space Trivia*. Burlington, Ontario: Apogee Books, 2003.

Riabchikov, Evgeny. *Russians in Space*. New York: Doubleday, 1971.

Romanov, A. *Spacecraft Designer: The Story of Sergei Korolev*. Moscow: Novosti, 1976.

Sacknoff, Scott, ed. *In Their Own Words: Conversations with the Astronauts and Men Who Led America's Journey into Space*. Bethesda MD: Space Publications, 2003.

Schefter, James. *The Race: The Uncensored Story of How America Beat Russia to the Moon*. New York: Random House, 1999.

Schick, Ron, and Julia Van Haaften. *The View from Space: American Astronaut Photography 1962–1972*. New York: Clarkson N. Potter, 1988.

Schirra, Wally, and Richard Billings. *Schirra's Space*. Boston: Quinlan Press, 1988.

Scott, David, and Alexei Leonov. *Two Sides of the Moon*. New York: Simon & Schuster, 2004.

Shatalov, Vladimir, and Mikhail Rebrov. *Cosmonauts of the USSR*. Moscow: Prosveshcheniye, 1980.

Shayler, David J. *Disasters and Accidents in Manned Spaceflight*. Chichester UK: Praxis, 2000.

———. *Gemini: Steps to the Moon*. Chichester UK: Springer-Praxis, 2001.

Shelton, William. *Man's Conquest of Space*. Washington DC: National Geographic Society, 1968.

———. *Soviet Space Exploration: The First Decade*. London: Arthur Barker, 1968.

Shepard, Alan, and Deke Slayton. *Moon Shot: The Inside Story of America's Race to the Moon*. Atlanta: Turner, 1994.

Siddiqi, Asif. *Challenge to Apollo: The Soviet Union and the Space Race, 1945–1974*. Washington DC: NASA History Office, 2000.

———. *The Soviet Space Race with Apollo*. Gainesville: University Press of Florida, 2003.

———. *Sputnik and the Soviet Space Challenge*. Gainesville: University Press of Florida, 2003.

Slayton, Donald K., and Michael Cassutt. *Deke! U.S. Manned Space: From Mercury to the Shuttle*. New York: Forge Books, 1994.

Sobel, Lester A., ed. *Space: From Sputnik to Gemini*. New York: Facts on File, 1965.

Stafford, Thomas P., and Michael Cassutt. *We Have Capture*. Washington DC: Smithsonian Institution Press, 2002.

Syme, Anthony. *The Astronauts*. Sydney: Horwitz, 1965.

Thompson, Neal. *Light This Candle: The Life & Times of Alan Shepard, America's First Spaceman*. New York: Crown, 2004.

Trux, John. *The Space Race: From Sputnik to Shuttle*. London: New English Library, 1986.

United Press International. *Gemini: America's Historic Walk in Space*. New York: UPI, 1965.

Wagener, Leon. *One Giant Leap: Neil Armstrong's Stellar American Journey*. New York: Forge Books, 2004.

Wendt, Guenter, and Russell Still. *The Unbroken Chain*. Burlington, Ontario: Apogee Books, 2001.

West Point Military Academy. "Edward H. White," biography. In *Howitzer*, West Point Military Academy yearbook. New York: West Point Military Academy, 1952.

White, Frank. *The Overview Effect: Space Exploration and Human Evolution*. Wilmington MA: Houghton Mifflin, 1987.

Wilford, John Noble. *We Reach the Moon*. New York: W. W. Norton, 1969.

Wolfe, Tom. *The Right Stuff*. New York: Farrar, Straus & Giroux, 1979.

Zimmerman, Robert. *Genesis: The Story of Apollo 8*. New York: Four Walls Eight Windows, 1998.

———. *Leaving Earth: Space Stations, Rival Superpowers, and the Quest for Interplanetary Travel*. Washington DC: Joseph Henry, 2003.

Periodicals and Online Articles

"468 Miles Up—Two Men Catch Rocket." *The Sun* (Sydney), 19 July 1966.

"Astronauts Soar 474 Miles after Chase in Space." *Sydney Morning Herald*, 20 July 1966.

Charles, John B. "A Day in the Life of the American Astronaut Corps: January 27, 1967." *Quest* 12, no. 1 (2005): 23–25.

"Charles Conrad, Third Man to Walk on Moon, Dies in Accident." *Redlands Daily Facts*, 9 July 1999, A3.

Chriss, Nicholas. "After Tranquility, Astronauts' Lives Were Anything but Tranquil." *Houston Chronicle*, 16 July 1989.

Christy, Robert. "Soyuz 5 Commander." *Zarya: A Source of Information on Soviet and Russian Spaceflight*. http://www.zarya.info/Diaries/Soyuz4-5/Volynov.htm.

Conrad, Nancy, and D. C. Agle. "Voices from the Moon." *One Giant Leap for Mankind* (July 1994): 86–125.

"Cosmonaut Lands, Cold and Hungry." *Daily Mirror* (Sydney), 30 October 1968.

Cox, Billy. "Apollo 1: Heroes' Legacy, Readers Write of Their Apollo 1 Memories." *Florida Today*, 26 January 1997.

Demaret, Kent. "Group Think and Go Fever Brought the Shuttle Down, Says Ex-astronaut Donn Eisele," *People Weekly*, 24 March 1986.

"First Russian Spaceship Returns Safely." *Sydney Morning Herald*, 18 January 1969.

French, Francis. "Achieving the Impossible: Neil Armstrong and a Century of Adventure." *Spaceflight* 46, no. 2 (February 2004): 68–73.

———. "Apollo—The Best of Times: Dick Gordon and the Space Achievements of the Sixties." *Spaceflight* 44, no. 1 (2002): 26–30.

———. "Apollo 17—The End of the Beginning." *Spaceflight* 45, no. 9 (September 2003): 373–79.

———. "An Eye on the Earth—Dave Scott and Pure Test Piloting." *Spaceflight* 44, no. 9 (September 2002): 376–80.

———. "'I Worked With NASA, Not For NASA': An Interview with Wally Schirra." *Spaceflight* 43, no. 11 (November 2001): 471–75.

———. "Lost Faith: A Lone Rebel and Space Bureaucracy." *Spaceflight* 43, no. 9 (September 2001): 374–80.

———. "Return to the Moon: John Young on the Future of Space Exploration." *Spaceflight* 44, no. 2 (February 2002): 70–76.

———. "A Vital Link: Guenter Wendt and the Meaning of Responsibility." *Collect Space: The Source for Space History and Artifacts.* http://www.collectspace.com/news/news-052102a.html.

"Gemini Pilots Tell of Space Accident." *The Australian*, 27 July 1966.

"Gemini Was Near End of Tether." *Daily Telegraph* (Sydney), 31 August 1965.

Gottlieb, Walter. "Behind the Scenes: Red Moon Rising." *ICOM Film & Video Production and Postproduction*, January 2001. Archived at http://www.icommag.com/January-2001/jan-2001-page-1.html.

Graham, Bob. "Should Have Gone to the Moon." *Sunday Times* (London), no. 15, November 1998.

Grahn, Sven. "The Flight of Soyuz-4 and Soyuz-5." http://www.svengrahn.pp.se/trackind/soyuz45.html.

Izvekov, I., and I. Afanasyev. "How from a Failure Was 'Forged' the Next Victory." *Novosti kosmonautiki*, no. 23/24 (November 1998): 64–66.

Johnson, Adam. "Once Adored Soviet Cosmonaut Looks Back." *Johnson's Russia List*, 28 April 1999. Archived at http://www.cdi.org/russia/johnson/3264.html##4.

MacKnight, Nigel, and Eddie Pugh. "8 Days or Bust, Part I." *Space Flight News*, no. 21 (September 1987): 10–17.

———. "8 Days or Bust, Part II." *Space Flight News*, no. 22 (October 1987): 28–33.

Man Will Survive! Addresses at the Dedication of the Florida State Museum. University of Florida, Gainesville, 23 September 1971.

Mark, Ross. "Astronaut Donn Eisele Sued in First 'Space Divorce.'" *Daily Express* (London), 12 March 1969.

Milkus, Aleksandr. "Soyuz 5." *Komsomolskaya Pravda*, 10 April 1998.

"New Space Chase Under Way." *Daily Mirror* (Sydney), 19 July 1966.

Oberg, James. "Soyuz 5's Flaming Return." *Flight Journal: The Aviation Adventure—Past, Present, and Future* 7, no. 3 (June 2002): 56–60.

O'Toole, Thomas. "The Man Who Didn't Walk on the Moon." *New York Times*, 17 July 1994, 26–29.

Page, Eric. "Arthur F. Anders, 96, Hero Aboard U.S. Gunboat in 1937." *New York Times*, 31 August 2000.

Pugh, Eddie. "Space Patches." *Space Flight News*, no. 23 (November 1987): 44–45.

Rebrov, Mikhail. "A Difficult Re-Entry from Orbit." *Krasnaya zvezda*, 27 April 1996, 5.

"Russians Blast Man into Orbit of Earth." *Sydney Morning Herald*, 15 January 1969.

Russian Space History. Sotheby's Auction Catalogue, Sale 6516, New York, 11 December 1993.

"Russian Space-link Mystery." *Daily Telegraph* (Sydney), 27 October 1968.

Salakhutdinov, G. "Interview with Vasily Mishin." *Ogonyok*, no. 34 (18–25 August 1990): 4–5.

"Space Buddies' Record." *The Sun* (Sydney), 1 August 1968.

"Space Flight 750 Miles High." *Daily Mirror* (Sydney), 30 July 1968.

"Spaceships Link: Crew Transfers." *Daily Telegraph* (Sydney), 17 January 1969.

Swanson, Glen E. "In His Own Words." *Quest* 7, no. 4 (2000): 16–19.

"Swift Triumph by Space Pair." *Daily Mirror* (Sydney), 1 August 1968.

Weiss, Otis L., ed. *The United States Astronauts and Their Families*. World Book Encyclopedia Science Service, 1965.

West Point Military Academy. "Edward H. White" biography. *Assembly* (West Point Military Academy magazine), Summer 1971.

Woo, Elaine. "Astronaut Charles Conrad Embodied NASA's "Can-Do" Spirit." *Los Angeles Times*, 10 July 1999, A.

Interviews and Personal Communications

Aldrin, Buzz. Interviews by Francis French. Los Angeles, California. 7 December 2002 and 27 September 2003.

Anders, Bill. Interview by Francis French. San Diego, California. 20 March 2003.

———. NASA Oral History interview by Paul Rollins. 8 October 1997.

Anders, Valerie. Telephone interview by Francis French. 5 September 2003.

Armstrong, Neil. NASA Oral History interview by Stephen Ambrose and Douglas Brinkley. 19 September 2001.

Bassett, Jeannie. E-mail correspondence with Colin Burgess. 21–23 October 2002.

Beddingfield, Sam. Correspondence with Colin Burgess. 16 December 2003 and 8 January 2004.

Berry, Charles. NASA Oral History interview by Carol Butler. 29 April 1999.

Black, Susie Eisele. E-mail and mail correspondence with Francis French. 29 April 2004 to 27 August 2004.

Borman, Frank. NASA Oral History interview by Catherine Harwood. 13 April 1999.

Brown, W. O. E-mail correspondence with Colin Burgess. 31 July and 3 August 2003.

Cernan, Gene. Interview by Francis French. San Diego, California. 24 March 2001.

Collins, Michael. Correspondence with Francis French. 16 January 2001.

———. Interview by Francis French. Los Angeles, California. 27 September 2003.

———. NASA Oral History interview by Michelle Kelly. 8 October 1997.

Cooper, Gordon. Interviews by Francis French. Ventura, California. 25 April 2001. Santa Monica, California, 26 October 2002.

———. NASA Oral History interview by Roy Neal, 21 May 1998.

Cooper, Katherine (Apollo 1 Memorial Foundation). E-mail correspondence with Colin Burgess. January–March 2004.

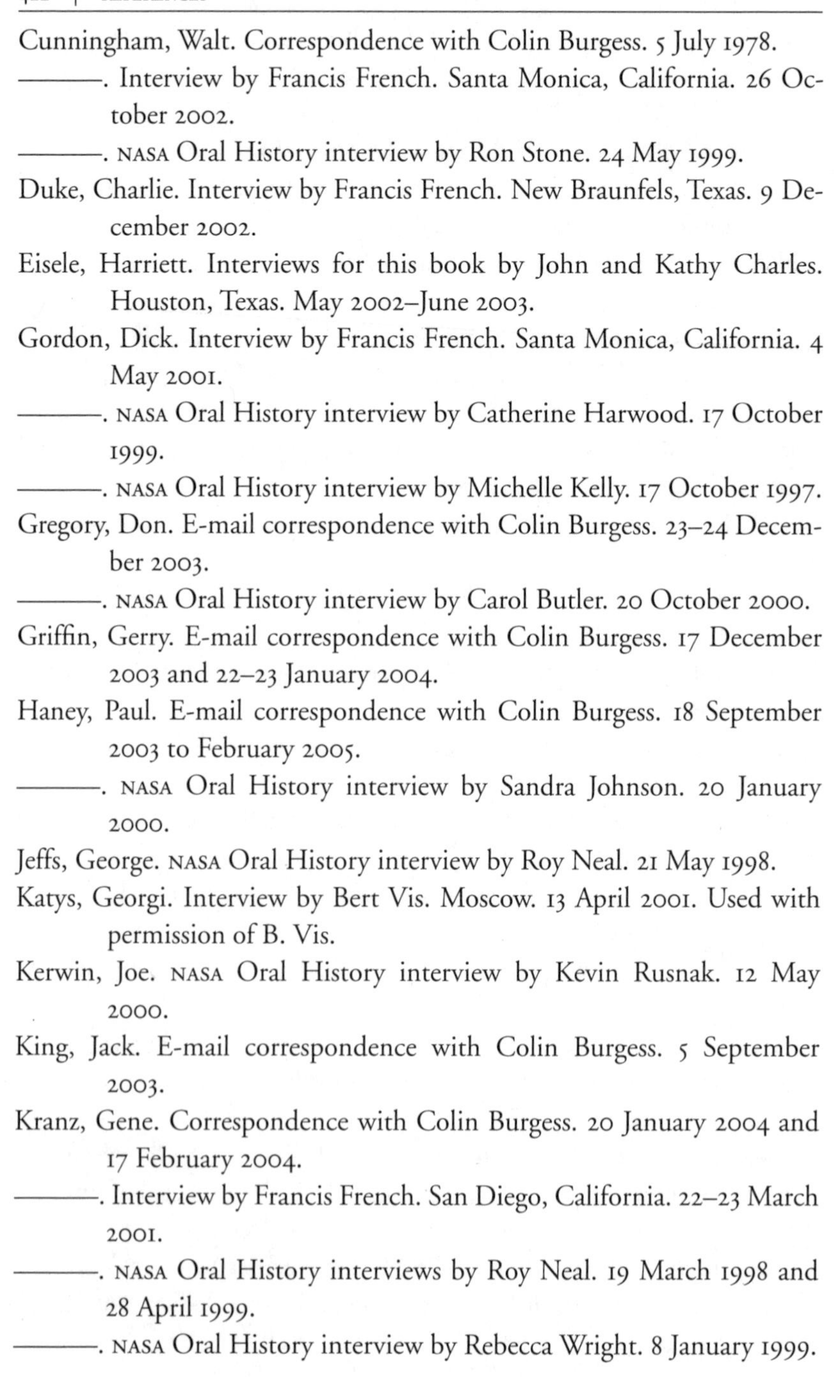

Cunningham, Walt. Correspondence with Colin Burgess. 5 July 1978.

———. Interview by Francis French. Santa Monica, California. 26 October 2002.

———. NASA Oral History interview by Ron Stone. 24 May 1999.

Duke, Charlie. Interview by Francis French. New Braunfels, Texas. 9 December 2002.

Eisele, Harriett. Interviews for this book by John and Kathy Charles. Houston, Texas. May 2002–June 2003.

Gordon, Dick. Interview by Francis French. Santa Monica, California. 4 May 2001.

———. NASA Oral History interview by Catherine Harwood. 17 October 1999.

———. NASA Oral History interview by Michelle Kelly. 17 October 1997.

Gregory, Don. E-mail correspondence with Colin Burgess. 23–24 December 2003.

———. NASA Oral History interview by Carol Butler. 20 October 2000.

Griffin, Gerry. E-mail correspondence with Colin Burgess. 17 December 2003 and 22–23 January 2004.

Haney, Paul. E-mail correspondence with Colin Burgess. 18 September 2003 to February 2005.

———. NASA Oral History interview by Sandra Johnson. 20 January 2000.

Jeffs, George. NASA Oral History interview by Roy Neal. 21 May 1998.

Katys, Georgi. Interview by Bert Vis. Moscow. 13 April 2001. Used with permission of B. Vis.

Kerwin, Joe. NASA Oral History interview by Kevin Rusnak. 12 May 2000.

King, Jack. E-mail correspondence with Colin Burgess. 5 September 2003.

Kranz, Gene. Correspondence with Colin Burgess. 20 January 2004 and 17 February 2004.

———. Interview by Francis French. San Diego, California. 22–23 March 2001.

———. NASA Oral History interviews by Roy Neal. 19 March 1998 and 28 April 1999.

———. NASA Oral History interview by Rebecca Wright. 8 January 1999.

Kubasov, Valery. Interview by Francis French. San Diego, California. 23 March 2001.

Leonov, Alexei. Interviews by Francis French. San Diego, California. 23–24 March 2001. Burbank, California. 2 September 2004.

Lovell, Jim. Interviews by Francis French. Los Angeles, California. 7 December 2002 and 25 September 2003.

———. NASA Oral History interview by Ron Stone. 25 May 1999.

Lunney, Glynn. NASA Oral History interview by Carol Butler. 8 February 1999.

———. NASA Oral History interview by Roy Neal. 26 April 1999.

McDivitt, Jim. NASA Oral History interview by Doug Ward. 29 June 1999.

Mitchell, Ed. E-mail correspondence with Francis French. 29 April and 25 August 2004.

Morrow, Lola. E-mail correspondence with Colin Burgess. 13–14 December 2003 and 10 March 2005.

Myers, Dale. NASA Oral History interviews by Carol Butler. 26 August 1998 and 5 March 1999.

O'Hara, Dee. Interview by Colin Burgess and Francis French. San Diego, California. 18 January 2003.

———. NASA Oral History interview by Rebecca Wright. 23 April 2002.

Pogue, William. NASA Oral History interview by Kevin Rusnak. 17 July 2000.

Popovich, Pavel. Interview by Rex Hall. Coventry, England. 22 November 2003.

———. Interview by Bert Vis. Ottawa, Canada. 30 September 1996. Used with permission of B. Vis.

Radimer, Milt. Correspondence with Colin Burgess. 17 March 2004.

Renshaw, Harvey. E-mail correspondence with Colin Burgess. 16 September 2003.

Schirra, Wally. Interviews by Francis French. San Diego, California. 19 February 2001 and 2 May 2001. Los Angeles, California. 26–27 September 2003.

———. NASA Oral History interview by Roy Neal. 1 December 1998.

Schmitt, Harrison. Interview by Francis French. Los Angeles, California. 7 December 2002.

Schweickart, Rusty. E-mail correspondence with Colin Burgess and Francis French. July 2002–August 2003.

———. Interview for this book by Bert Vis. The Hague, Netherlands. 5 August 2002.

———. NASA Oral History interview by Rebecca Wright. 19 October 1999 and 8 March 2000.

Shelly, Kelly (University of Houston–Clear Lake archivist). E-mail correspondence with Colin Burgess. June–August 2004.

Stafford, Thomas. Interview by Francis French. San Diego, California. 23 March 2001.

———. NASA Oral History interview by William Vantine. 15 October 1997.

Stevenson, Bob. Correspondence with Colin Burgess. 18 June 2001.

Taylor, Teddy. Correspondence with Colin Burgess. 12 August 2003.

Tinnirello, Al. Correspondence with Colin Burgess. 23 August 2003.

Volynov, Boris. Interview by Bert Vis. London. 16 May 2001. Used with permission of B. Vis.

Waddell, Hank. Correspondence with Colin Burgess. 28 July 2003.

Wendt, Guenter. NASA Oral History interview by Catherine Harwood. 25 February 1999.

———. NASA Oral History interview by Doyle McDonald. 16 January 1998.

Yardley, John. NASA Oral History interview by Chick Bergen. 29 June 1998.

Yeliseyev, Alexei. E-mail correspondence with Colin Burgess. 17 May 2004.

Other Sources

Apollo 7 Mission Report (MSC-PA R-68-15). Prepared by *Apollo 7* Mission Evaluation Team, NASA, Houston, Texas. December 1968.

Apollo 7 onboard voice transmissions (as recorded on spacecraft's onboard recorder/data storage equipment). NASA Manned Spacecraft Center, Houston, Texas. December 1968 (formerly classified document).

Apollo 7 Technical Debriefing (parts 1 and 2). NASA Manned Spacecraft Center, Houston, Texas. October 1968 (formerly classified documents).

Apollo 11 crew postflight press conference. NASA Manned Spacecraft Center, Houston, Texas. 12 August 1969.

Apollo 11 Technical Crew Debriefing. NASA Manned Spacecraft Center, Houston, Texas. 31 July 1969 (formerly classified document).

Cernan, Gene. Presentation at the San Diego Aerospace Museum, San Diego, California. 13 May 2004.

Cunningham, Walt. Personal Web site. http://www.waltercunningham.com.

———. Presentation at Reuben H. Fleet Science Center, San Diego, California. Hosted by Francis French. 18 October 2003.

Encyclopedia Astronautica. http://www.astronautix.com.

Freitag, Robert F. Robert F. Freitag to Shirley Thomas (University of Southern California), 7 November 1996. For International Astronautical Federation paper, *The Apollo Fire and Investigation: Facts Not Considered.* IAA-96-IAA.2.1.06.

Minzey, Josn. *Naval Academy Graduates in Gemini 6 and 7 Space Program.* USN archives release No. 22580. N.d.

NASA Mission Transcripts (CD-ROM). No. SP-2000-4602. Washington DC: NASA HQ, 2000.

Stafford, Thomas. Presentation at Reuben H. Fleet Science Center, San Diego, California. Hosted by Francis French. 23 March 2003.

Young, John W. Personal Web site, *Astronaut John W. Young, American and International Hero*, created by Dana Holland. http://www.nav.cc.tx.us/staff–pages/dana/jwy/bio/bio.htm.

———. Press conference, "Return to the Moon III." Lunar Development Conference, Las Vegas, Nevada. 19 July 2001.

In the Outward Odyssey: A People's History of Spaceflight Series

Into That Silent Sea
Trailblazers of the Space Era, 1961–1965
Francis French and Colin Burgess

In the Shadow of the Moon
A Challenging Journey to Tranquility, 1965–1969
Francis French and Colin Burgess

UNIVERSITY OF NEBRASKA PRESS

Also of Interest

Into That Silent Sea
Trailblazers of the Space Era, 1961–1965

By Francis French and Colin Burgess
With a foreword by Paul Haney

Through dozens of interviews and access to Russian and American official documents and family records, *Into That Silent Sea* captures the intimate stories of the men and women who made the space race their own and gave the era its compelling character.

ISBN: 978-0-8032-1146-9 (cloth)

Fallen Astronauts
Heroes Who Died Reaching for the Moon

By Colin Burgess and Kate Doolan, with Bert Vis
Foreword by Captain Eugene A. Cernan U.S. Navy (Ret.),
Commander, Apollo 17

This book enriches the saga of mankind's greatest scientific undertaking, Project Apollo, and conveys the human cost of the space race—by telling the stories of those sixteen astronauts and cosmonauts who died reaching for the moon.

ISBN: 978-0-8032-6212-6 (paper)

Teacher in Space
Christa McAuliffe and the Challenger Legacy

By Colin Burgess
Foreword by Grace George Corrigan

Christa McAuliffe's name is deeply entrenched in American history as the teacher who died when the *Challenger* exploded in January 1986. *Teacher in Space* explores and celebrates Christa's life and legacy and suggests that her goals of involving and educating children are being fulfilled even today.

ISBN: 978-0-8032-6182-2 (paper)